Scientific Workshop on Nuclear Fission and the Emission of Prompt Neutrons and Gamma Rays (Theory-4)

EPJ Web of Conferences Volume 169 (2018)

Varna, Bulgaria
20 – 22 June 2017

Editors:

F.-J. Hambsch
N. Carjan
I. Ruskov

ISBN: 978-1-5108-5946-3

Printed from e-media with permission by:

Curran Associates, Inc.
57 Morehouse Lane
Red Hook, NY 12571

Some format issues inherent in the e-media version may also appear in this print version.

Printed by Curran Associates, Inc. (2018)

For additional information, please contact EDP Sciences – Web of Conferences
at the address below.

EDP Sciences – Web of Conferences
17, Avenue du Hoggar
Parc d'Activité de Courtabœuf
BP 112
F-91944 Les Ulis Cedex A
France

Phone: +33 (0) 1 69 18 75 75
Fax: +33 (0) 1 69 28 84 91

contact-edps@webofconferences.org

Additional copies of this publication are available from:

Curran Associates, Inc.
57 Morehouse Lane
Red Hook, NY 12571 USA
Phone: 845-758-0400
Fax: 845-758-2633
Email: curran@proceedings.com
Web: www.proceedings.com

TABLE OF CONTENTS

STUDYING FISSION NEUTRONS WITH 2E-2V AND 2E 1
Al-Adili Ali, Jansson Kaj, Tarrío Diego, Hambsch Franz-Josef, Göök Alf, Oberstedt Stephan, Olivier Frégeau Marc, Gustavsson Cecilia, Lantz Mattias, Mattera Andrea, Prokofiev Alexander V., Rakopoulos Vasileios, Solders Andreas, Vidali Marzio, Österlund Michael, Pomp Stephan

PROMPT FISSION γ-RAY CHARACTERISTICS FROM NEUTRON-INDUCED FISSION ON ^{239}PU AND THE TIME-DEPENDENCE OF PROMPT γ-RAY EMISSION 7
Gatera Angélique, Göök Alf, Hambsch Franz-Josef, Moens André, Oberstedt Andreas, Oberstedt Stephan, Sibbens Goedele, Vanleeuw David, Vidali Marzio

PROMPT FISSION NEUTRON EMISSION IN THE REACTION 235U(N,F) 15
Göök Alf, Hambsch Franz-Josef, Oberstedt Stephan

THE 4-DIMENSIONAL LANGEVIN APPROACH TO LOW ENERGY NUCLEAR FISSION 27
Ivanyuk F. A., Ishizuka C., Usang M. D., Chiba S.

IMPLEMENTING AND TESTING THEORETICAL FISSION FRAGMENT YIELDS IN A HAUSER-FESHBACH STATISTICAL DECAY FRAMEWORK 35
Jaffke Patrick, Möller Peter, Stetcu Ionel, Talou Patrick, Schmitt Christelle

DETERMINATION OF THE NUCLEAR LEVEL DENSITIES AND RADIATIVE STRENGTH FUNCTION FOR 43 NUCLEI IN THE MASS INTERVAL 28≤A≤200 43
Knezevic David, Jovancevic Nikola, Sukhovoj Anatoly M., Mitsyna Ludmila V., Krmar Miodrag, Cong Vu D., Hambsch Franz-Josef, Oberstedt Stephan, Revay Zsolt, Stieghorst Christian, Dragic Aleksandar

NEW MEASUREMENTS ON ISOBARIC FISSION PRODUCT YIELDS AND MEAN KINETIC ENERGY FOR ^{241}PU THERMAL NEUTRON-INDUCED FISSION 49
Julien-Laferrière Sylvain, Kessedjian Grégoire, Serot Olivier, Chebboubi Abdelaziz, Bernard David, Blanc Aurélien, Köster Ulli, Litaize Olivier, Materna Thomas, Meplan Olivier, Rapala Michal, Sage Christophe

DETERMINATION OF THE FAST-NEUTRON-INDUCED FISSION CROSS-SECTION OF ^{242}PU AT NELBE 57
Kögler Toni, Beyer Roland, Junghans Arnd R., Schwengner Ronald, Wagner Andreas

MEASUREMENT OF THE ROT EFFECT IN THE NEUTRON INDUCED FISSION OF ^{235}U IN THE 0.3 EV RESONANCE AT A HOT SOURCE OF POLARIZED NEUTRONS 65
Kopatch Yuri, Novitsky Vadim, Ahmadov Gadir, Gagarsky Alexei, Berikov Daniyar, Danilyan Gevorg, Hutanu Vladimir, Klenke Jens, Masalovich Sergey

INVESTIGATION OF THE HEAVY NUCLEI FISSION WITH ANOMALOUSLY HIGH VALUES OF THE FISSION FRAGMENTS TOTAL KINETIC ENERGY 71
Khryachkov Vitaly, Goverdovskii Andrei, Ketlerov Vladimir, Mitrofanov Vecheslav, Sergachev Alexei

INFLUENCE OF PRIMARY FRAGMENT EXCITATION ENERGY AND SPIN DISTRIBUTIONS ON FISSION OBSERVABLES 77
Litaize Olivier, Thulliez Loïc, Serot Olivier, Chebboubi Abdelaziz, Tamagno Pierre

STUDY OF FISSION USING MULTI-NUCLEON TRANSFER REACTIONS 87
Nishio Katsuhisa, Hirose Kentaro, Vermeulen Mark, Makii Hiroyuki, Orlandi Riccardo, Tsukada Kazuaki, Asai Masato, Toyoshima Atsushi, Sato Tetsuya K., Nagame Yuichiro, Chiba Satoshi, Aritomo Yoshihiro, Tanaka Shouya, Ohtsuki Tsutomu, Tsekhanovich Igor, Petrache Costel M., Andreyev Andrei

PROMPT GAMMA RAYS FROM ^{252}CF(SF) AND THEIR ANGULAR DISTRIBUTIONS 97
Oberstedt Andreas, Billnert Robert, Gatera Angélique, Göök Alf, Oberstedt Stephan

ROLE OF THE EXCITATION ENERGY OF THE COMPOUND NUCLEUS IN BINARY DECAY PROCESSES 103
Pasca H., Andreev A. V., Adamian G. G., Antonenko N. V.

FISSION FRAGMENT MASS AND TOTAL KINETIC ENERGY DISTRIBUTIONS OF SPONTANEOUSLY FISSIONING PLUTONIUM ISOTOPES 109
Pomorski K., Nerlo-Pomorska B., Bartel J., Schmitt C.

MEASUREMENT OF FISSION YIELDS AND ISOMERIC YIELD RATIOS AT IGISOL 119
Pomp Stephan, Mattera Andrea, Rakopoulos Vasileios, Al-Adili Ali, Lantz Mattias, Solders Andreas, Jansson Kaj, Prokofiev Alexander V., Eronen Tommi, Gorelov Dimitri, Jokinen Ari, Kankainen Anu, Moore Iain D., Penttilä Heikki, Rinta-Antila Sami

PROMPT FISSION GAMMA-RAY EMISSION SPECTRAL DATA FOR 239PU(N,F) USING FAST DIRECTIONAL NEUTRONS FROM THE LICORNE NEUTRON SOURCE 127
Qi L., Wilson J. N., Lebois M., Al-Adili A., Chatillon A., Choudhury D., Gatera A., Georgiev G., Göök A., Laurent B., Maj A., Matea I., Oberstedt A., Oberstedt S., Rose S. J., Schmitt C., Wasilewska B., Zeiser F.

FISSION DYNAMICS WITH MICROSCOPIC LEVEL DENSITIES 133
Randrup Jørgen, Ward Daniel, Carlsson Gillis, Døssing Thomas, Möller Peter, Åberg Sven

FOURIER TRANSFORM OF SINGLE PARTICLE WAVE FUNCTIONS IN EXTREMELY DEFORMED NUCLEI: TOWARDS THE MOMENTUM DISTRIBUTION OF SCISSION NEUTRONS 143
Rizea M., Carjan N.

NEUTRON INDUCED FISSION OF ^{237}NP – STATUS, CHALLENGES AND OPPORTUNITIES 149
Ruskov Ivan, Goverdovski Andrei, Furman Walter, Kopatch Yury, Shcherbakov Oleg, Hambsch Franz-Josef, Oberstedt Stephan, Oberstedt Andreas

THE GEF MODEL: ASSESSMENT OF FISSION-FRAGMENT PROPERTIES OVER AN EXTENDED REGION 155
Schmitt Christelle, Schmidt Karl-Heinz, Jurado Beatriz

STAGGERING OF ANGULAR MOMENTUM DISTRIBUTION IN FISSION 165
Tamagno Pierre, Litaize Olivier

FISSION FRAGMENT YIELDS AND TOTAL KINETIC ENERGY RELEASE IN NEUTRON-INDUCED FISSION OF ^{235,238}U,AND ^{239}PU 173
Tovesson F., Duke D., Geppert-Kleinrath V., Manning B., Mayorov D., Mosby S., Schmitt K.

PARAMETERISATION OF THE RESIDUAL TEMPERATURE DISTRIBUTION BASED ON THE MODELLING OF SUCCESSIVE EMISSION OF PROMPT NEUTRONS 179
Tudora Anabella, Hambsch Franz-Josef

MEASUREMENT OF THE PROMPT FISSION γ-RAY SPECTRUM OF ^{242}PU 191
Urlass Sebastian, Beyer Roland, Junghans Arnd Rudolf, Kögler Toni, Schwengner Ronald, Wagner Andreas

FISSION OBSERVABLES FROM 4D LANGEVIN CALCULATIONS WITH MACROSCOPIC TRANSPORT COEFFICIENTS 201
Usang Mark D., Ivanyuk Fedir A., Ishizuka Chikako, Chiba Satoshi, Maruhn Joachim A.

STUDY OF PHOTON EMISSION WITH THE FISSION EVENT GENERATOR FREYA 207
Vogt Ramona, Randrup Jørgen

ANGULAR DISTRIBUTION OF SCISSION NEUTRONS STUDIED WITH TIME-DEPENDENT SCHRÖDINGER EQUATION 213
Wada Takahiro, Asano Tomomasa, Carjan Nicolae

STUDIES OF FISSION FRAGMENT YIELDS VIA HIGH-RESOLUTION γ -RAY SPECTROSCOPY 219
Wilson J. N., Lebois M., Qi L., Amador-Celdran P, Bleuel D., Briz J. A., Carroll R., Catford W., Witte H. De, Doherty D. T., Eloirdi R., Georgiev G., Gottardo A., Goasduff A., Hadyñska-Klek K., Hauschild K., Hess H., Ingeberg V., Konstantinopoulos T., Ljungvall J., Lopez-Martens A., Lorusso G., Lozeva R., Lutter R., Marini P., Matea I., Materna T., Mathieu L., Oberstedt A., Oberstedt S., Panebianco S., Podolyak Zs., Porta A., Regan P. H., Reiter P., Rezynkina K., Rose S. J., Sahin E., Seidlitz M., Serot O., Shearman R., Siebeck B., Siem S., Smith A. G., Tveten G. M., Verney D., Warr N., Zeiser F., Zielinska M.

Author Index

Preface

Franz-Josef Hambsch[1], *Ivan* Ruskov[2], and *Nicolae* Carjan[3]

[1] *European Commission, Joint Research Centre, Directorate G, 2440 Geel, Belgium*
[2] *Institute for Nuclear Research and Nuclear Energy of Bulgarian Academy of Sciences, 1784 Sofia, Bulgaria*
[3] *National Institute of Physics and Nuclear Engineering-Horia-Hulubei, Department of Theoretical Physics, 76900 Bucharest-Magurele, Romania*

On June 20–22, 2017 the Scientific Workshop on Nuclear Fission Dynamics and the Emission of Prompt Neutrons and Gamma Rays (THEORY–4) was held in Varna, Bulgaria. It represents the fourth workshop of a series initiated in 2010 in Sinaia, Romania and continued in 2012 (Biarritz, France) and 2014 (Opatija, Croatia). The workshop was co-organized by JRC-Geel-G.2 (Belgium), INRNE (Sofia, Bulgaria) and NIPNE (Bucharest, Romania). The organisers were assisted by a Programme Advisory Committee consisting of R. Capote (IAEA, Austria), S. Oberstedt (EC-JRC-G.2, Belgium), O. Serot (CEA-Cadarache, France), P. Talou (LANL, USA) and A. Tudora (Bucharest University, Romania).

The purpose of the workshop was to provide a comprehensive overview of the dynamical aspects of nuclear fission and their consequences on the fission fragment properties. The contributions have highlighted recent theoretical advancements in the field of nuclear fission and in particular for the emission of prompt neutrons and gamma rays.

Both invited presentations and normal contributions have been given. The success of the workshop must be attributed to the more than 30 contributions, many of which were post-doctoral fellows and PhD students.

The workshop was supported by the JRC-Geel-G.2. The general aim is to support the EU integration process by fostering collaboration with governmental organisations (or the like), which will have an institutional role for providing Science and Technology support to the implementation of policies, particularly through the associated scientific and technical organisations.

We would like to express our gratitude to the program advisory committee, for their help to choose key topics to debate and to the speakers for their carefully prepared presentations. A very special thank you goes to the perfect local organisational support of Carmen Cabanillas Platero.

edp sciences

Web of Conferences

Statement of Peer review

In submitting conference proceedings to *Web of Conferences*, the editors of the proceedings certify to the Publisher that

1. They adhere to its Policy on Publishing Integrity in order to safeguard good scientific practice in publishing.
2. All articles have been subjected to peer review administered by the proceedings editors.
3. Reviews have been conducted by expert referees, who have been requested to provide unbiased and constructive comments aimed, whenever possible, at improving the work.
4. Proceedings editors have taken all reasonable steps to ensure the quality of the materials they publish and their decision to accept or reject a paper for publication has been based only on the merits of the work and the relevance to the journal.

Title, date and place of the conference

Scientific Workshop on "Nuclear Fission Dynamics and the Emission of Prompt Neutrons and Gamma Rays" (Theory-4)
20-22 June 2017
Varna, Bulgaria

Proceedings editor(s):

F.-J. Hambsch, EC-JRC-Geel, Belgium

N. Carjan, NIPNE-HH, Bucharest, Romania

J. Ruskov, INRNE, Sofia, Bulgaria

Date and editor's signature

11.12.2017

Studying fission neutrons with 2E-2v and 2E

Ali Al-Adili[1,*], *Kaj* Jansson[1], *Diego* Tarrío[1], *Franz-Josef* Hambsch[2], *Alf* Göök[2], *Stephan* Oberstedt[2], *Marc Olivier* Frégeau[3], *Cecilia* Gustavsson[1], *Mattias* Lantz[1], *Andrea* Mattera[1], *Alexander V.* Prokofiev[1], *Vasileios* Rakopoulos[1], *Andreas* Solders[1], *Marzio* Vidali[2], *Michael* Österlund[1], and *Stephan* Pomp[1]

[1] *Department of Physics and Astronomy, Uppsala University, Sweden*
[2] *European Commission, Joint Research Centre, Directorate G-2, Geel, Belgium*
[3] *GANIL CEA/DRF-CNRS/IN2P3, Caen, France*

Abstract. This work aims at measuring prompt-fission neutrons at different excitation energies of the nucleus. Two independent techniques, the 2E-2v and the 2E techniques, are used to map the characteristics of the mass-dependent prompt fission neutron multiplicity, $\overline{\nu}(A)$, when the excitation energy is increased.
The VERDI 2E-2v spectrometer is being developed at JRC-GEEL. The Fission Fragment (FF) energies are measured using two arrays of 16 silicon (Si) detectors each. The FFs velocities are obtained by time-of-flight, measured between micro-channel plates (MCP) and Si detectors. With MCPs placed on both sides of the fission source, VERDI allows for independent timing measurements for both fragments. ^{252}Cf(sf) was measured and the present results revealed particular features of the 2E-2v technique. Dedicated simulations were also performed using the GEF code to study important aspects of the 2E-2v technique. Our simulations show that prompt neutron emission has a non-negligible impact on the deduced fragment data and affects also the shape of $\overline{\nu}(A)$. Geometrical constraints lead to a total-kinetic energy-dependent detection efficiency.
The 2E technique utilizes an ionization chamber together with two liquid scintillator detectors. Two measurements have been performed, one of ^{252}Cf(sf) and another one of thermal-neutron induced fission in ^{235}U(n,f). Results from ^{252}Cf(sf) are reported here.

1 Introduction

The evolution of the fragment-specific neutron-multiplicity, $\overline{\nu}(A)$, as a function of excitation energy, is an important observable. It reflects the sharing of the total excitation energy (TXE) and how nuclear shell effects wash out at higher excitation energy. Some works (theoretical as well as experimental) assume a higher $\overline{\nu}(A)$ for all fragments as excitation energy increases. However, the few available experimental data sets contradict this and instead point to an increase of $\overline{\nu}(A)$ only for heavy fragments [1]. Hence, more attention needs to be drawn to resolve this question and arrive at a proper description for the excitation energy dependence of $\overline{\nu}(A)$.

In this paper two different methods are being used with the aim to study this matter, in the first chance fission of ^{235}U(n,f). To realize this goal, ^{252}Cf(sf) is used for development and calibration

*e-mail: ali.al-adili@physics.uu.se

purposes. In the so-called "2E-2v" method, we measure both fragments' velocities and energies, simultaneously. By using these four observables, one can attempt to derive the fission fragment pre- and post-neutron emission masses. The average neutron multiplicity, $\overline{\nu}$(A), is thus obtained as the difference of the pre- and post-neutron emission masses on an event-by-event basis. In the "2E" method, both fragments' kinetic energies are measured simultaneously. To determine $\overline{\nu}$(A), neutron detectors (ND) are added to measure the rate of coincidences, between the fission chamber and the NDs. The experimental setups of both systems are sketched in Fig. 1.

2 Double-energy double-velocity method (2E-2v)

Currently, a few 2E-2v-instruments are being developed across the world, e.g., FALSTAFF, SPIDER, STEFF and VERDI [2–5]. One important goal is to suppress uncertainties in fission yield experiments, by enhancing the mass resolution. In principle, the method itself should be able to deliver a mass resolution of 1–2 mass units. The resolution is mainly limited by the neutron boil-off, but additional instrument-dependent effects apply (e.g. energy-losses in targets and foils and energy spreads in Si detectors and ionization chambers). The 2E-2v method has one obvious advantage, namely, that it does not need external input data regarding $\overline{\nu}$(A), Total Kinetic Energy (TKE) or incident neutron energy (E_{n}), during analysis. Such data are indispensable input to the 2E method.

2.1 VERDI

The VERDI (VElocity foR Direct particle Identification) spectrometer is developed at the JRC-GEEL. The instrument has two independent time-of-flight (TOF) sections, each hosting 16 NTD-Silicon (Si) detectors (PIPS). The velocity of each fragment, respectively its time-of-flight, is measured between either of two MCP and one of the Si detectors. The Si detectors also measure the kinetic energy of the fission fragments. The setup is schematically shown in Fig. 1a.

In a fission event, the released energy (Q) is divided between the TKE and the TXE. The TKE is shared between the nascent fragments according to momentum conservation. The fragments TOF thus decompose into two, overlapping, distributions. The FFs, after neutron emission, exit the target back-to-back and travel towards the Si-detectors. Some electrons are sputtered from the target and get deflected 90° by an electrostatic mirror towards respective MCP. A few electrons are enough to give rise to a start signal for the fragment TOF. The FFs traverse a distance of about 50 cm before reaching the Si detectors with a present size of 450 mm^2. The Si detectors are of double use: providing both the TOF stop signals and the FFs kinetic energies.

First results from VERDI, using spontaneous fission of ^{252}Cf, were published by Frégeau et al. [5]. Decent results were reported already using one MCP providing a common start signal. Another MCP was installed recently in order to enhance the TOF precision, and to improve the calibration procedure. After installation, a refined data analysis revealed limitations of two of the corrections needed, namely, the energy-calibration, responsible for the pulse-height-defect (PHD) and the plasma delay time (PDT) correction.

The energy-calibration was previously performed using the Schmitt calibration procedure [6]. However, the mean energy values turned out to be too high for ^{252}Cf. Two different methods were tested to remedy this. One was a newly developed iterative Schmidt procedure [7]. The other was simply an adaptation of the adjusted Schmidt values from Müller et al. [8].

The pre-neutron emission mass distributions, integral and as a function of TKE, of the updated VERDI instrument were promising. The mass yield showed an enhanced peak-to-valley ratio and a somewhat narrower distribution (see Fig. 2 for one selected Si detector pair), compared to reference

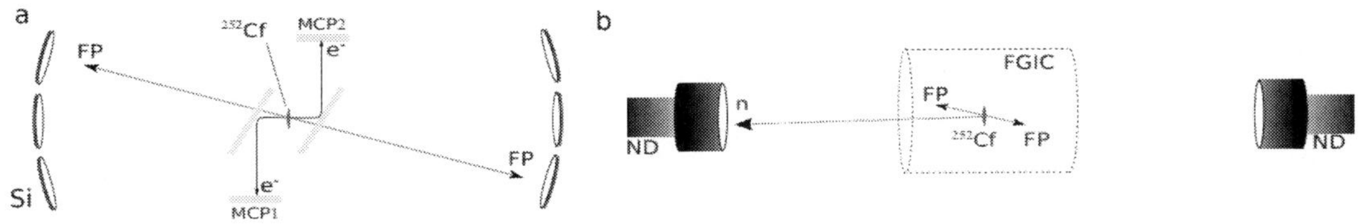

Figure 1. a) Schematics of the VERDI 2E-2v setup, with two arrays of 16 Si detectors and two Micro Channel Plate (MCP) assemblies as signal trigger. b) Schematics of the 2E setup with a Frisch-Grid Ionization Chamber (FGIC) and two liquid scintillator neutron detectors (ND).

2E data [9]. Calculation of the 2v mass yields will cancel some systematic errors, which is why these yields are rather robust against potential problems [10].

The mass-dependent neutron multiplicity shape differed from the 2E data [9], especially around mass symmetry. The more pronounced peaks were first interpreted as a sign of a better mass resolution compared to 2E. A set-back was however encountered in the $\overline{\nu}$(TKE) shape, which showed the opposite trend of what is expected. Since $Q = \text{TKE} - \text{TXE}$, the $\overline{\nu}$(TKE) should have a negative slope, as fewer neutrons are emitted for higher TKE values, i.e., lower TXE values. Our data exhibited a higher $\overline{\nu}$ as a function of TKE which was obviously wrong. The PDT was the prime suspect, as it is a delicate part of the analysis and which relies directly on the related 2E analysis from the silicon detectors.

The measured TOFs are delayed, due to the creation of a plasma in the PIPS detectors. In literature there are contradicting correction methods for this effect [10–12]. The simplified linear relationship ("scaling law") used by Velkovska et al. [11] was initially adopted for VERDI. The measured velocities were plotted as a function of the calculated velocities from the 2E energies. A conventional 2E analysis was performed on the Si energy data providing the "true" velocities. The velocity plot was linearly fitted to provide the "true" velocity as a function of the measured one. However, this method seems to contradict the PDT-systematics of Bohne et al. [12]. And indeed, the application of the scaling law produced unphysical results based on the VERDI data.

A preliminary analysis was recently done, using a new PDT-correction method. The PDT was parameterized as a function of pre-neutron mass and energy. Through an iterative procedure the parameters were improved. As a result from the new approach $\overline{\nu}$(TKE) showed a more reasonable trend. The obtained VERDI slope corresponds to 13 MeV/n, which agrees rather well with literature data of 12.6 MeV/n [9]. We plan to refine this method by parameterizing the PDT consistently with the systematics of Ref. [12].

2.2 GEF simulations for VERDI

In general, the 2E-2v data suffers from a limited solid angle coverage (1%). This does not only affect statistics, but could also introduce biases caused by the instrument geometry. We know that in about half of the acquired fission events, only one fragment is detected. The other fragment missed the complementary Si detector. To check the effect on fission observables, dedicated simulations were performed. The simulations are based on the GEF event-by-event output. From GEF we obtain information on A and Z for each fragment, their kinetic energies, and information on the emitted secondary particles. We sample the neutron emission isotropically in the center of momentum (CM) frame. The full relativistic kinematics of the fragments and emitted neutrons have been taken into

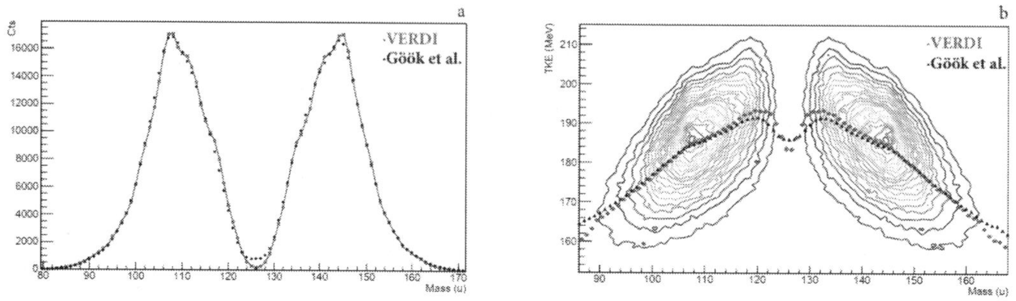

Figure 2. The ^{252}Cf mass (a) and mass vs. TKE (b) in red, for one Si pair, compared to Ref. [9] in black.

account. The relativistic calculations are necessary to prevent systematic errors in the order of 0.1 u compared to classical kinematics.

The simulations help to understand several features of the 2E-2v method. Firstly, the detection efficiency of coincidence events was about half, confirming the VERDI observation. Moreover, the detection efficiency is TKE-dependent as seen in Fig. 3a. For low TKE values, the efficiency becomes as low as 40% while for the highest TKE values it reaches 100%. This is due to neutron emission which increases for lower TKE, giving rise to larger fragment angular deviation from the fission axis. Secondly, a central assumption in 2E-2v is the one on average unchanged fragment velocity, after neutron emission. But since the fragment velocity is conserved only on average, and not for specific events, this could partially wash out fine structures in the mass distribution. This was observed in a pseudo-analysis of the input GEF data. Another consequence of the neutron emission are changes in the sawtooth shape as demonstrated in Fig. 3b; effects which could lead to an erroneous interpretation of the data. The input GEF data are shown in black. The data are analysed via the 2E-2v method (in red). Some deviations are encountered mainly around mass symmetry. The coincidence requirement (in blue) does not seem to be of major concern. Our simulations shed light on vital aspects of the 2E-2v technique itself and are subject of a dedicated publication under preparation.

3 Double-energy method (2E)

The 2E method is based on the analysis of the measured kinetic energies of both fragments, by momentum conservation. During the iterative analysis, known $\overline{\nu}(A)$ values must be used [1]. The data on ^{252}Cf are obtained by using a Frisch-grid ionization chamber (FGIC). The chamber is filled with P-10 counting gas, which is ionized when charged fragments are stopped. The total number of electrons is proportional to the deposited energy. Five electrodes are hosted in the chamber; two anodes, two grids and one common cathode where the target is placed. The Frisch-grids provide the emission angle of the fragments relative to the chamber axis. NE-213 equivalent neutron detectors (ND) are used in conjunction with the chamber. Both are put axially to the chamber symmetry axis, effectively reducing the need to determine the azimuthal angle of the fragments relative to the chamber axis. The neutron detectors have an intrinsic threshold for neutrons lying above 0.5 MeV in neutron energy. Both ND are cylindrical in shape and have a size of 102 mm × 51 mm (diameter × length). A digital data acquisition was used, storing all signal traces for off-line analysis.

During data analysis the neutron data treatment starts with a calibration both in TOF and pulse-heights (PH). The prompt fission γ peak is utilized for the TOF calibration and for time-walk corrections. The obtained timing resolution is 1.3 ns, determined as the full-width-at-half-maximum of the γ

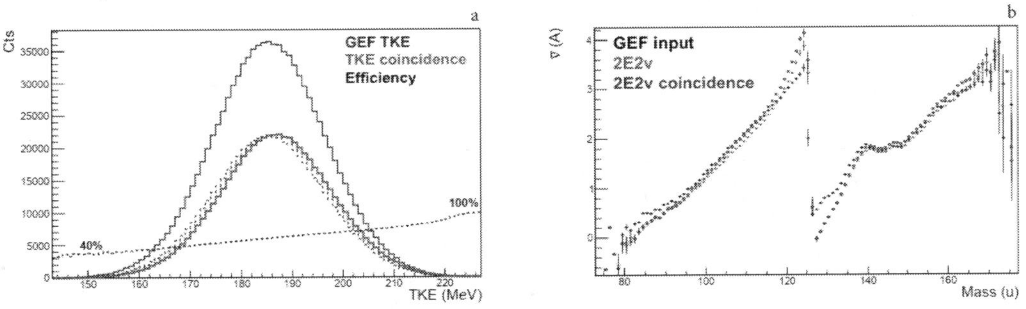

Figure 3. a) TKE distributions from GEF (blue) compared to the TKE for coincident events (red). The dashed line is the TKE from GEF, normalized to the integral of the coincidence data. The black dashed line depicts the ratio between the coincidence data and the input data scaled to 10^4, revealing a strong energy-dependent efficiency. b) $\bar{\nu}$(A) from our GEF-simulations. The GEF input data are depicted by the black symbols. The corresponding data obtained from the 2E-2v analysis on the simulated fission events are shown with red symbols, and those requiring coincidences of opposite Si-detector pairs are depicted with the blue symbols.

peak. Secondly, various cuts are applied to the neutron data to discriminate γ rays from neutrons and to correct for neutron scattering. At last, the neutron energy in the laboratory frame is determined and can be used to evaluate the detector intrinsic efficiency relative to the Mannhart ^{252}Cf evaluation (see Ref. [9] and references therein). The FF data are corrected for energy-losses after determining the angular distributions. Iterative calculations lead to the final pre-neutron emission FF mass yields, by using input $\bar{\nu}$(A) from Ref. [9]. The FF data are corrected for PHD and for the TKE dependence of $\bar{\nu}$ [13]. The fragment-neutron coincidence data, are corrected for the neutron momentum recoil effects. In addition, a cut in the CM system for neutron emission above $cos(\phi) = 0$ has been applied (where ϕ is the angle between the neutron momentum and the fission axis), in accordance with the approach presented in Ref. [9].

3.1 Results of 2E

Results on ^{252}Cf are shown in Figs. 4a–d. The neutron TOF data are plotted in Fig. 4a, in black after calibration and walk correction. After pulse-shape discrimination using the zero-crossing-method, the blue line shows a much reduced prompt fission γ peak. Further cuts on the TOF removed fully the γ peak as seen in the green curve. Final cuts were applied in the TOF-vs.-PH matrix considering the maximum proton recoil. The clean prompt neutron TOF spectrum is shown in red. The mass distribution is shown in Fig. 4b where the sample side (blue) and backing side (red) are plotted together with the average (green). The distribution is in good agreement with the literature data (in black) but shows a slightly degraded mass resolution. This is due to the ^{252}Cf target, which is of a lower quality (in terms of uniformity of target and backing) compared to the one used in Ref. [9]. In addition, the statistics obtained in this work is significantly lower, which affects the quality of the results. $\bar{\nu}$(A) is obtained by dividing the coincident mass distribution with the non-coincident one, after normalization for solid angle, total fission rate and proper scaling for $\bar{\nu}_{tot}$. In Fig. 4c the neutron sawtooth data are drawn, where our data (in red) are in decent agreement to the reference in black. The minimum around 130 u is well reproduced. Small differences can also be seen and could be caused by a lower target quality as well as lower resolution in the neutron TOF measurement. The $\bar{\nu}$(TKE) is shown in Fig. 4d

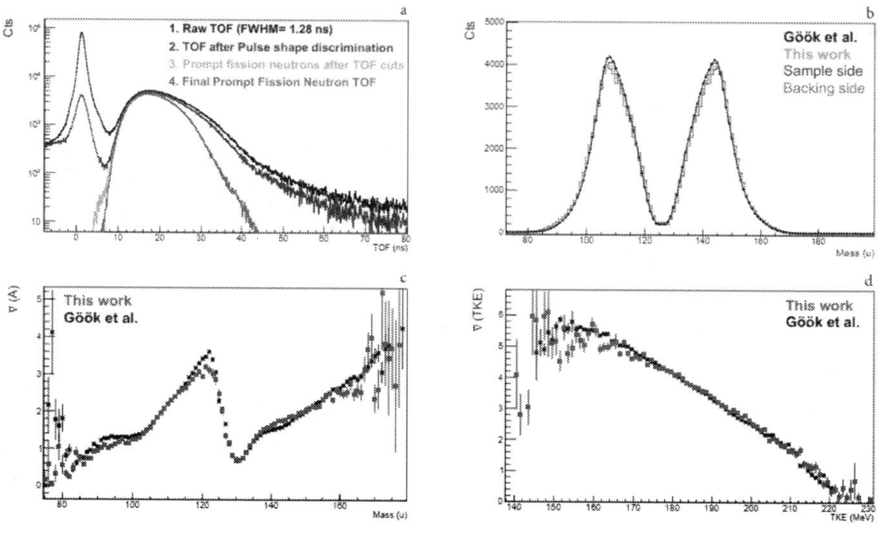

Figure 4. a) The TOF before and after applied cuts (see text for details). b) The 2E mass yields for ^{252}Cf(sf) are shown, for sample and backing sides and their average; c) $\bar{\nu}$(A) and d) $\bar{\nu}$(TKE). All were compared to Ref. [9].

and exhibits a slope very similar to Ref. [9]. This Cf experiment served as proof-of-principle and the results are quite satisfactory.

These projects were partially financed by EU FP7 CHANDA project, by granting A.A and K.J. scientific visits to the JRC-GEEL. The JRC-GEEL MONNET staff is acknowledged for technical support.

References

[1] A. Al-Adili et al., Nuclear Data Sheets **119**, 342 (2014)
[2] D. Doré et al., EPJ Web of Conferences **62**, 05005 (2013)
[3] J. Matarranz et al, Physics Procedia **47**, 76 (2013)
[4] K. Meierbachtol et al., Nucl. Instr. and Meth. A **788**, 59 (2015)
[5] M.O. Frégeau et al., Nucl. Instr. and Meth. A **817**, 35 (2016)
[6] H.W. Schmitt et al., Phys. Rev. **137**, B837 (1965)
[7] K. Jansson et al., EPJ Web of Conferences **146**, 04016 (2017)
[8] R.Müller et al., Phys. Rev. C **29**, 885 (1984)
[9] A.Göök et al., Phys. Rev. C **90**, 064611 (2014)
[10] K.Th. Brinkmann et al., Nucl. Instr. and Meth. A **276**, 557 (1989)
[11] J. Velkovska and R.L. McGrath, Nucl. Instr. and Meth. A **430**, 507 (1999)
[12] W. Bohne et al., Nucl. Instr. and Meth. A **240**, 145 (1985)
[13] A. Al-Adili et al., Phys. Rev. C **93**, 034603 (2016)

Prompt fission γ-ray characteristics from neutron-induced fission on ^{239}Pu and the time-dependence of prompt γ-ray emission

Angélique Gatera[1,2], *Alf* Göök[2], *Franz-Josef* Hambsch[2], *André* Moens[2], *Andreas* Oberstedt[3], *Stephan* Oberstedt[2,⋆], *Goedele* Sibbens[2], *David* Vanleeuw[2], and *Marzio* Vidali[2]

[1] *Ghent University, Department of Physics and Astronomy, Proeftuinstraat 86, 9000 Ghent, Belgium*
[2] *European Commission, DG JRC, Directorate G - Nuclear Safety and Security, G.2 SN3S, 2440 Geel, Belgium*
[3] *Extreme Light Infrastructure - Nuclear Physics (ELI-NP) / Horia Hulubei National Institute for Physics and Nuclear Engineering (IFIN-HH), 077125 Bucharest-Magurele, Romania*

Abstract. Recent years have seen an increased interest in prompt fission γ-ray (PFG) measurements motivated by a high priority request of the OECD/NEA for high precision data, mainly for the nuclear fuel isotopes ^{235}U and ^{239}Pu. Our group has conducted a PFG measurement campaign using state-of-the-art lanthanum halide detectors for all the main actinides to a precision better than 3%. The experiments were performed in a coincidence setup between a fission trigger and γ-ray detectors. The time-of-flight technique was used to discriminate photons, traveling at the speed of light, and prompt fission neutrons. For a full rejection of all neutrons below 20 MeV, the PFG time window should not be wider than a few nanoseconds. This window includes most PFG, provided that no isomeric states were populated during the de-excitation process. When isomeric states are populated, PFGs can still be emitted up to 1 μs after the instant of fission or later. To study these γ-rays, the detector response to neutrons had to be determined and a correction had to be applied to the γ-ray spectra.
The latest results for PFG characteristics from the reaction ^{239}Pu(n$_{th}$,f) will be presented, together with an analysis of PFGs emitted up to 200 ns after fission in the spontaneous fission of ^{252}Cf as well as for thermal-neutron induced fission on ^{235}U and ^{239}Pu. The results are compared with calculations in the framework of the Hauser-Feshbach Monte Carlo code CGMF and FIFRELIN.

1 Introduction

After almost 80 years since the discovery of nuclear fission the underlying process is still not understood in sufficient details. There are many fundamental questions still to be answered, touching both the nuclear energy as well as the basic science community. The first community is represented by the OECD Nuclear Energy Agency (NEA) [1], who expressed the need for more accurate fission cross-section and fragment yield data for safety assessments of Generation-IV reactor systems [2]. The second community, dealing with nuclear theory and the modeling of the fission process, are aiming at a detailed description of the nuclear fission process with predictive power for the observables, e.g.

⋆e-mail: stephan.oberstedt@ec.europa.eu

fragment yield and total kinetic energy distributions. Of particular interest is the understanding of the de-excitation mechanism of very neutron-rich isotopes that are produced in nuclear fission and how the excitation energy is shared between the two fission fragments.

The energy released in nuclear fission is distributed in kinetic and excitation energy of the two fragments. The excitation energy manifests itself in fragment deformation and intrinsic excitation energy. The first step of fission fragment de-excitation takes place at a very early stage after scission through the successive emission of neutrons and γ rays. In nuclear applications the prompt energy release by neutrons and γ rays accounts for about 5% of the total prompt heat set free in nuclear fission. Per definition prompt fission γ-rays may be emitted still after several microseconds, i.e. before the onset of β decay. Those isomeric γ-rays may be used to detect and identify actinides, that recently gained considerable attention for developing techniques to prevent illicit trafficking of fissile material.

The advances made in recent editions of various fission models presently on the market [3–9] require precise experimental data on prompt fission neutron and γ-ray emission as input parameters. For example multiplicity, average energy per particle and total dissipated energy per fission, preferably as function of fission-fragment mass and total kinetic energy, are key input to benchmark nuclear fission models attempting to describe the competition between prompt neutron and γ-ray emission.

Some years ago, a collaboration of scientists from JRC Geel (the former JRC IRMM) and other institutes took the lead in establishing a dedicated measurement programme on prompt fission neutron [10, 11] and fission γ-ray characteristics [12–17] responding to a high-priority data request of the OECD NEA. Here, we report about our most recent measurement of prompt fission γ-ray spectral PFGS characteristics from neutron-induced fission on ^{239}Pu. Those results made it necessary to verify previous results obtained for prompt fission γ-ray characteristics from neutron-induced fission on ^{235}U [14]. In addition, we present first results on time-dependent PFGS emission.

In the following we recall the essential information about our experimental technique and the applied data analysis, before we present and discuss our experimental results.

2 Experiment technique and data analysis

Experiments with spontaneously fissioning isotopes were performed at the JRC Geel, thermal-neutron induced fission at the cold-neutron beam at the Budapest Research Reactor. As detector for fission fragments a Frisch-grid ionization chamber (FGIC) is used providing a detection efficiency close to 100%. Counting gasses with a electron drift velocity of at least 10^5 m/s were used, CH_4 and Ar-CF_4(10%), to minimize the α-fission and the fission-fission pile-up rate. In combination with current sensitive pre-amplifiers the signal length was kept to 300 ns and 150 ns, respectively.

As γ ray detectors we used three cerium-doped lanthanum-bromide ($LaBr_3$:Ce) detectors of size 51 mm $\times$ 51 mm (diameter $\times$ length), which provided the best timing and energy resolution with sufficiently high detection efficiency. For the PFGS measurement on ^{235}U a single 76 mm $\times$ 76 mm (diameter $\times$ length) $LaBr_3$:Ce detector was used.

The data-acquisition system was based on 2-channel waveform digitizer cards with a sampling frequency of 400 MHz and a resolution of 14 bit [18].

Validation of our measurement setup and data analysis routines was done measuring prompt fission γ-rays from spontaneous fission of ^{252}Cf. We used a ^{252}Cf source with 11000 fission/s deposited on a 250 nm thick Ni-foil and placed in the FGIC on the central part of the cathode. In Table 1 we compare the average multiplicity, $\overline{M}_\gamma$, the mean photon energy, ϵ_γ, and the average total γ-ray energy per fission, $E_{\gamma,tot}$, with our previous measurements as well as with data from literature. The validation of our method reflects in a reproduction of our previous results for $\overline{M}_\gamma$, ϵ_γ and $E_{\gamma,tot}$ [12, 19] within 1%, 1.3% and 2.7%, respectively.

Table 1. Validation of our measurement setup and data analysis routines was done on spontaneous fission of ^{252}Cf. The previous results on $\overline{M}_\gamma$, ϵ_γ and $E_{\gamma,tot}$ [12, 19] are reproduced within 1%, 1.3% and 2.7%, respectively.

Results	Detector	$\overline{M}_\gamma$ (per fission)	ϵ_γ (MeV)	$E_{\gamma,tot}$ (MeV)	Energy range (MeV)
This work	$LaBr_3$:Ce (Q489)	8.32 ± 0.10	0.83 ± 0.02	6.89 ± 0.11	$0.1 - 7.0$
This work	$LaBr_3$:Ce (Q491)	8.41 ± 0.11	0.81 ± 0.02	6.80 ± 0.11	$0.1 - 7.0$
This work	$LaBr_3$:Ce (5414)	8.35 ± 0.10	0.81 ± 0.02	6.78 ± 0.10	$0.1 - 7.0$
This work	$LaBr_3$:Ce (5415)	8.44 ± 0.12	0.81 ± 0.02	6.82 ± 0.11	$0.1 - 7.0$
This work	Sum spectrum	8.38 ± 0.11	0.81 ± 0.02	6.82 ± 0.10	$0.1 - 7.0$
Billnert	$LaBr_3$:Ce	8.30 ± 0.06	0.80 ± 0.01	6.64 ± 0.07	$0.1 - 6.0$

The ^{239}Pu target was electrodeposited on 25 μm thick Al. The 2 cm large circular spot contained 430 μg high-purity ^{239}Pu (99.97%) leading to a fission rate of 4.4×10^4/s.

A thorough investigation of the detector response in a setup with a FGIC using the Geant4 code [21, 22] revealed an underestimation of the backscatter contribution to the response, possibly leading to an overestimation of the γ-ray yield below 300 keV. This required a verification of our results on PFGS from thermal-neutron induced fission on ^{235}U [14]. Therefore, we added a γ-ray detector to the SCINTIA array [11], an instrument dedicated to measure prompt-neutron spectral data on ^{235}U at thermal and resolved-resonance energies. The ^{235}U target was a circular 7 cm diameter spot of 67.2 μgU/cm^2 UF_4 (99.94% ^{235}U vacuum-evaporated onto a 27 μg/cm^2 polyimide foil plated with 50 μg/cm^2 Au. The average fission rate was about 4.5/s.

For the investigation of time-dependent PFG emission neutron - γ separation via their characteristic time-of-flight is no longer sufficient. It requires the determination of the detector response to fast neutrons. In our case this will manifest in the detection of γ-rays following the inelastic scattering of prompt fission neutrons essentially on ^{79,81}Br and ^{139}Ce from the scintillation material, and on ^{27}Al from the detector encapsulation. In a first approach the Geant4 simulation was performed time-integrated. As input to Geant4 we used prompt fission neutron spectra from the ENDF/B-VII.1 nuclear data library [20] keeping the same geometry as previously defined for the simulation of the γ response of our detectors. The simulated response is then subtracted from the time-integrated PFG spectrum and compared with the PFG spectrum as obtained within the prompt time window from -3 to 3 ns.

3 Experimental results and consequences for future evaluations

In Fig. 1 the emission spectra from the reaction ^{239}Pu(n_{th}, f) is shown in logarithmic scale for all three detectors together with the emission spectrum obtained from the sum of the three measured spectra. In the inset a zoom into the energy range below 800 keV is shown in linear scale. For the detector with serial No. 5415 a dip between 200 and 400 keV is visible, whose origin is not yet explained. However, this does not compromise the significance of the spectral characteristics. Our spectral data essentially confirms the PFGS characteristics of Verbinski et al. [23] and shows the deficiency of the spectral data from Ref. [24] at energies below 300 keV (see Fig. 2). For details we refer to Ref. [27].

Model calculations performed with the Monte-Carlo codes FIFRELIN [25] (green dashes) and CGMF [26] (black dots) are in very good agreement with our results.

With an average total prompt γ-ray energy released per fission, $E_{\gamma,tot}$, the result from Verbinski et al. [23] is confirmed within the uncertainty (1σ). This means that the observed excess in the γ heating does not originate from deficient PFGS data from thermal-neutron induced fission of ^{240}Pu*. All numerical details may be found in Table 1 of Ref. [27].

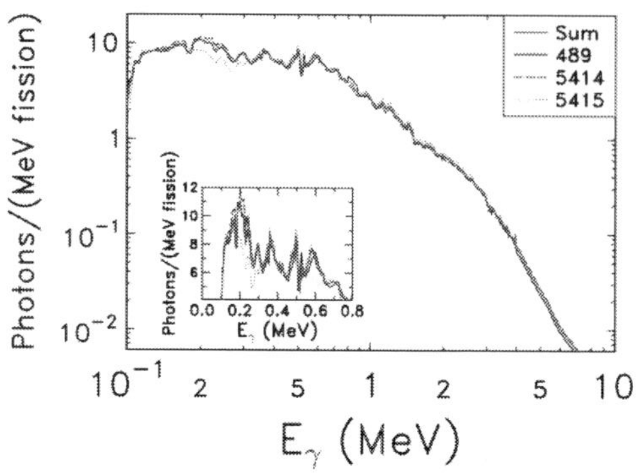

Figure 1. Prompt fission γ-ray spectra from thermal-neutron induced fission on ^{239}Pu shown in logarithmic scale, taken with three different 51 mm × 51 mm (diameter × length) cerium-doped (5%) $LaBr_3$ detectors (serial numbers given). Depicted are the individual emission spectra and the one drawn from the sum of the three measured spectra. The inset shows a zoom-in on the low-energy region in linear scale.

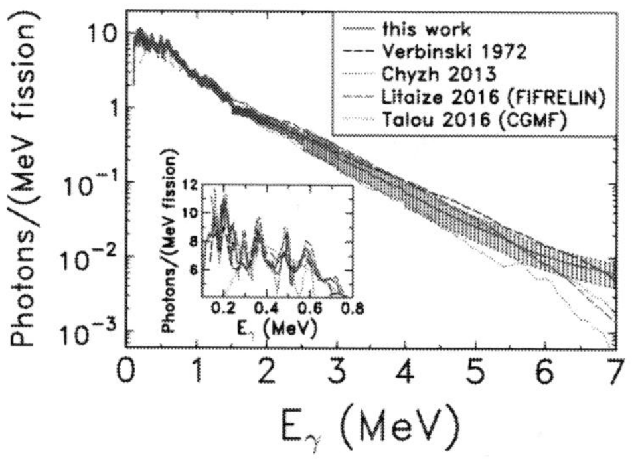

Figure 2. Prompt fission γ-ray spectra from thermal-neutron induced fission on ^{239}Pu (red line with error bars) compared with data from Verbinski et al. [23] and Chyzh et al. [24]. We also show recent calculations performed with the Monte-Carlo codes FIFRELIN [25] and CGMF [26]. The inset shows a zoom-in on the low-energy region in linear scale.

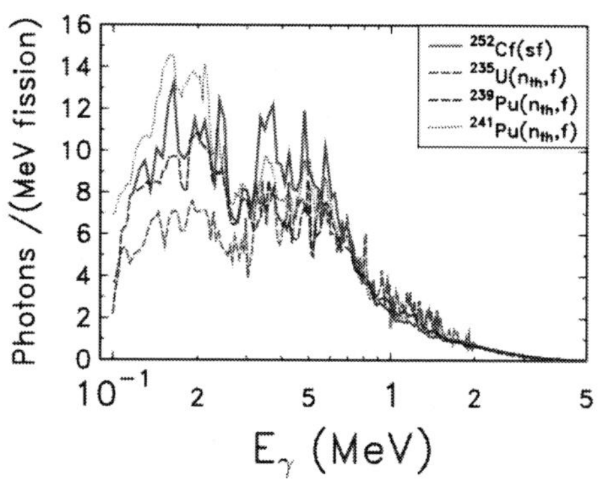

Figure 3. Summary of prompt fission γ-ray spectra for all fission reactions investigated by our collaboration (logarithmic energy scale). The spectral data for the reaction ^{235}U(n$_{th}$, f) are from this work. The changing yields at different γ energies reflect the variation in the pre-neutron fission fragment yield distributions.

The PFGS of our verification run on thermal-neutron induced fission on ^{235}U is shown in Fig. 3 as the green dashed line. Not only the photon yield below E_γ = 300 keV is here drastically reduced, also all depicted PFG spectra - taken from all fission reactions investigated by our collaboration - reflect consistently the variations in the corresponding pre-neutron fragment mass yield distributions confirming the validity of the present detector response matrix. The comparison of our spectra also teaches us that the PFG spectrum for the reaction ^{241}Pu(n$_{th}$, f), which was measured under the same conditions as the one for ^{235}U(n$_{th}$, f), needs to be unfolded with the new response matrix.

Last but not least, we present our first results on time-dependent emission of prompt fission γ-rays or *late-time emission*, here for the thermal-neutron induced fission on ^{239}Pu. As input for the

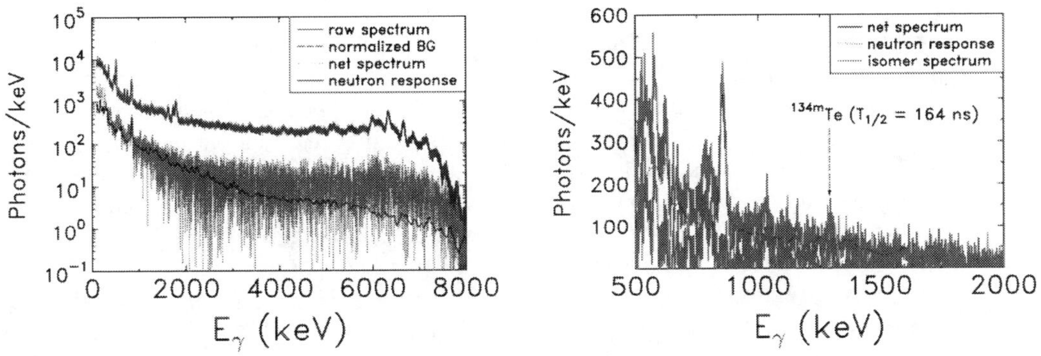

Figure 4. *Left part:* neutron response of a 51 mm × 51 mm (diameter × length) $LaBr_3$:Ce(5%) (black line); the dotted (red) line is the net spectrum obtained as difference between the raw and the normalized background (BG) spectrum; *right part:* part of the isomer spectrum integrated between 3 and 200 ns obtained as difference between the net spectrum (red) and the neutron response spectrum (black dotted line). The proof of principle is given by the γ transition from the 1st excited state to the ground-state, E_γ = 1279.11 keV, following the decay of the isomeric level in ^{134m}Te (E* = 1691.34 keV).

Table 2. Prompt fission γ-ray emission as a function of time relative to the instant of fission in the thermal-neutron induced fission of ^{240}Pu*; the data are for the detector with serial No. Q489. The cumulated multiplicity, $\overline{M}_\gamma$, is compared to calculations with the CGMF code [26].

time bin (ns)	$\overline{M}_\gamma$ (this work) (per fission)	$\overline{M}_\gamma$ [26] (per fission)
−3, 3	7.27 ± 0.11	7.10
−3, 240	7.66 ± 0.12	7.53

Geant4 simulations the prompt fission neutron spectrum from Ref. [20] was used. In the upper part of Fig. 4 the simulated detector response of our 51 mm × 51 mm $LaBr_3$:Ce detectors is shown (black line) together with the integral raw spectrum (green line), the normalized background spectrum (blue dashed line) and the background-corrected total prompt fission γ-ray spectrum. The transition from the 1st excited to the ground state in ^{27}Al is visible at E_γ = 843 keV.

In the lower part of Fig. 4 the net spectrum (red) is shown together with the neutron response (dotted black) and the difference of the two, i.e. the isomer spectrum (green). The figure shows a γ energy range, which includes the transition from the isomer level in the fission product ^{134m}Te, that has a half-life $T_{1/2}$ = 164 ns.

From this spectrum the isomeric, or late-time, contribution to the PFGS characteristics is deduced. First, tentative, numerical values for the cumulated PFG multiplicity as a function of time after fission are summarized in Table 2 and compared to results from model calculations [26]. Our data lies systematically higher, but a difference of 3% may, at the present stage, be considered as reasonable agreement.

4 Summary and outlook

During the last decade we conducted prompt fission γ-ray spectral measurements on a series of major and minor actinide samples, i.e. ^{235}U(n$_{th}$,f), ^{239,241}Pu(n$_{th}$,f) and ^{240,242}Pu(sf), respectively. Our measurement technique and data analysis was validated by several measurements of the spontaneous fission of ^{252}Cf samples employing targets with different fission rates and various lanthanide-halide detectors. We noticed that the determination of the detector response is crucial for obtaining the correct spectral shape at low γ-ray energies. First measurements on thermal-neutron induced fission of ^{236}U* were repeated under different experimental conditions. The new spectral data shows that the previous data needs to be reanalyzed using a more realistic response function.

With the PFGS data from neutron-induced fission of ^{240}Pu* we finalized the OECD/NEA HPRL requests for thermal energies. We do not find significant differences of PFGS characteristics compared to e.g. Verbinski's data. We may, therefore, conclude that the observed underestimation of the prompt γ heating is due to either fast-neutron induced fission or photo-fission induced by γ-rays from neutron capture in the construction material around the nuclear core.

We started to investigate so-called *late-time emission* of prompt fission γ-rays, i.e. isomeric decay in a time window adjacent to the prompt window, $\Delta t = \pm 3$ ns, extending up to 240 ns in case of ^{240}Pu*. A first global assessment of the neutron response of our LaBr$_3$:Ce(5%) detectors was done using Geant4. A comparison with the CGMF code shows global agreement.

In future we plan dedicated experiments in order to determine the time-dependent PFG emission in greater detail.

References

[1] Nuclear Data High Priority Request List of the NEA (Req. ID: H.3, H.4), `http://www.nea.fr/html/dbdata/hprl/hprlview.pl?ID=421` and `http://www.nea.fr/html/dbdata/hprl/hprlview.pl?ID=422`

[2] G. Rimpault, in Proceedings of the Workshop on Nuclear Data Needs for Generation IV, April 2005, edited by P. Rullhusen (World Scientific, Singapore, 2006), p. 46

[3] B. Becker, P. Talou, T. Kawano, Y. Danon, I. Stetcu, Phys. Rev. C **87**, 014617 (2013)

[4] O. Litaize, D. Regnier, O. Serot, Phys. Proc. **59**, 89 (2014)

[5] P. Talou, T. Kaweno, I. Stetcu, R. Vogt, J. Randrup, Nucl. Data Sheets **118**, 227 (2014)

[6] O. Serot, O. Litaize, D. Regnier, Phys. Proc. **59**, 132 (2014)

[7] I. Stetcu, P. Talou, T. Kawano, M. Jandel, Phys. Rev. C **90**, 024617 (2014)

[8] K.-H. Schmidt, B. Jurado, Ch. Amouroux, JEFF-Report 24, Data Bank, Nuclear Energy Agency, OECD (2015)

[9] R. Vogt, J. Randrup, Nucl. Data Sheets **118**, 220 (2014)

[10] N. Kornilov, F.-J. Hambsch, I. Fabry, S. Oberstedt, T. Belgya, Z. Kis, L. Szentmiklosi, S. Simakov, Nucl. Sci. Eng. **165**, 117 (2010)

[11] A. Göök, F.-J. Hambsch, S. Oberstedt, Eur. J. Phys. Web of Conf. **111**, 05001 (2016)

[12] R. Billnert, F.-J. Hambsch, A. Oberstedt, S. Oberstedt, Phys. Rev. C **87**, 024601 (2013)

[13] M. Lebois, J.N. Wilson, P. Halipré, A. Oberstedt, S. Oberstedt, P. Marini, C. Schmitt, S.J. Rose, S. Siem, M. Fallot, A. Porta, A.-A. Zakari, Phys. Rev. C **92**, 034618 (2015)

[14] A. Oberstedt, T. Belgya, R. Billnert, R. Borcea, T. Bryś, W. Geerts, A. Göök, F.-J. Hambsch, Z. Kis, T. Martinez, S. Oberstedt, L. Szentmiklosi, K., Takàcs, and M. Vidali, Phys. Rev. C. **87**, 051602(R) (2013)

[15] S. Oberstedt, A. Oberstedt, A. Gatera, A. Göök, F.-J. Hambsch, A. Moens, G. Sibbens, D. Vanleeuw, and M. Vidali, Phys. Rev. C **93**, 054603 (2016)
[16] S. Oberstedt, R. Billnert, F.-J. Hambsch, M. Lebois, A. Oberstedt, and J.N. Wilson, Eur. Phys. J. A **51**, 178 (2015)
[17] S. Oberstedt, R. Billnert, T. Belgya, T. Bryś, W. Geerts, C. Guerrero, F.-J. Hambsch, Z. Kis, A. Moens, A. Oberstedt, G. Sibbens, L. Szentmiklosi, D. Vanleeuw, and M. Vidali, Phys. Rev. C **90**, 024618 (2014)
[18] `http://spdevices.com/index.php/adq214`
[19] A. Oberstedt, R. Billnert, F.-J. Hambsch, and S. Oberstedt, Phys. Rev. C **92**, 014618 (2015)
[20] ENDF/B-VII.1 Evaluated Nuclear Data File, MF = 5, MT = 18 (2011), `http://www.nndc.bnl.gov/exfor/endf00.jsp`
[21] S. Agostinelli and the Geant4 collaboration, Nucl. Inst. Meth. A **506**, 250 (2003)
[22] PENELOPE11 Computer code, `http://www.oeced-nea.org/tools/abstract/detail/nea-1525`
[23] V.V. Verbinski, H.Weber, and R.E. Sund, Phys. Rev. C **7**, 1173 (1973)
[24] A. Chyzh, C.Y. Wu, E. Kwan, R.A. Henderson, J.M. Gostic, T.A. Bredeweg, R.C. Haight, A.C. Hayes-Sterbenz, M. Jandel, J.M. O'Donnell, and J.L. Ullman, Phys. Rev. C **85**, 021601 (2012)
[25] O. Litaize and O. Serot, FIFRELIN ver. 2016 (private communication)
[26] P. Talou, T. Kaweno, I. Stetcu, J.P. Lestone, E. McKigney, and M.B. Chadwick, Physical Review C **94**, 064613 (2016)
[27] A. Gatera, T. Belgya, W. Geerts, A. Göök, F.-J. Hambsch, M. Lebois, B. Maróti, A. Moens, A. Oberstedt, S. Oberstedt, F. Postelt, L. Qi, L. Szentmiklosi, G. Sibbens, D. Vanleeuw, M. Vidali, and F. Zeiser, Phys. Rev. C **95**, 064609 (2017)

Prompt fission neutron emission in the reaction ^{235}U(n,f)

Alf Göök*, *Franz-Josef* Hambsch, and *Stephan* Oberstedt

European Commission, Joint Research Centre, Directorate G - Nuclear Safety and Security, Unit G.2, Retieseweg 111, 2440 Geel, Belgium

Abstract. Experimental activities at JRC-Geel on prompt fission neutron (PFN) emission in response to OECD/NEA nuclear data requests are presented in this contribution. Specifically, on-going investigations of PFN emission from the reaction ^{235}U(n,f) in the region of the resolved resonances, taking place at the GELINA facility, are presented. The focus of this contribution lies on studies of PFN correlations with fission fragment properties. The experiment employs a scintillation detector array for neutron detection, while fission fragment properties are determined via the double kinetic energy technique using a position sensitive twin ionization chamber. This setup allows us to study several correlations between properties of neutron and fission fragments simultaneously. Results on PFN correlations with fission fragment properties from the present study differ significantly from earlier studies on this reaction, induced by thermal neutrons.

1 Introduction

Experimental investigations of prompt fission neutrons and fission fragment properties in resonance neutron induced fission on ^{235}U are taking place at the GELINA facility of the JRC-Geel. Improved knowledge of the properties of prompt fission neutrons (PFN), their multiplicities, as well as their energy and angular distributions are not only of interest for questions related to the neutron emission itself, but also to questions relevant to the formation of the fission fragments, the sharing of excitation energy among them and the time scale of the process. In recent years large efforts have been put into the modeling of PFN emission in fission [1]. The efforts are quite successfu in both consistency between the different approaches and reproducing available experimental data. However, in the case of ^{235}U(n,f) difficulties have been encountered [2], especially regarding the dependency of $\overline{\nu}$ on the fragment total kinetic energy (TKE). In Ref. [3] possible deficiencies in the experimental data were pointed out. For these reasons it is interesting to revisit PFN and fragment correlations in ^{235}U(n,f) experimentally.

2 Experimental setup

The experiment has been performed at the GELINA facility of the JRC-Geel site. The experimental setup, illustrated in Fig. 1, is located at a distance of 9.2 m from the GELINA neutron source. Fission fragments are detected in a position sensitive twin ionization chamber (IC). This detector has been

*e-mail: alf.gook@ec.europa.eu

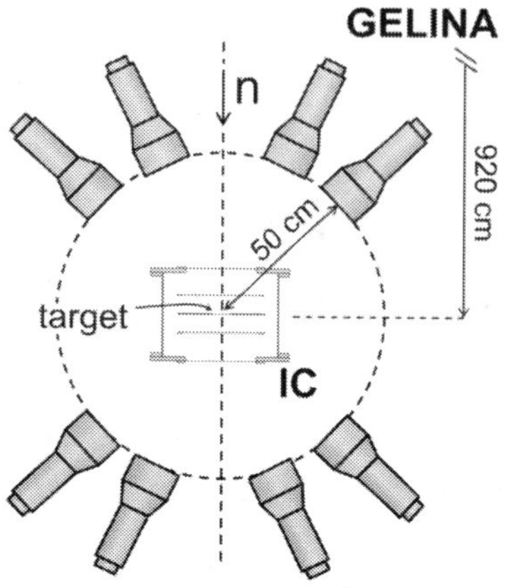

Figure 1. Schematic representation of the experimental setup located at the GELINA facility. The ^{235}U-target is placed inside the twin ionization chamber (IC), 9.2 m away from the neutron producing target. The neutron detectors are placed in a quasi spherical geometry around the IC at a distance of about 50 cm from the center of the ^{235}U-target.

developed at JRC-Geel, and is described in Ref. [4] in detail. Essentially, it is a standard twin Frisch grid ionization chamber with the anode plates replaced by position sensitive readouts. It is capable of determining the fission axis orientation in space with a resolution of 7°, simultaneously with the fission fragment masses and energies.

Neutrons are detected in an array of proton recoil scintillators. The array consist of 19 NE213 equivalent BC-501 liquid scintillators, 3 p-terphenyl and 1 stilbene crystal scintillators. The detectors are placed in a quasi-spherical geometry around the center of the ^{235}U-target at distance of about 50 cm. The actual position of the detectors relative to the IC has been determined with high precision using a ROMER arm[1] measuring device.

The data acquisition is based on wave-form digitizers, sampling the detector signals at 400 MSample/s with 14-bit resolution. A fission trigger derived from the charge induced on the central cathode of the IC, triggers the data acquisition. For every fission trigger digital wave-forms of all neutron detectors as well as the IC's electrodes are stored on disk, together with time-stamp information, for off-line analysis. The time stamp is reset by every beam pulse from the GELINA and used to determine the incident neutron time-of-flight (ToF).

3 Data analysis

3.1 PFN data

Events in the scintillation detectors corresponding to neutrons are selected by means of pulse-shape discrimination (PSD). The PFN energy is measured by means of the ToF technique, with a resolution of ~1 ns (FWHM). In addition to PFNs, the fission is also accompanied by prompt fission γ-rays, which might be detected in the scintillation detectors as well. Most of the γ-emission takes place within a few ns after the instant of fission. Therefore, the range of high-energy neutrons is most sensitive to false events induced by γ-rays. The PSD is not able to give a clean discrimination between γ-rays and neutron events for pulse heights corresponding to a proton recoil energy smaller than ~1 MeV. In order to reduce the background caused by the γ-emission a dynamic light-threshold [5] is applied.

Each detector in the SCINTIA array has been characterized using the PFNs from ^{252}Cf(sf), with experimental setup and procedures described in Ref. [6]. To correct for the energy dependent detection

[1]`http://www.hexagonmi.com/products/portable-measuring-arms/romer-absolute-arm`

efficiency and multiple scattering of neutrons on the surrounding materials, the setup is modeled with GEANT4. The simulation uses experimentally determined proton light output functions, for other recoil-particle species literature data are used [7, 8]. The Monte-Carlo model has been validated against the standard PFN spectrum [9] from the spontaneous fission decay of ^{252}Cf.

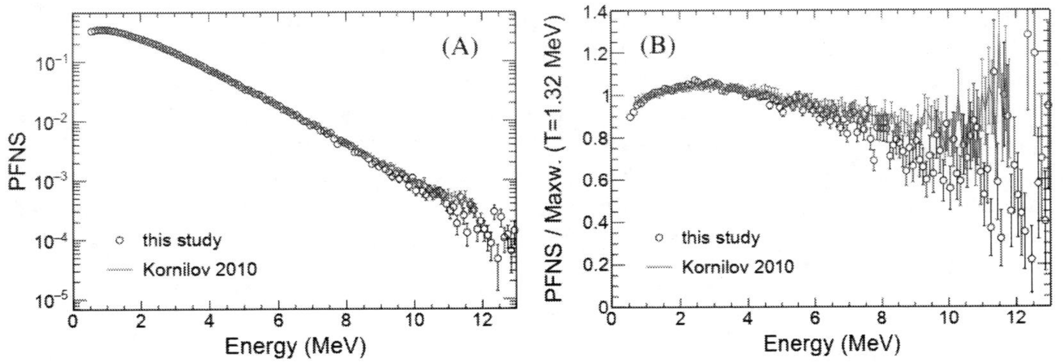

Figure 2. (A) The prompt fission neutron spectrum in the laboratory frame compared to data from cold-neutron induced fission from Kornilov et al. [10]. (B) The prompt fission neutron spectrum represented as a ratio to a Maxellian with a temperature of 1.32 MeV.

In Fig. 2 the ^{235}U(n,f) PFN spectrum observed in this study, when selecting the incident neutron energy range [0.3 eV, 45 keV], is compared to the spectrum from cold neutron induced fission determined by Kornilov et al. [10]. The two spectra are in quite close agreement, although the spectrum observed in the resonance region is slightly softer.

3.2 Determination of fission fragment properties

The fission fragment mass and kinetic energy before neutron emission is obtained by means of the well-established double kinetic energy (2E) technique, which relates the masses $m^*_{1,2}$ and energies $E^*_{1,2}$ before neutron emission in a binary fission event according to

$$m^*_{1,2} = m_{cn} \cdot \frac{E^*_{2,1}}{E^*_1 + E^*_2}, \tag{1}$$

where m_{cn} is the mass of the compound nucleus undergoing fission. Under the assumption of isotropic neutron emission from fully accelerated fragments, the energies before neutron emission E^* are related to the energies after neutron emission E according to

$$E^* = E \cdot \frac{m^*}{m^* - \overline{\nu}(m^*, \mathrm{TKE})}, \tag{2}$$

where $\overline{\nu}$ is the number of neutrons emitted by the fragment. The dependence of $\overline{\nu}$ on mass and TKE can only be derived from the data once the 2E-analysis is completed. As initial assumption we have used the evaluated data on $\overline{\nu}(m^*)$ from Wahl [11] and the parameterization

$$\overline{\nu}(m^*, \mathrm{TKE}) = \overline{\nu}(m^*) + \frac{\overline{\nu}(m^*)}{\overline{\nu}(m^*) + \overline{\nu}(m_{cn} - m^*)} \cdot \Delta_{\mathrm{TKE}},$$

$$\Delta_{\mathrm{TKE}} = \frac{\overline{\mathrm{TKE}}(m^*) - \mathrm{TKE}}{E_{\mathrm{sep}}}, \tag{3}$$

where $E_{\mathrm{sep}} = 8.6$ MeV/n is the average energy necessary to emit a neutron [12]. The analysis was later repeated using the results on $\overline{\nu}(m^*)$ and $E_{\mathrm{sep}} = 8.51$ MeV/n derived from the data. No significant changes in the results were observed between the two analyses. In the 2E-technique the main contribution to the mass resolution is the neutron evaporation, since Eq. (2) only holds on average. In this work the mass resolution is 4–5 u (FWHM), determined by comparing the measured thermal mass yield to data from Ref. [13]. This mass resolution is close to what can be expected considering only the contribution from the neutron evaporation [14]. For events where neutrons and fission fragments are detected in coincidence, an additional correction [15] for the recoil energy imparted to the fragment is added to Eq. (2).

4 Experimental results

All results on correlations between fission fragment properties and PFN properties presented here are from the selected incident neutron energy range [0.3 eV, 45 keV]. Some of the results, presented in this section and originating from a subset of the data from the present experiment, has been reported upon earlier [16]. The main difference between results reported by us in Ref. [16] and reported here is, aside from the improved statistical accuracy, that the absolute neutron detection efficiency was re-evaluated. This effectively raised the mean total neutron multiplicity from 2.3 to 2.4 neutrons/fission.

4.1 Prompt fission neutron angular distributions in the laboratory frame

Most model calculations of the properties of PFNs are based on the assumption that the neutrons are emitted from the fragments after they have reached their terminal velocity. However, theoretical arguments have been raised asserting that at least a fraction of the neutrons are emitted during the scission process [17] and/or during the acceleration of the fragments [18, 19]. A large number of experiments have been devoted to investigate the source of prompt neutrons in detail. However, considering the variation of experimental results it is difficult to draw definitive conclusions. The experimental method used here, as well as in many earlier studies [20–22] of PFN emission in ^{235}U(n,f), assumes that the neutrons are emitted from the fully accelerated fragment. Hence, it is of great importance to investigate to what extent the experimental data support this assumption. In order to do so, we follow a procedure similar to that of Vorobyev et al. [23]. The energy–angle distribution of neutrons emitted from moving fragments when observed in the laboratory frame is

$$N(E_n, \theta) = N_L(\eta_L, \vartheta_L)\sqrt{\frac{E_n}{\eta_L}} + N_H(\eta_H, \vartheta_H)\sqrt{\frac{E_n}{\eta_H}}, \tag{4}$$

where $N_{L,H}(\eta_{L,H}, \vartheta_{L,H})$ are the center-of-mass (c.m.) energy–angle distributions of neutrons from the light and heavy fragments, respectively. The neutron energies in the c.m. frame $\eta_{L,H}$ can be derived from experimental observables in the laboratory frame; light (L) and heavy (H) fragment energies

$E_{L,H}$ and masses $m_{L,H}$, neutron energy E_n and angle θ between the neutron's and the light fragment's directions of motion

$$\eta_L = E_n + E_L \frac{m}{m_L} - 2\sqrt{E_n E_L \frac{m}{m_L}} \cos\theta, \tag{5}$$

$$\eta_H = E_n + E_H \frac{m}{m_H} + 2\sqrt{E_n E_H \frac{m}{m_H}} \cos\theta, \tag{6}$$

where m is the mass of the neutron. For the purpose of the model a single fragmentation is used,

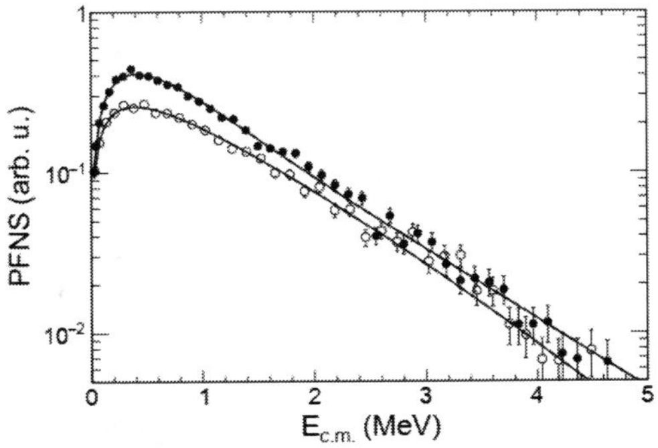

Figure 3. The c.m. energy spectra of prompt neutrons derived for the light (•) and heavy (○) fragment groups.

with $E_L \frac{m}{m_L}$ = (1.02 ± 0.01) MeV and $E_H \frac{m}{m_H}$ = (0.491 ± 0.01) MeV taken as averages from the experimental data. Under the assumption of isotropic emission from the fully accelerated fragments, only the neutron spectra in the c.m. frame are unknown. They are determined from the experimental data by selecting small angles in the laboratory frame, where the contribution of neutrons from the complementary fragment is the smallest. Neutrons detected at angles smaller than 12° relative to the motion of the light and heavy fission fragments are selected. The observed laboratory spectra for these selections are transformed into the c.m. frame of the respective fragment and corrected for the small contribution of neutrons from the complementary fragment, the result is displayed in Fig. 3. Each of the spectra has been fitted with a linear superposition of a Maxwellian and a Watt spectrum shape, represented by the full black lines. The fitted shapes are then used to calculate the spectrum as a function of the angle between the light fission fragment direction and the detected neutron, according to Eq. (4). The result of the calculation is presented as a contour plot in Fig. 4. As can be seen, the contour lines of the model do follow the contour lines of the experimental data quite well. In Fig. 5 the zeroth (A) and the first (B) moments of the calculated and experimentally observed spectra as a function of $\cos\theta$ are compared. The agreement is fair, although a slight underestimation of the experimental data is evident. The underestimation amounts to 2.5% of the total neutron yield. The description of the experimental data could be improved by assuming a small anisotropy in the c.m. frame [23]. However, we did not include an anisotropy term in the evaluation of the laboratory data as the angular distribution observed in the c.m. frame does not allow for this degree of freedom, see Sect. 4.2.3.

4.2 Experimental results in the fission fragment rest-frame

The complete experimental determination of all relevant kinematic parameters allows the transformation from the laboratory frame of reference into the rest frame of the fully accelerated fission fragments. This transformation is, however, complicated by the fact that for each laboratory angle only the

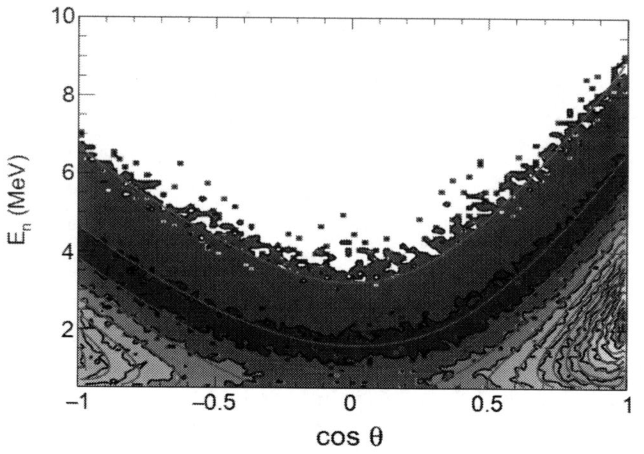

Figure 4. The laboratory energy-angle distribution of prompt neutrons. The experimental data are represented by a contour-plot with black contour lines on top of a color-coded scatter-plot. The red contour lines represent the model calculation according to Eq. (4) with adjusted parameters, see text for details.

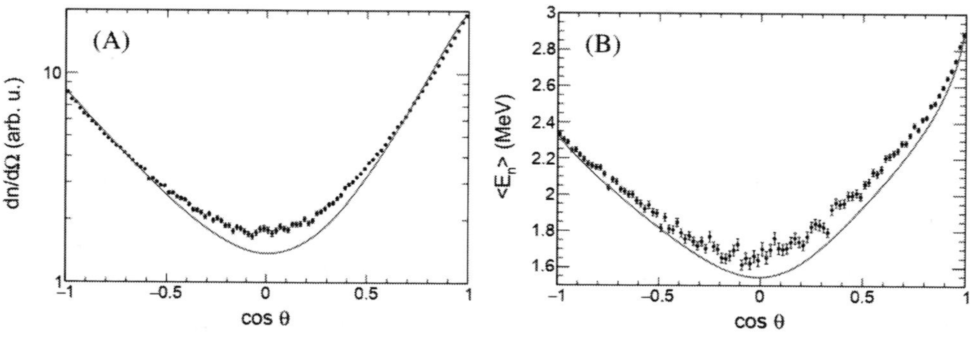

Figure 5. (A) Projection of the data in Fig. 4 onto cos θ. (B) The Average neutron energy as a function of cos θ. The red lines represent the result of the model calculation.

sum of contributions from the two fission fragments is observed. Thus, an unknown contribution must first be subtracted from any observed distribution. Fortunately, the contribution of neutrons emitted from the fragment flying away from the neutron detector is small. Therefore, in a first approximation it is assumed that all neutrons that are detected with a c.m. angle smaller than 90°originate from the fragment directed towards the neutron detector. The resulting distribution is then used to calculate the disturbing component from the complementary fragment. The magnitude of the correction as a function of fission fragment mass is shown in Fig. 6.

4.2.1 Angular distribution

Using the procedures discussed above the angular distribution of prompt neutrons in the rest frame of the fully accelerated fragments has been determined. Figure 7 shows the angular distribution integrated over all c.m. neutron energies and all fission configuration, i.e. no selection in mass or TKE has been made. The result has been fitted with a second order Legendre polynomial. Obviously, the result supports isotropic emission in the c.m.-frame. This does, however, not exclude emission from non

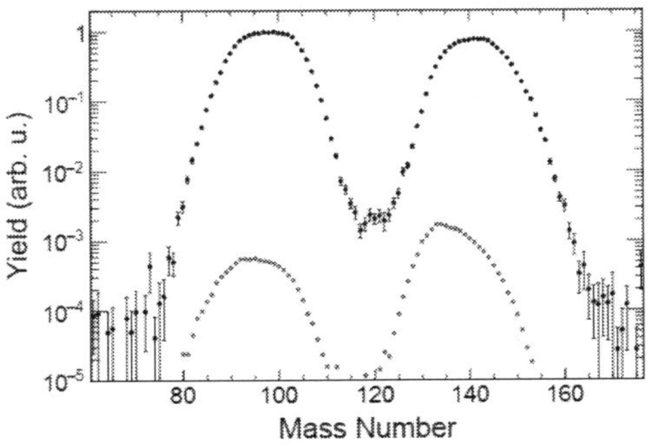

Figure 6. Fission neutron yield for $\theta_{c.m.} > 90°$ as a function of the fission fragment mass. The red points represent the calculated contribution of neutrons from the complementary fission fragment.

fully accelerated fragments due to the selection of events with $\theta_{c.m.} > 90°$. The selection implies that events with laboratory neutron energies smaller than the fragment energy per nucleon are not taken into account.

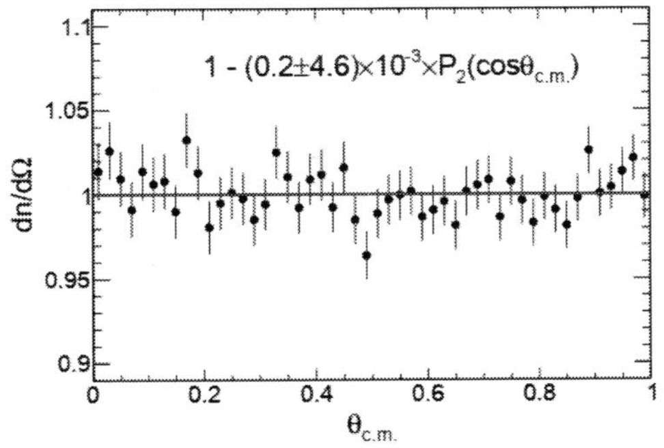

Figure 7. The angular distribution of prompt neutrons in the c.m.frame integrated over all neutron energies and fission configurations.

4.2.2 Average neutron multiplicity

The average neutron multiplicity as a function of the fragment mass is shown in Fig. 8 (A). For comparison, experimental data from Refs. [20, 21] and the evaluation from Ref. [11] are also shown in the figure. The general shape is reproduced in this work, however the minima around mass number ~80 u for the light fragments and ~130 u for the heavy fragments appear more pronounced in the present data. The average neutron multiplicity per fission is shown as a function of the heavy fragment mass in Fig. 8 (B). A pronounced minimum close to heavy fragment mass A_H ~132 u observed. The shape of the present data agrees very well with the data from Maslin et al. [22].

In Fig. 9 the average neutron multiplicity per fission is plotted as a function of the fragment TKE. As expected, a close to linear dependence is observed, except for TKE≲145 MeV. A least square fit, indicated by the full black line in Fig. 9(A), results in $-\partial \mathrm{TKE} / \partial \bar{\nu}$ = 12.0 MeV/n. This value is substantially lower than the values 16.7 – 18.5 MeV/n determined in earlier studies, performed at thermal incident neutron energies [20–22, 24]. Data from Refs. [20–22, 24] are also included in Fig. 9(A). By studying the fission fragment TKE distributions from Refs. [20–22] it has been found that these studies suffer from substantial resolution broadening. The broadening also affects the correlation with detected neutrons, which explains the strong difference between the present data and

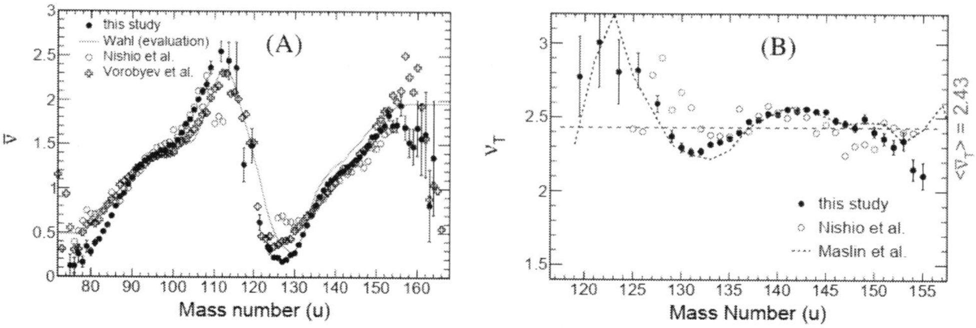

Figure 8. (A) The average neutron multiplicity per fragment as a function of the fragment mass from this study compared to data from Refs. [20, 21] and the evaluation by Wahl [11]. (B) The average neutron multiplicity per fission as a function of the heavy fragment mass from this study compared to data from Refs. [20, 22]. The dashed red line indicates the mean multiplicity per fission.

the literature data. This conclusion was already drawn form a subset of the present data, and a more detailed discussion of this has been published in Ref. [16].

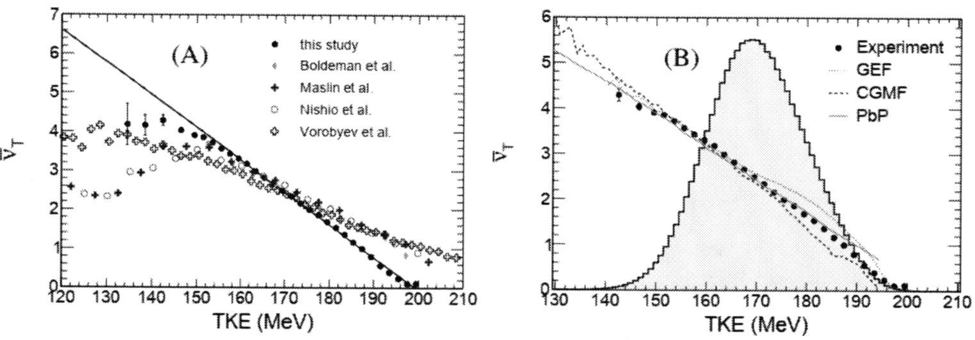

Figure 9. The average neutron multiplicity per fission as a function of the TKE. (A) The result from this study is compared to experimental data from Refs.[20–22, 24]. The black line represent a straight line with an inverse slope of 12.0 MeV/n, obtained from a least square fit of the data from this study. (B) The result from this study is compared to model calculations [1, 25]. The fission yield as a function of TKE determined in this study is also shown.

A comparison of the experimental result for average neutron multiplicity per fission $\overline{\nu}_T$(TKE) with different model calculations, taken from a recent review of the available fission fragment deexcitation models [1, 25], is shown in Fig. 9(B). The different models presented in Ref. [1] agree well with each other. Therefore, not all of them are included in the figure. It is clear that the model calculations are in much better agreement with the present data set than with earlier experiment.

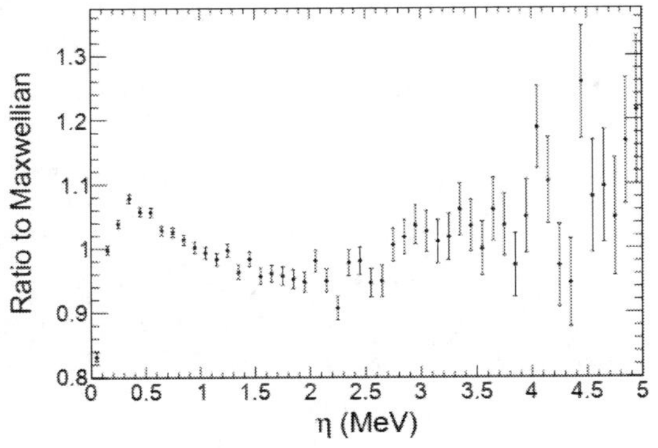

Figure 10. The prompt neutron spectrum in the c.m. frame, integrated over all fragmentations, presented as a ratio to a Maxwellian with temperature 0.826 MeV.

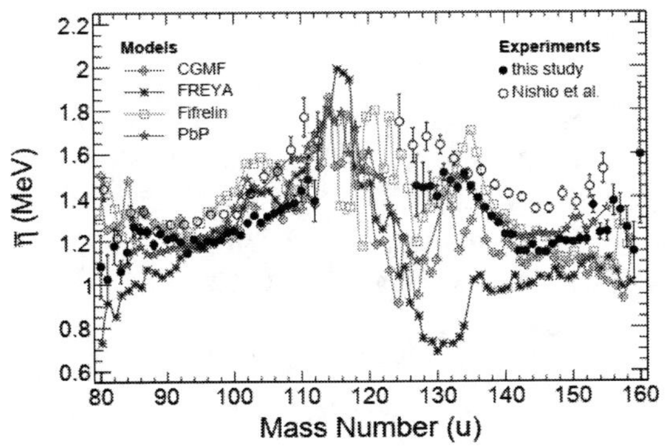

Figure 11. The average c.m. neutron energy $\overline{\eta}$ as a function of the fission fragment mass, results from this experiment are compared to model predictions from Refs. [1] and experimental data from Ref. [20].

4.2.3 Prompt neutron energy

The neutron energy in the c.m. frame η has also been evaluated on an event-by-event basis. The integral spectrum is presented as a ratio to a Maxwellian with a fitted temperature of 0.826 MeV in Fig. 10. The average neutron energy as a function of the fission fragment mass is compared to theoretical calculations and an earlier experiment in Fig. 11. The average values for the light- and heavy fragment groups are 1.23 MeV and 1.26 MeV, respectively. The average c.m. neutron energy is an important validation parameter for neutron emission models as it relates to both the excitation energy partition between the fragments as well as the distribution of temperatures in the sequential emission of neutrons. The quantity is also important for application as it has a direct influence on the laboratory neutron spectrum. It is evident from Fig. 11 that $\overline{\eta}(A)$ is almost symmetric with respect to symmetric mass division, in contradiction to $\overline{\nu}(A)$, which shows the well know sawtooth shape. This is very similar to earlier findings for the prompt fission neutrons from ^{252}Cf(sf) [26–28]. The different model predictions are in general in good agreement with the experimental data. The FREYA model does however predict a pronounced sawtooth shape which is not observed in the experiment. All models predict a more or less pronounced minimum in the region below A=130, while the experimental data shows only a flattening of the trend when approaching this region from heavier masses. Considering the strong slope of the mass yield curve and the experimental mass resolution it is expected that sharp changes in the properties of neutrons as a function of mass are washed out.

5 Conclusion

A multi-parameter experiment on correlations between the properties of fission fragments and PFNs in the reaction ^{235}U(n,f) for an incident neutron energy range [0.3 eV,45 keV] has been presented. The laboratory angular distribution of prompt neutrons have been compared with a model calculation based on isotropic emission from fully accelerated fragments. The model agrees fairly well with the experimental data, however the model calculation does underestimate the neutron yield at large angles with the fission axis. The underestimation amounts to 2.5% of the total neutron yield. Results on average neutron multiplicities in correlation with fission fragment mass and TKE show significant differences to earlier studies on this reaction, induced by thermal neutrons. The sawtooth shape of the average neutron multiplicity per fragment show more pronounced minima at $A \approx 130$ and $A \approx 80$. The TKE dependence of the neutron multiplicity per fission shows an inverse slope $\partial \mathrm{TKE} / \partial \overline{\nu}$ approximately 35% weaker than observed in earlier studies [20–22, 24]. The difference can be attributed to improved fission fragment TKE resolution in the present experiment. The present result for the average neutron multiplicity as a function of TKE is in good agreement with model calculations [1, 25].

References

[1] R. Capote, Y.-J. Chen, F.-J. Hambsch, N.V. Kornilov, J.P. Lestone, O. Litaize, B. Morillon, D. Neudecker, S. Oberstedt, T. Ohsawa, N. Otuka, V.G. Pronyaev, A. Saxena, O. Serot, O.A. Shcherbakov, N.-C. Shu, D.L. Smith, P. Talou, A. Trkov, A.C. Tudora, R. Vogt, A.S. Vorobyev, Nucl. Data Sheets **131**, 1 (2016)

[2] S. Lemaire, P. Talou, T. Kawano, M. Chadwick, D. Madland, Phys. Rev. C **72** (2005), cited By 71

[3] N.V. Kornilov, F.-J. Hambsch, A.S. Vorobyev, Nucl. Phys. A **789**, 55 (2007)

[4] A. Göök, W. Geerts, F.-J. Hambsch, S. Oberstedt, M. Vidali, S. Zeynalov, Nucl. Instr. and Meth. A **830**, 366 (2016)

[5] N.V. Kornilov, Tech. rep., IAEA, INDC(USA)-108 (2015)

[6] N.V. Kornilov, I. Fabry, S. Oberstedt, F.-J. Hambsch, Nucl. Instr. and Meth. A **599**, 226 (2009)

[7] V.V. Verbinski, W.R. Burrus, T.A. Love, W. Zobel, N.W. Hill, R. Textor, Nucl. Instr. and Meth. **65**, 8 (1968)

[8] M. Tajik, N. Ghal-Eh, G.R. Etaati, H. Afarideh, Nucl. Instr. and Meth. A **704**, 104 (2013)

[9] W. Mannhart, *IAEA-TECDOC-410* (Leningrad, 1986), p. 158

[10] N.V. Kornilov, F.-J. Hambsch, I. Fabry, S. Oberstedt, T. Belgya, Z. Kis, L. Szentmiklosi, S. Simakov., Nucl. Sci. and Eng. **165**, 117 (2010)

[11] C. Wahl, Atomic Data and Nuclear Data Tables **39**, 56 (1988)

[12] H. Nifenecker, C. Signarbieux, R. Babinet, J. Poitou, 3rd IAEA Symp. on the Physics and Chemistry of Fission **2**, 117 (1973)

[13] P. Geltenbort, F. Gönnenwein, A. Oed, Radiation Effects **93**, 57 (1986)

[14] J. Terrell, Phys. Rev. **127**, 880 (1962)

[15] A. Gavron, Nucl. Instr. and Meth. **115**, 99 (1974)

[16] A. Göök, F.-J. Hambsch, S. Oberstedt, EPJ Web of Conf. ND 2016 Int. Conf. on Nucl. Data for Sci. and Tech. (2016)

[17] M. Rizea, N. Carjan, Nucl. Phys. A **909**, 50 (2013)

[18] V.P. Eismont, Jour. Nucl. Energy A/B **20**, 875 (1965)

[19] V.P. Eismont, Atomnya Energiya **19**, 113 (1965)

[20] K. Nishio, Y. Nakagome, H. Yamamoto, I. Kimura, Nucl. Phys. A **632**, 540 (1998)
[21] A.S. Vorobyev, O. Shcherbakov, A. Gagarski, G. Val'ski, G. Petrov, EPJ Web of Conf. **8**, 03004 (2010)
[22] E.E. Maslin, A.L. Rodgers, W.G.F. Core, Phys. Rev. **164**, 1520 (1967)
[23] A.S. Vorobyev, O.A. Shcherbakov, Yu.S. Pleva, A.M Gagarski, G.V. Val'ski, G.A. Petrov, V.I. Petrova, T.A. Zavarukhina, Nucl. Instr. and Meth. A **598**, 795 (2009)
[24] J.W. Boldeman, A.R. de L. Musgrove, R. L. Walsh, Aust. J. Phys. **24**, 821 (1971)
[25] K.H. Schmidt, B. Jurado, C. Amouroux, C. Schmitt, Nucl. Data Sheets **131**, 107 (2016)
[26] A. Göök, F.-J. Hambsch, M. Vidali, Phys. Rev. C **90**, 064611 (2014)
[27] C. Budtz-Jørgensen, H.H. Knitter, Nucl. Phys. A **490**, 307 (1988)
[28] H.R. Bowman, S.G. Thompson, J.C.D. Milton, W.J. Swiatecki, Phys. Rev. **126**, 2120 (1962)

The 4-dimensional Langevin approach to low energy nuclear fission

F.A. Ivanyuk[1,2,*], C. Ishizuka[2], M.D. Usang[2], and S. Chiba[2]

[1] *Institute for Nuclear Research, Kiev, Ukraine*
[2] *Institute of Innovative Research, Tokyo Institute of Technology, Tokyo, Japan*

Abstract. We applied the four-dimensional Langevin approach to the description of fission of ^{235}U by neutrons and calculated the dependence of the excitation energy of fission fragments on their mass number. For this we have fitted the compact just-before-scission configuration obtained by the Langevin calculations by the two separated fragments and calculated the intrinsic excitation and the deformation energy of each fragment accurately taking into account the shell and pairing effects and their dependence on the temperature and mass of the fragments. For the sharing of energy between the fission fragments we have used the simplest and most reliable assumption - the temperature of each fragment immediately after the neck rupture is the same as the temperature of mother nucleus just before scission. The calculated excitation energy of fission fragments clearly demonstrates the saw-tooth structure in the dependence on fragment mass number.

1 Introduction

One of the most successful approaches to the description of fusion-fission reactions and the fission process is the approach based on the Langevin equations for the shape degrees of freedom. In this approach instead of co-ordinates of individual nucleons one introduces the parametrization of the shape of nuclear surface and solves the dynamical equations for the time evolution of deformation parameters and conjugated momenta. The internal state of the system in such calculations is assumed not to deviate much from the thermal equilibrium. Up to now the shape of fissioning system was described mainly by three deformation parameters that specify the main fission degrees of freedom - the total elongation, the mass asymmetry and the neck radius of the fissioning system. In some works the fourth degree of freedom was introduced for the axial asymmetry or for the orientation of approaching ions. Meanwhile, it was pointed out a long ago by V. Pashkevich [1] that the deepest valley on the deformation energy surface of heavy nuclei corresponds to the configuration when one "fragment" is almost spherical, and another - very elongated. Such configuration corresponds to very low energy due to the large shell correction of almost spherical part (double magic ^{132}Sn) and relatively small Coulomb interaction between spherical and elongated parts. Such configuration is impossible to describe within the three-parametric shape parametrization.

That is why recently [2] we took into account one more deformation parameter. Namely, we describe the shape of fissioning system by the two-center shell model (TCSM) parametrization in which

*e-mail: ivanyuk@kinr.kiev.ua

the deformation parameters δ_1 and δ_2 are considered as independent dynamical variables. In addition, the mean field of Nilsson type was replaced by the Woods-Saxon potential that better reproduces the single-particle energies of heavy nuclei, especially at large deformations. Within this approach we have managed to reproduce rather accurately the mass and total kinetic energy distribution of fission fragments. We have also found a very clear correlation between the neutron multiplicity and the deformation of fission fragment just before scission.

The existing descriptions of excitation energy immediately after scission [3–6] are based on very different assumptions and approximations that even contradict to each other. So, the reason of the saw-tooth structure of the excitation energy and neutron multiplicity at present is not clear. In present work we use the results of the Langevin calculations in order to examine the dependence of fragments excitation energy on the its mass number and, thus, clarify the reason for the saw-tooth structure of neutron multiplicity.

2 The Langevin approach

We describe the fission process within *four-dimensional* Langevin approach [2, 7]. I.e., we solve the system of equations for time-dependent deformation parameters q_μ and conjugated momenta p_μ,

$$\frac{dq_\mu}{dt} = \left(m^{-1}\right)_{\mu\nu} p_\nu, \quad \frac{dp_\mu}{dt} = -\frac{\partial V(q,T)}{\partial q_\mu} - \frac{1}{2}\frac{\partial}{\partial q_\mu}\left(m^{-1}\right)_{\nu\sigma} p_\nu p_\sigma - \gamma_{\mu\nu}\left(m^{-1}\right)_{\nu\sigma} p_\sigma + g_{\mu\nu}R_\nu(t), \tag{1}$$

where the collective coordinates, $\{q_\mu\} = \{z_0/R_0, \delta_1, \delta_2, \alpha\}$ fix the shape of a fissioning nucleus within the two-center shell model shape parametrization [8]. Here z_0 is the elongation parameter of the nucleus, R_0 the radius of spherical compound nucleus, δ_1 and δ_2 are related to the deformation of fragments, and α is the mass asymmetry. The neck parameter ϵ is kept constant, $\epsilon = 0.35$. The detailed description of the two-center shell model shape parametrization can be found in [9, 10].

For the tensors of mass and friction in four-dimensional calculations we used the macroscopic quantities - Werner-Wheeler method [11] for the mass tensor $m_{\mu\nu}$ and reduced ($k_s = 0.27$) wall-and-window formula [12, 13] for the friction tensor $\gamma_{\mu\nu}$. The Langevin random force $g_{\mu\nu}R_\nu(t)$ is the product of white noise $R_\nu(t)$ and the temperature dependent strength factors $g_{\mu\nu}$. The factors $g_{\mu\nu}$ are related to the temperature and friction tensor via the modified Einstein relation [14],

$$\sum_\sigma g_{\mu\sigma} g_{\sigma\nu} = T^* \gamma_{\mu\nu}, \qquad \text{with } T^* = \frac{\hbar\varpi}{2}\coth\frac{\hbar\varpi}{2T}, \qquad \text{and } \hbar\varpi = 2 \text{ MeV}. \tag{2}$$

The temperature T in this context is related to total excitation energy $E_x^{(tot)}$ and intrinsic energy E_{int},

$$E_{int} = E_x^{(tot)} - \frac{1}{2}\left(m^{-1}\right)_{\mu\nu} p_\mu p_\nu - V(q, T = 0) = aT^2, \tag{3}$$

where a is the level density parameter.

The potential, $V(q,T)$ used here was calculated with the deformed Woods-Saxon potential by the code of Pashkevich [1]. The shell corrections δE_{shell} and δE_{pair} were calculated for $T = 0$. For the temperature dependence of the shell and pairing effects we adopt the approximation suggested in [15].

The Langevin calculations start at either the potential minimum (the ground state shape) or at the second minimum. As long as the results presented in this work are concerned, both choices gave identical results. Initially, the momentum p_μ's were set to be zero, and Langevin motions are initiated by the conservative and random forces. Such calculations are continued until the trajectories reach the scission points, which are defined as points where the neck radius becomes zero. Calculations were repeated typically from 10^5 to 10^6 times, the final phase-space data were stored on a disk, and various quantities were calculated after all the Langevin calculations were finished.

3 The deformation and the intrinsic excitation energy

One of the quantities of interest for this work is the average value of the deformation parameter δ_1 (or δ_2). An example of dependence of the average value of fragment deformation $\langle\delta\rangle$ on the mass asymmetry is shown in left part of Fig. 1. Please, note that within the TCSM shape parametrization the parameters δ_1 and δ_2 fix the deformation of the outer parts of "fragments" only, from the tips of nucleus to the thickest place. The deformation of the inner part, from the thickest place till the neck, is regulated by all 5 deformation parameters. At the scission point the nucleus is very elongated and the inner part is elongated too. In order to make the fragment as a whole more spherical, the outer part should be somewhat oblate, see the right part of Fig. 1. Thus, negative δ within TCSM shape parametrization does not mean that the whole fragment is oblate.

For the comparison we show also in the left part of Fig. 1 the neutron multiplicities $\overline{\nu}$ from works [16, 17]. The average value $\langle\delta\rangle$ was shifted up by 0.25 in order to see better the similarity between $\overline{\nu}$ and $\langle\delta\rangle$. Indeed, one can see a surprising correlation between $\langle\delta\rangle$ and $\overline{\nu}$. Such similarity was noticed already in [18]. However, the similarity between the fragment deformation and $\overline{\nu}$ does not explain much. It is the fragment excitation energy that should be compared with $\overline{\nu}$, not the fragment's deformation. In order to calculate the fragment excitation energy the information on its shape is not sufficient. One should know also the sharing of excitation energy between fission fragments.

To clarify this question we calculate in present work the excitation energy of *each* fragment separately. This is not so trivial since from the Langevin calculations we can get only the excitation energy of the *whole* nucleus just before the scission. Very often it is assumed that the excitation energy is shared between fragments proportionally (or inversely proportionally) to their masses. Here we will not use none of such assumption but will calculate the excitation energy of each fragment directly, assuming that the temperature of both fission fragments immediately after scission is exactly the same as the temperature of mother nuc leus just before scission.

In order to do this we fit the compact shape (with zero or finite neck) just before scission by two separate fragments. An example of such fit is shown in the right part of Fig. 1. As one can see from the figure, the fit can be done quite accurately.

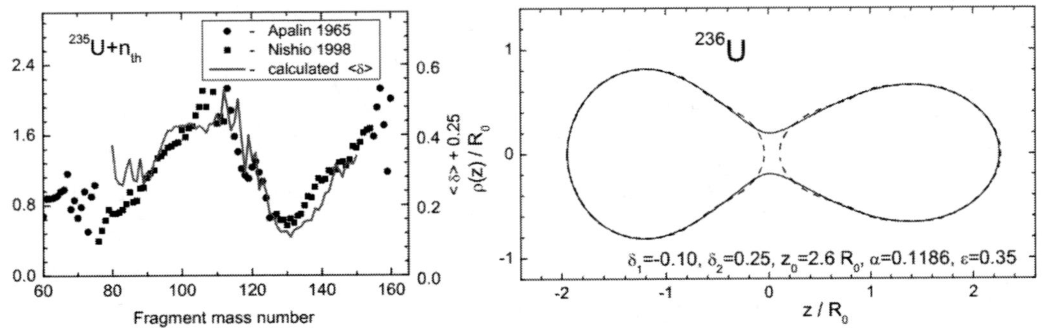

Figure 1. *Left*: the average value $\langle\delta\rangle$ of the fragment deformation (right scale) and the average neutron multiplicities $\overline{\nu}$ [16, 17] as function of mass asymmetry (fragment mass number). *Right*: an example of the shape of fissioning system just before scission and its fit by the two separate fragments.

To check the accuracy of the fit we compare the liquid drop energies and the shell corrections calculated for the compact shape just before scission and for the fit by two separate fragments. As one can see from Fig. 2, both the liquid drop energy and the shell corrections calculated for the compact shape and the two separated fragments are very close to each another. This means that the replacement of the shape just before scission by separate fragments is quite accurate.

In case of compact shape the number of particles to the right or the left from the neck ($z = 0$) is not necessarily an integer number. After the fit the number of particles in each fragment must be integer. Thus, the Z/N ratio in the compact shape and fragments is slightly different. This explains the wiggles of red curve in left part of Fig. 2. Due to the same reason the fragments may not be even-even nuclei (the mother nucleus ^{236}U is even-even) and the odd-even effects are seen in the shell corrections of fitted fragments.

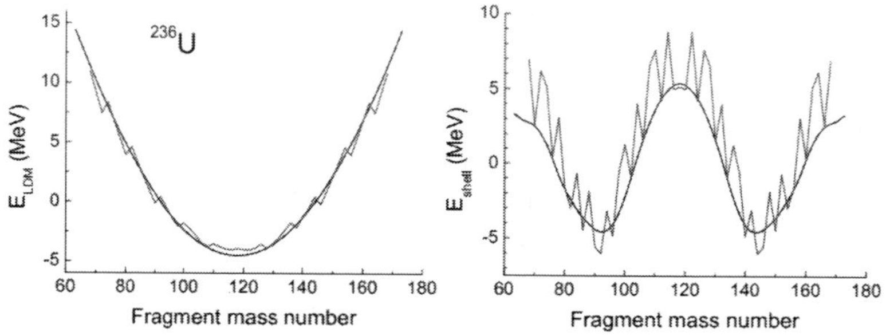

Figure 2. The liquid drop energy (left) and the shell correction at $T = 0$ (right) calculated for the compact shape just before scission (black lines) and the fit by two separate fragments (red) as function of mass asymmetry (fragment mass number).

To calculate the excitation energy, especially the shell correction, at finite but small temperature one should be very cautious. The popular relations like $E_x = aT^2$ or $E_{shell}(T) = E_{shell}(T = 0) \cdot \Phi(T)$ works somehow "on average" at large enough temperatures. At small temperature, say below $T =$ 1 MeV both relations are not reliable. Their accuracy is hard to estimate.

In order to avoid the inaccuracy caused by the above relations we will not use them but will calculate the energy at finite temperature directly. I.e. we define the energy at finite temperature as the sum of the energy within independent particle model $E_{ipm}(T)$ and the pairing energy E_{pair} defined within BCS approximation,

$$E(T) = E_{ipm}(T) + E_{pair}(T), \tag{4}$$

with

$$E_{ipm}(T) = 2\sum_k \epsilon_k n_k(\mu, T), \text{with } n_k(,\mu, T) = \frac{1}{1 + e^{(\epsilon_k - \mu)/T}}, \text{and } N = 2\sum_k n_k(\mu, T), \tag{5}$$

and

$$E_{pair}(T) = 2\sum_{k=N_1}^{N_2} \epsilon_k \left[n_k(\Delta, \lambda, T) - n_k(\mu, T)\right] - \Delta^2/G, \tag{6}$$

with

$$n_k(\Delta, \lambda, T) = \frac{1}{2}\left(1 - \frac{\epsilon_k - \lambda}{E_k} \tanh\frac{E_k}{2T}\right), \sum_{k=N_1}^{N_2} n_k(\Delta, \lambda, T) = \sum_{k=N_1}^{N_2} n_k(\mu, T), \sum_{k=N_1}^{N_2} \frac{1}{E_k} = \frac{2}{G}. \tag{7}$$

In (6)–(7) G is the strength of pairing interaction, N_1 and N_2 are the limits of the pairing window and E_k is the quasiparticle energy $E_k = \sqrt{(\epsilon_k - \lambda)^2 + \Delta^2}$.

For convenience one may add and subtract from (4) the Strutinsky averaged quantities $\widetilde{E}_{ipm}(T=0)$ and $\widetilde{E}_{pair}(T=0)$ and eventually apply the Strutinsky renormalization, i.e., replace the $\widetilde{E}_{ipm}(T=0)+\widetilde{E}_{pair}(T=0)$ by the liquid drop energy E_{LDM}. Then the energy (4) attains the form

$$E(T) = E_{LDM} + [E_{ipm}(T) - \widetilde{E}_{ipm}(T=0)] + [E_{pair}(T) - \widetilde{E}_{pair}(T=0)]\,. \tag{8}$$

The intrinsic excitation energy $E_x(T)$ is defined in a usual way

$$E_x(T) = E(T) - E(T=0)\,. \tag{9}$$

One can see that the liquid drop energy and average $\widetilde{E}_{ipm}(T=0)$ and $\widetilde{E}_{pair}(T=0)$ do not contribute to the excitation energy. That is only the temperature dependent $E_{ipm}(T)$ and $E_{pair}(T)$ that matter. These quantities are completely defined by Eqs. (5)–(7) and can be calculated for any temperature.

The dependence of the excitation energy on the temperature and the mass number is shown in Fig. 3. One can see that dependence of E_x on temperature T differs substantially from the popular approximation $E_x = aT^2$. Even more important is the dependence of E_x on the mass number at fixed temperature. Only at very high temperature, above $T = 3$ MeV, the excitation energy is proportional to the mass of nucleus. Below $T = 1$ MeV the dependence is more complicated. Say, at $T = 0.5$ MeV the excitation energy is almost zero around double magic ^{132}Sn and much larger away from $A = 132$, both at larger and *smaller A*. Thus, the fragment with mass number *smaller* than $A = 132$ can have *larger* intrinsic excitation energy (compared with $A = 132$) *even if the temperature of both is the same*. This circumstance has an important effect on the mass dependence of the excitation energy, see below.

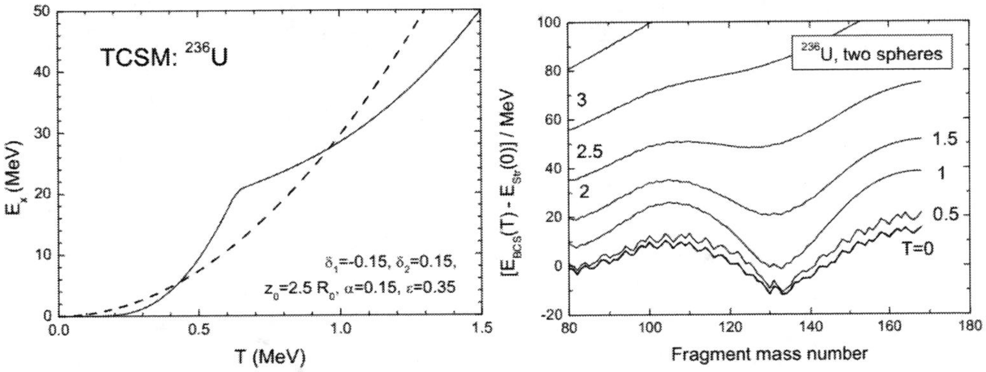

Figure 3. *Left*: the excitation energy (9) versus temperature for the scission shape given by $z_0 = 2.5R_0, \delta_1 = -0.15, \delta_2 = 0.15, \alpha = 0.15, \epsilon = 0.35$. The dash curve - the approximation for the excitation energy $E_x = a\,T^2$, with $a = A/8$. *Right*: the dependence of energy (8) (minus E_{LDM}) on the mass of *spherical* fragment. The Z/N ratio of the fragment is chosen the same as for ^{236}U.

4 The total kinetic energy and the neutron multiplicity

Having at our disposal the results shown in Fig. 3 we are ready to calculate the excitation energy of fission fragments as a function of their mass number. Because of random force in Langevin equations the calculations are repeated many times. Some trajectories reach the scission line, some not. Each trajectory reaching the scission line corresponds to a certain shape and temperature. These results are shown by the points in the left-upper part of Fig. 4. The solid line represent the average value of temperature at fixed mass asymmetry.

For each point in the left-upper part of Fig. 4 (the shape and temperature) we fit the pre-scission shape by two separate fragments, like it is shown in Fig. 1, and calculate the excitation energy (9) of each fragment for the temperature given by the output of Langevin calculations. The average value of intrinsic excitation energy (9) (red) is shown in right-upper part of Fig. 4 as function of mass asymmetry. It may seems unexpected, but this curve demonstrates the saw-tooth structure *without assuming that the light fragment is hotter*. This feature can be understood looking at Figs. 1, 3. The value of excitation energy at given temperature depends very much on the density of states around Fermi energy. Around $A = 132$ the shape of heavy fragment is close to spherical, see right part of Fig. 1. The density of states around Fermi energy is very low (because ^{132}Sn is a double magic nucleus). Thus, the excitation energy around $A = 132$ is the smallest. The fragment with $A \approx 120$ is a middle-shell nucleus, and it is very deformed. The density of states around Fermi energy is high in this case and the excitation energy is large. This explains the sudden drop of the excitation energy from $A \approx 120$ to $A \approx 132$ *at constant temperature*. Besides, the average temperature for symmetric fission is slightly larger compared with asymmetric, see left-upper part of Fig. 4.

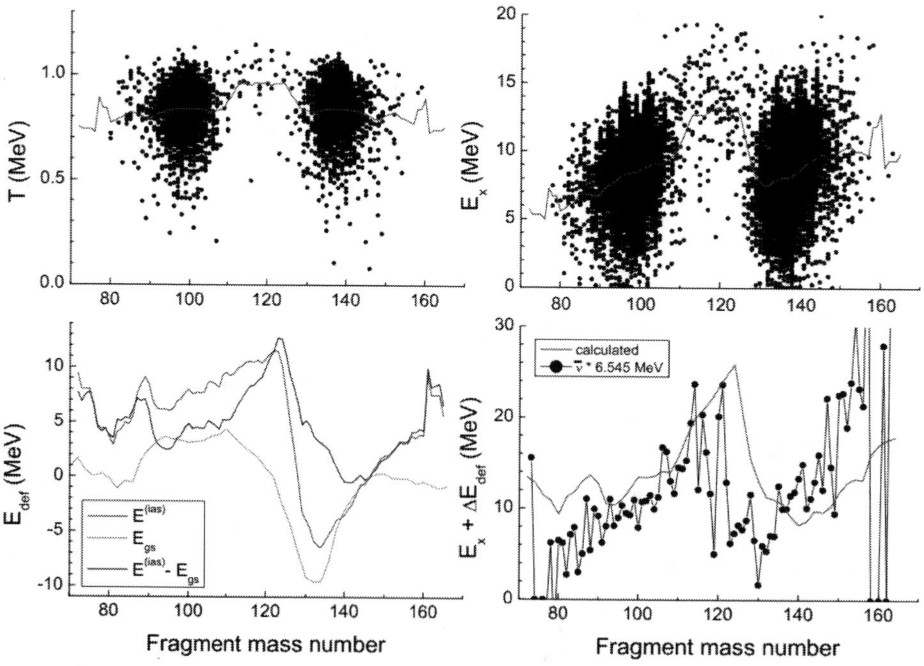

Figure 4. *Upper left*: distribution of temperature just before scission of ^{236}U for $E_n = 5$ MeV. The solid curve is the average temperature for given mass asymmetry. *Upper right*: the intrinsic excitation energy (9). *Bottom left*: the deformation energy immediately after scission $E^{(ias)}$ (red), the ground state deformation energy E_{gs} (green) and the extra deformation energy $\Delta E_{def} = E^{(ias)} - E_{gs}$ (blue). *Bottom right*: the total excitation energy $E_x + \Delta E_{def}$ (red) and the neutron multiplicity [19] multiplied by the neutron separation energy $S_n = 6.545$ MeV.

Another contribution to the excitation energy comes from the extra deformation energy $\Delta E_{def} = E^{(ias)} - E_{gs}$, i.e. the difference between the energy $E^{(ias)}$ immediately after scission configuration (at $T = 0$) and that in the ground state E_{gs}. The fragment deformation energy $E^{(ias)}$ is shown in the left-bottom part of Fig. 4. Similar to the intrinsic excitation energy it exhibits the saw-tooth behaviour. Around $A \approx 132$ it is small (because the fragment is almost spherical) and away from $A \approx 132$ it is large (because the fragment is very deformed). The problem with the deformation energy is that the ground state energy is also minimal at $A = 132$. So, the extra deformation energy ΔE_{def} does not show much of saw-tooth structure.

Still, the sum of intrinsic excitation energy and extra deformation energy $E_x + \Delta E_{def}$ clearly shows the saw-tooth behaviour very similar to experimental results for $\bar{\nu}$, see the right-bottom part of Fig. 4.

5 Summary

The saw-tooth structure in the fragments excitation energy is a natural consequence of shell effects in the deformation and intrinsic excitation energies at low excitations (temperatures). This conclusion is based on the results of four-dimensional Langevin calculations that lead to the realistic configuration at the scission point - the heavy fragment is almost spherical and light - very deformed. The only assumption is: just before scission the temperature is constant over the whole volume of a fissioning nucleus; immediately after neck rupture each fragment has *the same* temperature as it was just before scission. The calculated sum of the two contributions to the total excitation energy (deformation and excitation at scission) for ^{236}U as a function of the fragment mass clearly shows a saw-tooth behaviour in qualitative agreement with the measured dependence of the prompt neutron multiplicity.

This contribution includes the results of "Comprehensive study of delayed-neutron yields for accurate evaluation of kinetics of high-burn up reactors" entrusted to the Tokyo Institute of Technology by the Ministry of Education, Culture, Sports, Science and Technology of Japan (EXT). We appreciate very much the informative discussions with Dr. C. Schmitt and Prof. A. Tudora.

References

[1] V.V. Pashkevich, Nucl. Phys. A **169**, 275 (1971)
[2] C. Ishizuka, M.D. Usang, F.A. Ivanyuk, J.A. Maruhn, K. Nishio, S. Chiba (submitted to PRC)
[3] K.-H. Schmidt and B. Jurado, Phys. Rev. Lett. **104**, 212501 (2010)
[4] K.-H. Schmidt and B. Jurado, Phys. Rev. C **83**, 061601(R) (2011)
[5] C. Morariu, A. Tudora, F.-J. Hambsch et al., J. Phys. G. Nucl. Part. Phys. **39**, 055103 (2012)
[6] A. Tudora, F.-J. Hambsch, I. Visana, G. Giubega, Nucl. Phys. A **940**, 242 (2015)
[7] Y. Abe, S. Ayik, P.-G. Reinhard and E. Suraud, Phys. Rep. **275**, 49 (1996)
[8] J. Maruhn and W. Greiner, Z. Phys. **251**, 431 (1972)
[9] F.A. Ivanyuk, S. Chiba, and Y. Aritomo, Phys. Rev. C **90**, 054607 (2014)
[10] M.D. Usang, F.A. Ivanyuk, C. Ishisuka, and S. Chiba, Phys. Rev. C **94**, 044602 (2016)
[11] K.T.R. Davies, A.J. Sierk, J.R. Nix, Phys. Rev. C **13**, 2385 (1976)
[12] J. Blocki, Y. Boneh, J.R. Nix et al., Ann. Phys. **113**, 330 (1978)
[13] A.J. Sierk and J.R. Nix, Phys. Rev. C **21**, 982 (1980)
[14] H. Hofmann, D. Kiderlen, Int. Jour. Mod. Phys. E **7**, 243 (1998)
[15] J. Randrup and P. Moller, Phys. Rev. C **88**, 064606 (2013)
[16] V.F. Apalin, Yu.N. Gritsyuk, I.E. Kutikov et al., Nucl. Phys. **71**, 553 (1965)

[17] K. Nishio, Y. Nakagome, H. Yamamoto, I. Kimura, Nucl. Phys. A **632**, 540 (1998)
[18] B.D. Wilkins, E.P. Steinberg, and R.R. Chasman, Phys. Rev. C **14**, 1832 (1976)
[19] R. Müller, A.A. Naqvi, F. Käppeler, and F. Dickmann, Phys. Rev. C **29**, 885 (1984)

EPJ Web of Conferences **169**, 00006 (2018) https://doi.org/10.1051/epjconf/201816900006
Theory-4

Implementing and testing theoretical fission fragment yields in a Hauser-Feshbach statistical decay framework

Patrick Jaffke[1,⋆], *Peter* Möller[1], *Ionel* Stetcu[1], *Patrick* Talou[1], and *Christelle* Schmitt[2,3]

[1] *Theoretical Division, Los Alamos National Laboratory, Los Alamos NM, 87545, USA*
[2] *Grand Accélérateur National d'Ions Lourds, CEA/DSM-CNRS/IN2P3, Caen, France*
[3] *Institut Pluridisciplinaire Hubert Curien, 23 rue du Loess, BP. 28, 67037 Strasbourg Cedex 2, France*

Abstract. We implement fission fragment yields, calculated using Brownian shape-motion on a macroscopic-microscopic potential energy surface in six dimensions, into the Hauser-Feshbach statistical decay code CGMF. This combination allows us to test the impact of utilizing theoretically-calculated fission fragment yields on the subsequent prompt neutron and γ-ray emission. We draw connections between the fragment yields and the total kinetic energy TKE of the fission fragments and demonstrate that the use of calculated yields can introduce a difference in the $\langle$TKE$\rangle$ and, thus, the prompt neutron multiplicity $\bar{\nu}$, as compared with experimental fragment yields. We deduce the uncertainty on the $\langle$TKE$\rangle$ and $\bar{\nu}$ from this procedure and identify possible applications.

1 Introduction

Recent advances in theoretical models and computational abilities have allowed one to calculate many fission fragment properties at scission. Several models exist: macroscopic nuclear shape models [1–3], self-consistent microscopic models using effective two-body nuclear forces [4–6], or macroscopic-microscopic models of the potential energy surfaces with various treatments of the dynamics on these surfaces, such as Brownian shape motion or hydrodynamical models of the inertia and friction tensors in the equations of motion [7–9]. All of these can be used to determine fission fragment yields in charge Z, or mass A, or both. Certain models are capable of calculating the total kinetic energy TKE of the fission fragments and even the nuclear deformation at scission as well. These methods become powerful tools towards a predictive model of fission, where one could calculate the scission configurations for compound nuclei that are experimentally impossible to measure.

Combining theoretical fragment yields with a Hauser-Feshbach [10] statistical decay framework, such as CGMF [11], provides one with a method to evaluate the impact of using these calculated fragment yields instead of experimental ones on the prompt neutron and γ-ray data. Primary fission fragment data already provide a direct comparison with calculated yields, but the addition of prompt neutron and γ-ray data strengthens the constraints on the theoretical yields models. It is this secondary step we illustrate here for the fission reactions ^{239}Pu(n,f) and ^{241}Pu(n,f) with neutron energies E_n from thermal to 10 MeV. These cases are used to estimate the uncertainties in using theoretical yields for plutonium isotopes in the range $^{236-244}$Pu.

⋆e-mail: pjaffke@lanl.gov

Our process is as follows: we first detail the necessary input for the Hauser-Feshbach calculations, including our average total kinetic energy ⟨TKE⟩ parameterization, the assumed spin distribution and its dependence on E_n, the multi-chance fission probabilities, and the distributions of TKE parameters with fragment mass. Next, we present a simple comparison of fragment yields for ^{239}Pu(n,f) for which we have experimental data at various incident neutron energies [12]. Finally, we conduct the full CGMF calculations to determine the sensitivity of fission observables to the use of theoretically-calculated fission fragment yields, as compared with the experimental ones.

2 Input

The fission fragment yields are calculated from a macroscopic-microscopic (macro-micro) model that uses six shape parameters to determine $Y(Z, A)$ [13]. Previous studies have shown good agreement between the calculated charge yields and the experimental ones for a variety of compound nuclei and incident neutron energies [7, 14]. In our CGMF calculations, we take the mass yields from the macro-micro model and the charge distribution follows that of Wahl [15]. One could use the charge distributions directly from the calculated $Y(Z, A)$, but this would require an increase in computation time to achieve reasonable statistics in the wings of the charge distributions. As the calculated charge distributions are similar to those of Wahl, we use the systematics instead. The ⟨TKE⟩ and its dependence on mass ⟨TKE⟩(A) are taken from experimental data. For example, at thermal energies values of 177.76 MeV [16–18] and 178.89 MeV [19–21] are used for ^{239}Pu(n,f) and ^{241}Pu(n,f), respectively. The slopes $\partial\langle\mathrm{TKE}\rangle/\partial E_n$ are taken from a variety of sources [12, 19, 22]. The ⟨TKE⟩(A) are taken from Vorobyeva [23, 24] for the various plutonium isotopes, but we have also used other sources [25–27] and found minimal impact on our results. The sampled TKE for each event follows a Gaussian centered about ⟨TKE⟩(A) with a variance $\sigma^2_{\mathrm{TKE}}(A)$, taken from Ref. [18, 28]. For isotopes that did not have data on the distribution parameters, ⟨TKE⟩(A) and $\sigma^2_{\mathrm{TKE}}(A)$, we take the data of the nearest available plutonium isotope and allow for an overall scaling in magnitude. We determine the ⟨TKE⟩ for plutonium isotopes without data with a local $Z^2/A^{1/3}$ fit, similar to the Viola systematics [29]. We note that this simple two-point fit of the above data is quite inadequate, but could be improved with additional data. When no data is available on the slope $\partial\langle\mathrm{TKE}\rangle/\partial E_n$, we use the average over all plutonium isotopes with data.

The spin distributions of the fragments follows that of previous papers [30, 31] and the average spin $\langle J\rangle$ has an exponential form with E_n. The value at thermal neutron energies is fitted such that average prompt neutron multiplicity $\bar{\nu}$ matches the experimental data, when available, and the asymptote is derived from fits to ^{238}U(n,f) prompt γ-ray multiplicities.

Multi-chance fission probabilities are calculated with the CoH code [32]. We note that this parameter is highly dependent on the fission barrier heights and level density parameters, which are not experimentally bound very well. We take fission barrier heights and curvatures from JENDL-4.0 [33]. Our procedure is to fit the total fission cross-section from ENDF/B-VII.1 [34] with the above barrier heights and curvatures. The level density parameters are adjusted to obtain the best fit and we use the resulting multi-chance probabilities.

With the above prescription, the only remaining input are the fission fragment yields themselves. We fit the theoretically predicted mass yields with 3 Gaussians. The parameters of the Gaussians, weight w, mean μ, and variance σ^2, are dependent on the incident neutron energy E_n. This is done to provide a smooth function of the mass yields at any E_n, rather than an interpolation of tabulated values. They have the following form:

$$w_i(E_n) = \frac{w_i}{1 + e^{(E_n - x_i)/y_i}}, \qquad \mu_i(E_n) = m_i E_n^2 + n_i E_n + p_i, \qquad \sigma_i(E_n) = s_i E_n^2 + t_i E_n + r_i \tag{1}$$

The weights of the Gaussians follow a Fermi-function with E_n and the variances and means are allowed to be quadratic in E_n. Obviously, the mean for the symmetric Gaussian is $\mu_3 = A_C/2$, where A_C is the mass of the fissioning nucleus, and one of the weights can be constrained via $2w_1 + 2w_2 + w_3 = 2$, reducing the number of fitted parameters. We have found that using a linear function for $\sigma_i(E_n)$, instead of the quadratic one in Eq. (1), produces very good fits as well. The 3 Gaussian fit smoothly transitions to higher excitation energies and when multi-chance fission occurs, the proper compound nucleus' Gaussian parameters are selected, as we have calculated yields for $^{236-244}$Pu, instead of a shift in the means [35].

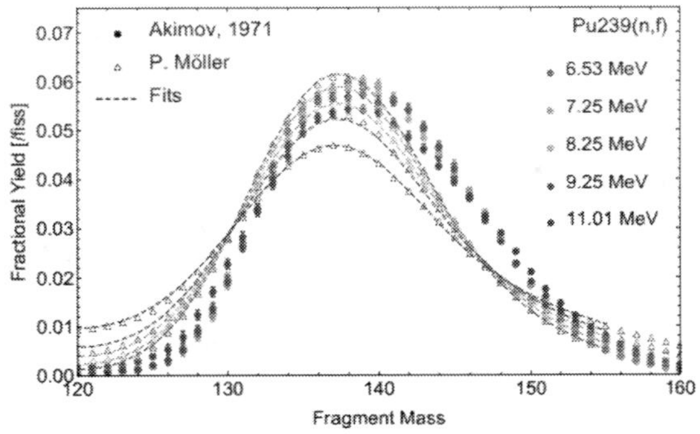

Figure 1. Comparison of the mass yields for the neutron-induced fission of ^{239}Pu from Ref. [12] (circles) and the calculated yields via Ref. [13] (triangles). Listed are the excitation energies, assuming a binding energy of $E_b = 6.53$ MeV for ^{239}Pu. Thus, the 6.53 MeV (red) symbols and curves denote thermal fission. The dashed lines are the 3 Gaussian fits. We have included a 4.5 amu mass resolution in the calculated yields. While none of the listed excitation energies in this plot are above second-chance fission, we included this effect in our study.

The mass yields of Ref. [13] are compared with the data of Ref. [12] at a variety of excitation energies. We have applied a mass resolution of 4.5 amu to the theoretical yields. One can see in Fig. 1 that there appears to be a shift in average heavy fragment mass $\langle A_h \rangle$ and the calculated yields appear to show a more rapid increase of the symmetric region with excitation energy. Overall, the calculated yields reproduce the general trends of the data, but the under-prediction of the mass yields around $A \sim 145$ causes a shift in $\langle A_h \rangle$ of about 1.15 amu relative to data. The impact of the excitation energy on the symmetric region is also more subtle in Akimov. The effect of these differences on the prompt particle emission is studied next.

3 Prompt particle emission

We perform Hauser-Feshbach fission fragment decay calculations for neutron-induced fission reactions on various plutonium isotopes. Most instructive is the case of ^{239}Pu(n,f) as it contains experimental data on the primary fragment yields over a range of incident energies. We also show the results for ^{241}Pu(n,f), where we only have the theoretical yields. In Fig. 2, we illustrate the total prompt neutron multiplicity as a function of incident neutron energy, which we note includes pre-scission neutrons. The comparison shows three data sources [34, 36, 37] and three computed cases. Case *A* corresponds to a calculation that uses the *Y*(*A*) from Ref. [12] and no adjustments have been made to the TKE input parameters described above. Case *B* swaps out the experimental *Y*(*A*) with the calculated values and maintains the same TKE input parameters as case *A*. Case *D* adjusts the TKE parameters so as to match the $\bar{\nu}(E_n)$ data.

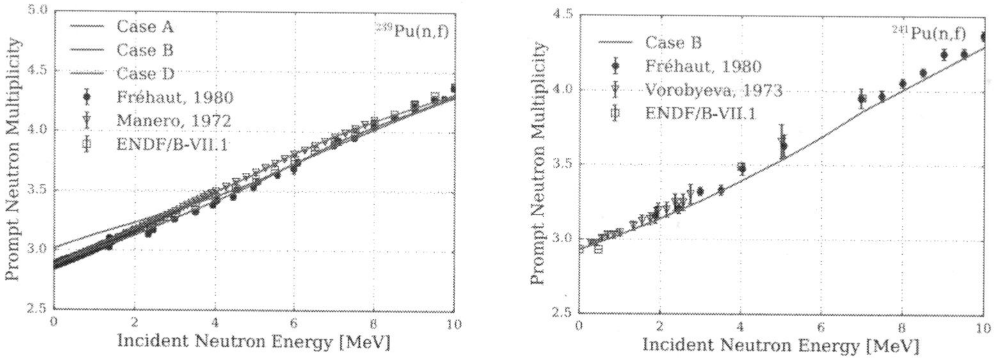

Figure 2. Comparison of the total prompt neutron multiplicity $\bar{\nu}$ (including pre-scission neutrons) from three data sources [34, 36, 37] and three calculated cases. The cases correspond to using experimental $Y(A)$ [12] (A), using calculated yields [13] and the same TKE parameters (B), and using calculated yields and adjusting the TKE parameters (D). We note that the change in TKE parameters is necessary in order to match the $\bar{\nu}(E_n)$ from experiment, but this produces strong tensions with ⟨TKE⟩ data, as seen in Fig. 3. Only case B is considered for ^{241}Pu(n,f) and is compared to Ref. [34, 37, 38], with good agreement.

We can see that a simple swap of the $Y(A)$ results in a change in $\bar{\nu}$, when we require the ⟨TKE⟩ to be equal in both cases A and B. This is precisely caused by the difference in $\langle A_h \rangle$ between the experimental yields and the calculated ones. The shape of the ⟨TKE⟩(A) distributions are kept constant, but we must scale the magnitude by a factor f, such that ⟨TKE⟩ are equal between cases A and B. The calculated yields have a lower $\langle A_h \rangle$, so $f_B < f_A$. This shift to an overall lower ⟨TKE⟩(A) in case B generates an overall larger total excitation energy via energy conservation and, thus, case B shows a larger $\bar{\nu}$. The change in slopes between cases A and B reflect the differences in the evolution of $\langle A_h \rangle$ with incident neutron energy for the two yields methods.

Case D is the result of altering both the ⟨TKE⟩ and its slope in E_n in order to match the shape of the experimental $\bar{\nu}$ [34]. We can see in Fig. 2 that these changes restore the agreement for ^{239}Pu(n,f). This change in the ⟨TKE⟩ slope acts to counter the difference generated from the change in $\langle A_h \rangle$ with E_n between the experimental and calculated yields. The benefit of using a correlated fission event method, such as `CGMF`, is that the connection between ⟨TKE⟩ and $\bar{\nu}$ is preserved. Thus, we can see the impact of this change in the TKE parameters in Fig. 3 and determine an uncertainty from the use of calculated yields. We find that $\delta_{\bar{\nu}} \sim 0.13\,\nu/f$, $\delta_{\partial\bar{\nu}/\partial E_n} \sim 0.05\,\nu/f/\mathrm{MeV}$, $\delta_{\langle \mathrm{TKE} \rangle} \sim 1.0\,\mathrm{MeV}$, and $\delta_{\partial\langle \mathrm{TKE} \rangle/\partial E_n} \sim 0.18$. With an average energy cost of $\sim 8.0\,\mathrm{MeV}$ per neutron, the values of $\delta_{\langle \mathrm{TKE} \rangle} \sim 1.0\,\mathrm{MeV}$ are in agreement with the inferred $\delta_{\bar{\nu}} \sim 0.13\,\nu/f$. The error bars for ⟨TKE⟩(E_n) are significantly larger in the ^{241}Pu case, where we find that the calculated yields can reproduce the ⟨TKE⟩ and $\bar{\nu}$ data simultaneously. This is most likely due to the larger experimental error bars and, thus, a larger acceptable range for ⟨TKE⟩, but may also indicate that this issue is a local problem instead of a global issue with the calculated yields. Further studies of uranium isotopes are underway to examine this.

The majority of this work has focused on a comparison of prompt neutron data, and its connection to the TKE of the fragments. We have also done a comparison of the prompt gamma observables, mainly the average total gamma energy $\langle E_\gamma^T \rangle$ and average total multiplicity $\langle M_\gamma \rangle$ as a function of incident neutron energy. These values are very sensitive to detector thresholds [40] and very limited

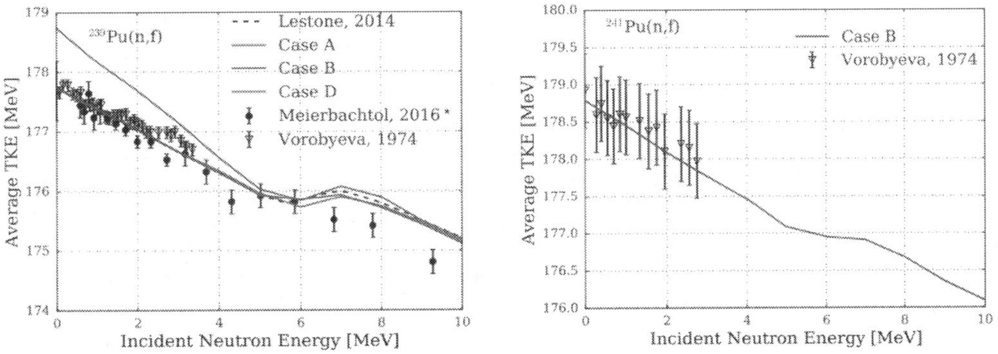

Figure 3. Comparison of the average total kinetic energy of the fission fragments ⟨TKE⟩ from two data sources [19, 22], a parameterized model [39], and the `CGMF` calculations under the same case conditions as above. The data and parameterized model have been adjusted to have the same value for thermal neutrons. We note that the Meierbachtol data [22] corresponds to the post-neutron emitted kinetic energies. The shape beyond second chance fission implies a lower ⟨TKE⟩ for the fissioning nucleus of ^{239}Pu than the one used in our simple $Z^2/A^{1/3}$ fit. One can clearly see that altering the TKE parameters to match the $\bar{\nu}(E_n)$ in Fig. 2 with the calculated yields results in an incompatible ⟨TKE⟩ (case *D*). Again, only case *B* is considered for ^{241}Pu(n,f) and is compared to Ref. [23], finding good agreement, although the error bars are large.

data is available across our range of incident neutron energies. Nevertheless, we observe a very small impact of the yields choice on the $\langle E_\gamma^T \rangle$ and $\langle M_\gamma \rangle$. This is because any extra excitation energy induced through the changes in yields is dominantly exhausted through prompt neutron emission, especially at the early stages of the Hauser-Feshbach de-excitation. The resulting errors are $\delta_{\langle M_\gamma \rangle} \sim 0.01\,\gamma/f$ and $\delta_{\langle E_\gamma^T \rangle} \sim 0.09\,\text{MeV}$ across the incident neutron energies studied here. This result indicates that these prompt γ-ray observables are less sensitive to the use of calculated versus experimental fission yields. More specific measurements, such as the intensity of low-energy γ-ray peaks, are more sensitive to the choice of $Y(A)$, on the order of $\sim 5\%$. A more detailed analysis of changes to the γ-ray spectrum is underway.

4 Conclusion

We have successfully implemented theoretically-predicted yields from the macroscopic-microscopic model into a Hauser-Feshbach statistical decay formalism. This allows us to infer the feasibility of using theoretically-calculated mass yields to determine the prompt neutron and γ-ray observables for unknown fissioning systems. First, we presented a direct comparison of the mass yields for ^{239}Pu(n,f) with available data. The average heavy fragment masses differed by 1.15 amu and the change with E_n in the symmetric region was more dramatic in the theoretical yields. These differences were propagated to the prompt neutron and γ-ray observables using the statistical Monte Carlo Hauser-Feshbach decay code `CGMF`.

The difference in $\langle A_h \rangle$ induced a difference in the ⟨TKE⟩ between the experimental and theoretical mass yields. This, in turn, resulted in a difference in the prompt neutron multiplicity $\bar{\nu}$. We have demonstrated that the experimental data can simultaneously reproduce the $\bar{\nu}$ and ⟨TKE⟩ trends with incident neutron energy. The calculated yields, however, introduce tension between these two observables. A resulting uncertainty of $\delta_{\bar{\nu}} \sim 0.13$ and $\delta_{\langle \text{TKE} \rangle} \sim 1.0\,\text{MeV}$ for the thermal values and

$\delta_{\partial\bar{\nu}/\partial E_n} \sim 0.05$ and $\delta_{\partial\langle \mathrm{TKE}\rangle/\partial E_n} \sim 0.18\,\mathrm{MeV}$ for the slopes was determined for the calculated yields. These uncertainties are mainly derived from the ^{239}Pu(n,f) reaction. Good simultaneous agreement between $\bar{\nu}$ and $\langle\mathrm{TKE}\rangle$ in the ^{241}Pu(n,f) reaction was found when using the calculated yields. The uncertainties on the prompt γ-ray observables are much smaller: $\delta_{\langle M_\gamma\rangle} \sim 0.01$ and $\delta_{\langle E_\gamma^T\rangle} \sim 0.09\,\mathrm{MeV}$ across the 10 MeV range of incident neutron energies.

With these derived uncertainties, we identify the current applications of the theoretically calculated yields. In criticality studies, which often require uncertainties on $\bar{\nu}$ to be much less than 1% [35], calculations with theoretically calculated yields would be difficult and the more-precise experimental values must be used. However, one could use the calculated yields to estimate the contribution to a prompt neutron multiplicity distribution from low-level impurities in a sample during active or passive neutron interrogation. In addition, the derived uncertainties are very encouraging for fission calculations in r-process simulations [41–43]. For estimates of the γ-ray heating due to fission of fissiles without experimental data, we find that the induced error from these calculated yields are low. Additional studies will be conducted on more actinides and super-heavy nuclei to extend our analysis.

The authors would like to thank T. Kawano for useful discussions on the calculations of the multi-chance fission probabilities. The work of P.J., P.M., I.S., and P.T. was conducted under the auspices of the National Nuclear Security Administration of the U.S. Department of Energy at Los Alamos National Laboratory under Contract DEAC52-06NA25396. The work of C.S. acknowledges support through FUSTIPEN (French-U.S. Theory Institute for Physics with Exotic Nuclei) under DOE grant number DE-FG02-10ER41700. This publication is LA-UR-17-27821.

References

[1] M.G. Mustafa, P.A. Baisden, H. Chandra, Physical Review C **25**, 2524 (1982)
[2] A.V. Karpov, P.N. Nadtochy, D.V. Vanin, G.D. Adeev, Physical Review C **63**, 054610 (2001)
[3] P. Nadtochy, E. Ryabov, A. Gegechkori, Y.A. Anischenko, G. Adeev, Physical Review C **85**, 064619 (2012)
[4] W. Younes, D. Gogny, Physical Review Letters **107**, 132501 (2011)
[5] N. Schunck, D. Duke, H. Carr, A. Knoll, Physical Review C **90**, 054305 (2014)
[6] A. Bulgac, P. Magierski, K.J. Roche, I. Stetcu, Physical Review Letters **116**, 122504 (2016)
[7] J. Randrup, P. Möller, Physical Review Letters **106**, 132503 (2011)
[8] T. Ichikawa, A. Iwamoto, P. Möller, A.J. Sierk, Physical Review C **86**, 024610 (2012)
[9] Y. Aritomo, S. Chiba, K. Nishio, Progress in Nuclear Energy **85**, 568 (2015)
[10] W. Hauser, H. Feshbach, Physical Review **87**, 366 (1952)
[11] P. Talou, T. Kawano, I. Stetcu, LA-CC-13063, Los Alamos Nat. Lab. (2013)
[12] N. Akimov, V. Vorob'eva, V. Kabenin, Soviet Journal of Nuclear Physics-USSR **13**, 484 (1971)
[13] P. Möller, T. Ichikawa, Eur. Phys. J. A **51**, 173 (2015)
[14] P. Möller, C. Schmitt, Eur. Phys. J. A **53**, 7 (2017)
[15] A.C. Wahl, Systematics of fission-product yields, Los Alamos Nat. Lab. (2002)
[16] H. Thierens, A. De Clercq, E. Jacobs, D. De Frenne, P. D'hondt, P. De Gelder, A. Deruytter, Physical Review C **23**, 2104 (1981)
[17] C. Wagemans, E. Allaert, A. Deruytter, R. Barthélémy, P. Schillebeeckx, Physical Review C **30**, 218 (1984)
[18] P. Schillebeeckx, C. Wagemans, A. Deruytter, R. Barthelemy, Nuclear Physics A **545**, 623 (1992)

[19] V. Vorobyeva, N. Dyachenko, N. Kolosov, B. Kuzminov, A. Sergachev, Soviet Journal of Nuclear Physics-USSR **19**, 489 (1974)
[20] E. Allaert, C. Wagemans, G. Wegener-Penning, A. Deruytter, R. Barthélémy, Nuclear Physics A **380**, 61 (1982)
[21] H. Thierens, E. Jacobs, P. D'hondt, A. De Clercq, M. Piessens, D. De Frenne, Physical Review C **29**, 498 (1984)
[22] K. Meierbachtol, F. Tovesson, D. Duke, V. Geppert-Kleinrath, B. Manning, R. Meharchand, S. Mosby, D. Shields, Physical Review C **94**, 034611 (2016)
[23] V. Vorobyeva, N. Dyachenko, N. Kolosov, B. Kuzminov, A. Sergachev, V. Surin, Soviet Journal of Nuclear Physics-USSR **19**, 621 (1974)
[24] V. Vorobyeva et al., Mass yields and kinetic energy of fragments for fission of plutonium isotopes, in Conf. on Neutron Physics **3**, 270 (1973)
[25] V. Surin, S. AI, B. Kuzminov, R. NI, Soviet Journal of Nuclear Physics-USSR **14**, 523 (1972)
[26] K. Nishio, Y. Nakagome, I. Kanno, I. Kimura, Journal of Nuclear Science and Technology **32**, 404 (1995)
[27] C. Tsuchiya, Y. Nakagome, H. Yamana, H. Moriyama, K. Nishio, I. Kanno, K. Shin, I. Kimura, Journal of Nuclear Science and Technology **37**, 941 (2000)
[28] J. Neiler, F. Walter, H. Schmitt, Physical Review **149**, 894 (1966)
[29] V. Viola, K. Kwiatkowski, M. Walker, Physical Review C **31**, 1550 (1985)
[30] P. Talou, B. Becker, T. Kawano, M. Chadwick, Y. Danon, Physical Review C **83**, 064612 (2011)
[31] B. Becker, P. Talou, T. Kawano, Y. Danon, I. Stetcu, Physical Review C **87**, 014617 (2013)
[32] T. Kawano, Los Alamos National Laboratory, unpublished (2003)
[33] K. Shibata, O. Iwamoto, T. Nakagawa, N. Iwamoto, A. Ichihara, S. Kunieda, S. Chiba, K. Furutaka, N. Otuka, T. Ohsawa et al., Journal of Nuclear Science and Technology **48**, 1 (2011)
[34] M. Chadwick, M. Herman, P. Obložinskỳ, M.E. Dunn, Y. Danon, A. Kahler, D.L. Smith, B. Pritychenko, G. Arbanas, R. Arcilla et al., Nuclear Data Sheets **112**, 2887 (2011)
[35] R. Vogt, J. Randrup, J. Pruet, W. Younes, Physical Review C **80**, 044611 (2009)
[36] F. Manero, V. Konshin, Atomic Energy Review **10**, 637 (1972)
[37] M. Soleilhac, J. Frehaut, J. Gauriau, G. Mosinski, Data EXFOR **20** (1980)
[38] V. Vorobyeva, N. Dyachenko, B. Kuzminov, Soviet Journal of Nuclear Physics-USSR **36**, 406 (1974)
[39] J. Lestone, T. Strother, Nuclear Data Sheets **118**, 208 (2014)
[40] I. Stetcu, P. Talou, T. Kawano, M. Jandel, Physical Review C **90**, 024617 (2014)
[41] I. Panov, E. Kolbe, B. Pfeiffer, T. Rauscher, K.L. Kratz, F.K. Thielemann, Nuclear Physics A **747**, 633 (2005)
[42] G. Martínez-Pinedo, D. Mocelj, N.T. Zinner, A. Kelić, K. Langanke, I. Panov, B. Pfeiffer, T. Rauscher, K.H. Schmidt, F.K. Thielemann, Progress in Particle and Nuclear Physics **59**, 199 (2007)
[43] S. Goriely, A. Bauswein, H.T. Janka, The Astrophysical Journal Letters **738**, L32 (2011)

EPJ Web of Conferences **169**, 00007 (2018) https://doi.org/10.1051/epjconf/201816900007
Theory-4

Determination of the nuclear level densities and radiative strength function for 43 nuclei in the mass interval 28≤A≤200

David Knezevic[1,5], *Nikola* Jovancevic[1,*], *Anatoly M.* Sukhovoj[2], *Ludmila V.* Mitsyna[2], *Miodrag* Krmar[1], *Vu D.* Cong[2], *Franz-Josef* Hambsch[4], *Stephan* Oberstedt[4], *Zsolt* Revay[3], *Christian* Stieghorst[3], and *Aleksandar* Dragic[5]

[1] *University of Novi Sad, Faculty of Science, Department of Physics, Trg Dositeja Obradovica 3, 21000 Novi Sad, Serbia*
[2] *Joint Institute for Nuclear Research, 141980 Moscow region, Dubna, Russia*
[3] *Technische Universität München, Forschungsneutronenquelle Heinz Maier-Leibnitz (FRM II), Lichtenbergstr. 1, 85747 Garching, Germany*
[4] *European Commission, Joint Research Centre, Directorate G – Nuclear Safety and Security, Unit G.2, Retieseweg 111, 2440 Geel, Belgium*
[5] *Institute of Physics Belgrade, Pregrevica 118, 11080 Zemun, Serbia*

Abstract. The determination of nuclear level densities and radiative strength functions is one of the most important tasks in low-energy nuclear physics. Accurate experimental values of these parameters are critical for the study of the fundamental properties of nuclear structure. The step-like structure in the dependence of the level densities ρ on the excitation energy of nuclei E_{ex} is observed in the two-step gamma cascade measurements for nuclei in the $28 \leq A \leq 200$ mass region. This characteristic structure can be explained only if a co-existence of quasi-particles and phonons, as well as their interaction in a nucleus, are taken into account in the process of gamma-decay. Here we present a new improvement to the Dubna practical model for the determination of nuclear level densities and radiative strength functions. The new practical model guarantees a good description of the available intensities of the two step gamma cascades, comparable to the experimental data accuracy.

1 Introduction

The development of theoretical models of nuclear structures requires a set of experimental information of the excited levels density, ρ, (with given quantum numbers) and of the values of the partial width (radiative strength function), Γ, of all possible decay channels. Correct interpretation of the dynamics of the nuclear transitions, in a broad variety from the simple low-lying levels (e.g., quasi-particle or phonon structure) to the very complex compound-states is possible by the theoretical calculations if those experimental data are available. One of the most suitable techniques for determination of required nuclear mater parameters (ρ and Γ) is the two-step gamma cascades methods based on measurement of gamma coincidences following neutron capture [1].

*e-mail: nikola.jovancevic@df.uns.ac.rs

Based on the experimental data collected by two-step gamma cascades experiment a model for description the gamma-decay of neutron resonance was developed at JINR, Dubna [2, 3]. In this model the level density ρ of quasi-particles in any nucleus is defined using the known model of n-quasi-particle levels. Here we presented the improved version of this model taking into account shell inhomogeneities of the single-particle level spectra and their influence on the functions: $\rho=\varphi(E_{ex})$ and $\Gamma=\psi(E_1)$, where E_{ex} is the excitation energy and E_1 is primary transition energy. The experimental results of two step gamma cascades intensity for 43 nuclei in the $28 \leq A \leq 200$ mass region were fitting by this model. This provide us possibility to extract parameters of nuclear structure such as breaking thresholds of the second and the third Cooper pairs, ratio of the collective level density to the total one or level parity.

2 Dubna two-step gamma cascades method

The two-step gamma-cascades method for obtaining information about the nuclear structure parameters following the thermal neutron captures was developed at FLNP, JINR, DUBNA [2, 3]. From amount of gamma-gamma coincidences the method allows to choose registration events of full energy of two-gamma transition cascade with a sufficiently low background. And the experimental intensity distributions of cascades to the final levels of compound-nucleus with excited energy below ~500–800 keV are obtained from these coincidences. Using the nuclear spectroscopy procedures allows decomposing the initial spectrum on primary and secondary transmission components of cascades with an acceptable uncertainty [2, 3].

The basic idea of this method comes from specific dependence of the two-step gamma- cascade intensity on the partial radiative width Γ and the density of excited levels:

$$I_{\gamma\gamma} = \sum_{\lambda,f}\sum_{i}\frac{\Gamma_{\lambda i}}{\Gamma_{\lambda}}\frac{\Gamma_{if}}{\Gamma_{i}} = \sum_{\lambda,f}\frac{\Gamma_{\lambda i}}{\langle\Gamma_{\lambda i}\rangle m_{\lambda i}}n_{\lambda i}\frac{\Gamma_{if}}{\left\langle\Gamma_{if}\right\rangle m_{if}} \tag{1}$$

where $\Gamma_{\lambda i}$ and Γ_{if} are the partial radiative widths corresponding to the primary and to the secondary transitions; $n_{\lambda i} = \rho\Delta E_i$ is the number of the excited intermediate levels in a certain interval of the excitation energy ΔE_i; $\langle\Gamma_{\lambda i}\rangle$ and $\langle\Gamma_{if}\rangle$ are the average values of the corresponding intervals of the nucleus excitation energy widths; $m_{\lambda i}$ and m_{if} are the number of levels in the same intervals. When this method was developed for the first time it was based on an interactive calculation. Using iterative process with "randomly" chosen functions ρ and Γ, it is possible to obtain the most probable values of level density and radiative width (or radiative strength function).

3 Model of the gamma-decay of neutron resonance

Here we present improved version of the model for the gamma-decay of neutron resonance [2] which can explain the experimental data based on combination of phenomenological and theoretical representations.

The level density, described by an expression for density ρ_l of Fermi levels, was taken from the model of density Ω_n of n-quasi-particle states [4]:

$$\rho_l = \frac{(2J+1)\exp\left(-(J+1/2)^2/2\sigma^2\right)}{2\sqrt{2\pi}\sigma^3}\cdot\Omega_n(E_{ex}), \Omega_n(E_{ex}) = \frac{g^n(E_{ex}-U_l)^{n-1}}{((n/2)!)^2(n-1)!} \tag{2}$$

Here J is the spin quantum number, $g = 6a/\pi^2$ is the density of the single-particle states near Fermi-surface, σ is the cut-off factor (a and σ values were taken from the back-shifted Fermi-gas model

[5]), and U_l, is the energy of the l-th Cooper pair breaking threshold. The effect of the collective enhancement was also included in this model by the coefficient C_{col} of the collective enhancement of the vibrational level density (or both vibrational and rotational ones for deformed nuclei). For a given excitation energy, E_{ex}, the phenomenological coefficient is determined by a theoretical description that can be found in Ref. [3]:

$$C_{coll} = A_l \exp(\sqrt{(E_{ex} - U_l)/E_\nu} - (E_{ex} - U_l)/E_\mu) + \beta \tag{3}$$

where A_l are parameters of density for the vibrational levels above the breaking point for each l-th Cooper pair, E_μ and E_ν determine the change in the nuclear entropy and the change of the quasi-particles excitation energies, respectively. Coefficients A_l for different pairs are fitted independently, as it was done in Ref. [2]. Coefficient β is used for a description of the rotation level density.

Radiative strength functions for $E1$- and $M1$-transitions are determined in this model by Ref. [6]:

$$k(E1, E_\gamma) + k(M1, E_\gamma) = w \frac{1}{3\pi^2\hbar^2c^2A^{2/3}} \frac{\sigma_G \Gamma_G^2 (E_\gamma^2 + \kappa 4\pi^2 T^2)}{(E_\gamma^2 - E_G^2)^2 + E_\gamma^2 \Gamma_\gamma^2} + $$
$$+ P\delta^- \exp(\alpha_p (E_\gamma - E_p)) + P\delta^+ \exp(\beta_p (E_p - E_\gamma)) \tag{4}$$

with fitting normalization parameter w and coefficient κ; thermodynamic temperature T; the location of the center of the giant dipole resonance E_G, with width Γ_G and cross section σ_G in the maximum for each nucleus. For description of experimental data of Ref. [3] it is necessary to add one or several narrow peaks to the strength function is based on the data of Ref. [3]. The second summand of Eq. (5) corresponds to the left slope of the peak (energies below the maximum), and the third summand is the right slope (energies above the maximum). Position E_p in the energy scale, amplitudes $P\delta^+$ and $P\delta^-$ and slope parameters α_p and β_p are fitted for each peak independently. At $E_1 \approx B_n$ the fitted ratios Γ_{M1}/Γ_{E1} of $E1$- and $M1$-strength functions are normalized to known experimental values, and their sum Γ_λ is normalized to the total radiation width of the resonance.

The influence of the shell correction δE on the density of the quasi-particle levels were tested in this work. It was done by using the $a(A)$ value, which depends on the excitation energy, included linearly in the parameter of the single-particle density g (see Eq. (2)). For a nucleus with mass A and excitation energy E_{ex}, $a(A)$ is expressed, as [3]:

$$a(A) = \tilde{a}(1 + ((1 - exp(\gamma E_{ex}))\delta E / E_{ex})) \tag{5}$$

where asymptotic value is $\tilde{a} = 0.114 \cdot A + 0.162 \cdot A^{2/3}$ and $\gamma = 0.054$. The δE values slightly varied relative to their evaluations [3] in order to keep an average spacing between neutron resonances (see [2]).

In our model the set of common parameters for fitting (see Eqs. (2, 3)) were:

1) the break up thresholds energies U_l up to l=4,

2) the E_μ and E_ν parameters, which are common for all Cooper pairs

3) the mutually independent parameters A_l of the density of vibrational levels above the break up threshold U_l

4) the coefficients w, κ and β

5) the ratio r of negative parity and the total level density.

Those parameters were used for the description of the intensity $I_{\gamma\gamma}(E_1)$ for 43 nuclei, in the framework of the proposed model.

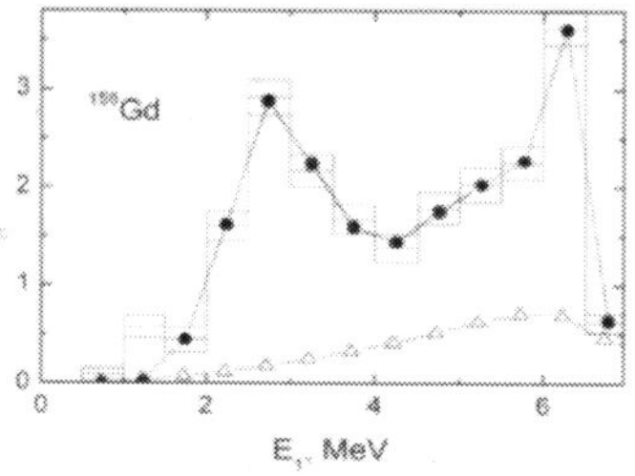

Figure 1. Histogram - experimental cascade intensity and its uncertainties for ^{156}Gd as function of primary cascade quanta E_1. Points - the best fit of the presented practical model; triangles - a calculation of $I_{\gamma\gamma}$ using models of Ref. [5, 6]. Recorded threshold for cascade gammas is E_γ = 520 keV.

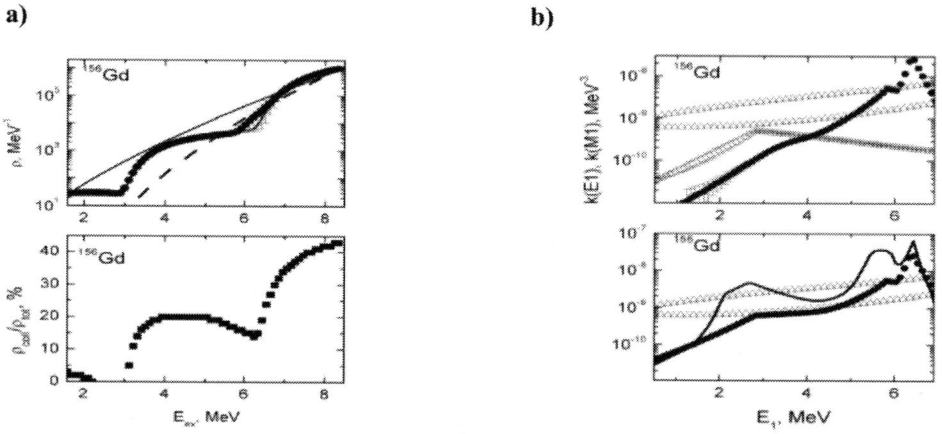

Figure 2. a) Level density of ^{156}Gd. *Top:* points are the best fit of level density (uncertainties – scatter of fits for different sets of initial parameters); dashed and solid lines are the level density calculated using the model of Ref. [5], with taking into account the shell correction δE (6) and without δE, correspondingly. *Bottom:* fitted ratio of density of collective levels to the total level density. **b)** Strength function for ^{156}Gd. *Top:* solid points are the best fit of the strength function of $E1$-transitions; open points are the best fit of the strength function of $M1$-transitions. *Bottom:* solid points are a sum of $E1$- and $M1$- strength functions; dash line is the sum of strength functions multiplied by ρ_{mod}/ρ_{exp} ratio (Ref. [7]). Calculations using the model of Ref. [6] (lower triangles) and using the model of Ref. [8] (upper triangles) were fulfilled with $k(M1)$= const.

4 Results and discussion

A solution of the system of Eq. (1) is performed by the Monte-Carlo method. The nonlinearity of the strongly correlated equations of the system (1) produces an uncertainty of extracting the ρ and Γ parameters from $I_{\gamma\gamma}$ intensities.

Experimental data on $I_{\gamma\gamma}(E_1)$ are usually obtained with a small total uncertainty and averaged over 500 keV energy intervals. The results for ^{156}Gd are shown, in more detail, in Figs. 1–2. The best fits to $I_{\gamma\gamma}(E_1)$, as well as the fitted level densities and strength functions, are compared to corresponding values calculated using the statistical model. The results and corresponding calculations of level density and radiative strength function for the rest of the investigated nuclei will not be shown in

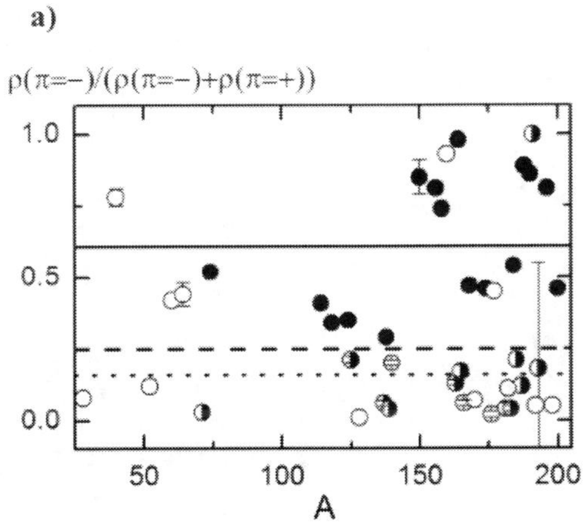

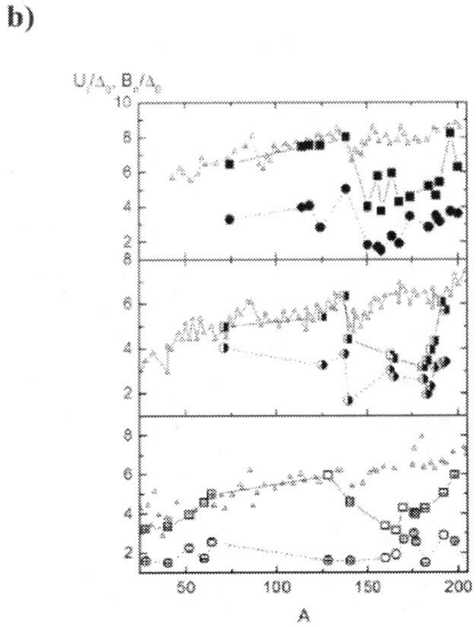

Figure 3. a) A-dependence of the ratios U_l/Δ_0, for the second (points) and the third (squares) Cooper pairs. Full points – even-even, half-open points are even-odd and open points are odd-odd compound nuclei. Triangles – the mass dependence of B_n/Δ_0 ratio. **B)** Mass dependence of the ratio of the level density with negative parity to the total level density at the upper energy border of the E_d and their averages for even-even nuclei (solid lines), even-odd (dashed lines) and odd-odd nuclei (dotted lines). Full points – even-even, half-open points – even-odd and open points – odd-odd compound nuclei.

this publication. However, we are presented here obtained results for some of parameters of nuclear structure.

One important parameter is the breaking thresholds for Cooper pairs. In the present analysis was confirmed the previous results about the connection between the shape of the investigated nucleus and the breaking thresholds. That was established for the first time in our prior analysis [3]. As the breaking thresholds differ for nuclei with various nucleon parities and depend on the average pairing energy (Δ_0) of the last nucleon, the mass dependencies for the ratios of the break up thresholds of the second and the third Cooper pairs to Δ_0, as well as the mass dependence of the binding energy to Δ_0, are presented in Fig. 3. As it can be seen in Fig. 3, there is a noticeable difference in U_2/Δ_0 and U_3/Δ_0 ratios for spherical and deformed nuclei in contrast to B_n/Δ_0.

In this work it was also obtained information about levels parity. For determination of the part $r = \rho(\pi-)/(\rho(\pi-) + \rho(\pi+))$ of levels $\rho(\pi-)$ with negative parity, a linear extrapolation for r value was applied in the $E_d \leq E_{ex} \leq B_n$ energy interval. At that, in the B_n point we use generally accepted assumption, that $\rho(\pi-) = 0.5(\rho(\pi-) + \rho(\pi+))$, and $\rho(\pi-)$ value in this energy point was fixed, and at the E_d energy the $\rho(\pi-)$ value varied.

The calculated ratios of density of the levels with negative parity to the total level density are shown in Fig. 3. The averages of these ratios are 0.61(22), 0.25(28) and 0.16(16) for even-even, even-odd and odd-odd nuclei, respectively (and for odd-even ^{177}Lu it is 0.65(1)). Hence, the behavior of the gamma-decay process is different for nuclei of various nucleon parities.

5 Conclusion

In this work we presented new variant of model for gamma decay of neutron resonance, taking into account shell inhomogeneities of the single-particle level spectra. We used this model for fitting the experimental intensity of two-step gamma cascades and to obtain information about parameters of nuclear structure.

The data on Cooper pair break-up energies, obtained with a high accuracy, are sufficient to conclude that the dynamics of interaction between superfluid and normal phases of a nucleus depends on its' shape. Our model allows for a separate determination of the density of vibrational levels between the breaking thresholds of the Cooper pairs.

Unfortunately, an existence of the sources of uncertainties of the sought ρ and Γ functions is a fundamental problem, and it is inevitable for any nuclear model used for experimental data analysis and for predictions of the spectra and cross sections. There are also fluctuations of the intensities of gamma-transitions in different nuclei, which has a contribution to the systematical error. Nevertheless, the practical model showed one possibility to describe the data of the two-step experiments with the accuracy that exceeds the statistical one.

For future development of reliable model of cascade gamma decay new experimental data are necessary. Because of that, ^{108}Ag, ^{110}Ag, ^{104}Rh and ^{56}Mn nuclei will be investigated by two step gamma cascade method.

References

[1] V.G. Soloviev, Nuclear Physics A **586**(2), 265 (1995)
[2] A.M. Sukhovoj, Phys. Atom. Nucl. **78**, 230 (2015)
[3] A.M. Sukhovoj, L.V. Mitsyna, N. Jovancevic, Phys. Atom. Nucl. **79**, 313 (2016)
[4] V.M. Strutinsky, in Proceedings of the International Congress on Nuclear Physics, Paris, France, p. 617 (1958)
[5] W. Dilg, W. Schantl, H. Vonach, and M. Uhl, Nucl. Phys. A **217**, 269 (1973)
[6] S.G. Kadmenskij, V.P. Markushev and W.I. Furman, Sov. J. Nucl. Phys. **37**, 165 (1983)
[7] N. Jovancevic, A.M. Sukhovoj, W.I. Furman, and V.A. Khitrov, in Proceedings of XX ISINN, Preprint E3-2013-22, p. 157 (Dubna, 2013); `http://isinn.jinr.ru/past-isinns.html`
[8] P. Axel, Phys. Rev. **126**, 671 (1962)

EPJ Web of Conferences **169**, 00008 (2018) https://doi.org/10.1051/epjconf/201816900008
Theory-4

New measurements on isobaric fission product yields and mean kinetic energy for ^{241}Pu thermal neutron-induced fission

Sylvain Julien-Laferrière[1,2,*], *Grégoire* Kessedjian[2,**], *Olivier* Serot[1,***], *Abdelaziz* Chebboubi[1], *David* Bernard[1], *Aurélien* Blanc[3], *Ulli* Köster[3], *Olivier* Litaize[1], *Thomas* Materna[4], *Olivier* Meplan[2], *Michal* Rapala[4], and *Christophe* Sage[2]

[1] *CEA, DEN, DER, SPRC, 13108 Saint-Paul-Lez-Durance, France*
[2] *LPSC, Université Grenoble-Alpes, CNRS/IN2P3, 38026 Grenoble, France*
[3] *ILL, 38042 Grenoble, France*
[4] *CEA, DSM, IRFU, SPhN, 91191 Gif-sur-Yvette, France*

Abstract. Nuclear fission yields data measurements for thermal neutron induced fission of ^{241}Pu have been carried out at the Institut Laue Langevin (ILL) in Grenoble, using the Lohengrin mass spectrometer. Mass, isotopic and isomeric yields have been extracted for the last measurements. A focus is given in this document to the mass yield results which are obtained for almost the entire heavy peak and most of the light high yields masses, along with the covariance matrix. The mean kinetic energy as a function of the fission product mass has also been extracted from the measurements. The total mean kinetic energy pre and post neutron emission have been assessed and compared to other works showing a rather good agreement.

1 Introduction

The isobaric fission product yield, or mass yield $Y(A)$, is the production rate by fission of a nucleus of mass A including all the possible nuclear charge Z. The fission product yield knowledge for thermal neutron induced fission, condensed in the recent evaluated nuclear data libraries (JEFF-3.1.1, ENDF/B-VII.1, JENDL-4.0 ...), is one of the major contributor to the uncertainty on computed reactor physics quantities (see for example [1]). Furthermore, the uncertainty propagation can not be properly done since the yield variance-covariance matrices are missing from the evaluated data libraries, despite recent efforts to address this issue [2–5]. Additionally, some discrepancies between the major libraries exist and need to be understood. For these reasons, a collaboration of the LPSC, the CEA and the ILL focuses on producing precise measurements of fission yields for major actinides with their related experimental variance-covariance matrice [6]. In this document, the ^{241}Pu thermal neutron induced fission isobaric yields and fission product kinetic energy distribution as function of the fission product mass will be addressed. The experimental setup will be presented in Sect. 2, followed by the analysis procedure to extract the isobaric yields in Sect. 3 along with the experimental

*e-mail: sylvain.julien-laferriere@cea.fr
**e-mail: kessedjian@lpsc.in2p3.fr
***e-mail: olivier.serot@cea.fr

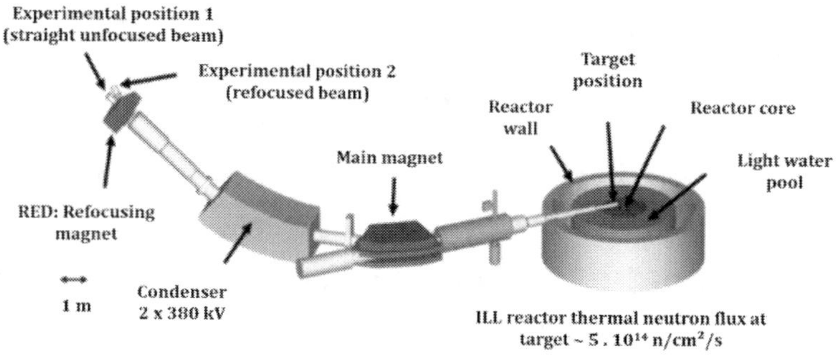

Figure 1. The Lohengrin mass spectrometer at the ILL, Grenoble. The ILL high neutron flux compensates the small geometrical acceptance of the instrument.

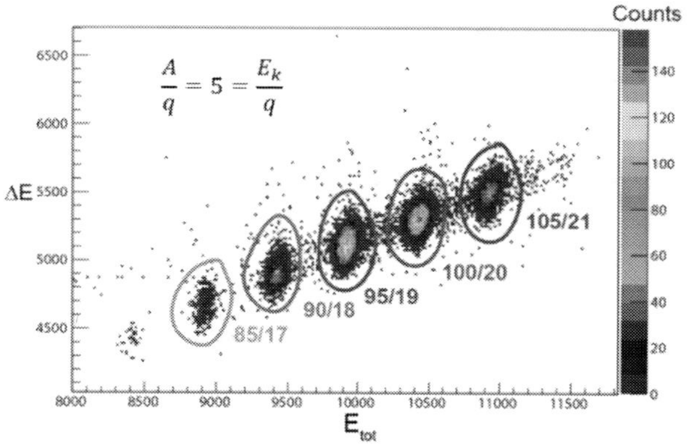

Figure 2. Raw results from the ionization chamber for the ratio $A/q = 5$ and $E_k/q = 5$. The total count rate for each mass can be measured unambiguously with sufficient statistics.

results obtained. Finally, the mean kinetic energy as a function of the fission product mass will be displayed and the corrections applied to get the mean kinetic energy before neutron emission will be detailed in Sect. 4.

2 Experimental setup

The Lohengrin recoil mass spectrometer [7, 8] of which a descriptive figure is presented Fig. 1, has been built in the 70's at the ILL and is since a major instrument for nuclear data measurements in particular due to its high mass resolution ($\Delta A/A \sim 1/400$). A ^{241}Pu target is placed close to the reactor core under a high thermal neutron flux ($\sim 5 \cdot 10^{14}$ n/s/cm^2 at target position) and undergoes a significant amount of fissions. Fission products which enter in the apparatus (solid angle $\leq 3.2{\cdot}10^{-5}$ sr) are deflected through a magnetic and an electric fields. The ion beam is selected according to the A/q and E_k/q ratios, where A is the fission product mass, E_k its kinetic energy and q its ionic charge. The ratios A/q and E_k/q can be achieved with different triplet (A, E_k, q) leading to a possible degeneracy. Two experimental positions exist, at position 1, the ion beam directly enters in an ionization chamber. This position is used to measure isobaric yields, $Y(A)$. A double ionization chamber with a Frisch grid is used, giving access to a bidimentional map of the deposited energy as it can be seen Fig. 2. From Fig. 2 one can also see that the previous degeneracy is well resolved, since each spot can be assigned to a single mass unambiguously. At position 2, the ion beam is refocused and implanted on a moving tape. Two clovers of four high purity germanium crystals each and two broad energy germanium detectors are placed in the vicinity of the tape and detect γ-rays resulting from the fission products β^- disintegration, allowing to measure isotopic yields $Y(A, Z)$.

3 Isobaric yields

Since a beam is selected with Lohengrin according to the mass A, the ionic charge q and the kinetic energy E_k of the fission product, one can measure with the ionization chamber the count rate $N(A, E_k, q)$. The proper definition of the total mass count rate is then:

$$N(A) = \int_{E_k} \sum_{q} N(A, E_k, q)\, dE_k \tag{1}$$

Since the mass yields are normalized to 1 for the heavy peak (H), the yield of the mass A_i will be:

$$Y(A_i) = \frac{N(A_i)}{\sum_{j \in H} N(A_j)} \tag{2}$$

3.1 Analysis procedure

Due to a limited beam time, it is not possible to measure the entire kinetic energy distribution for every ionic charge state. Thus, the measurements have to be optimised in order to avoid biases while performing as much scans as possible in a limited time. For each mass, at least three scans of the kinetic energy distribution at three different ionic charges q_{ref} and one scan of the ionic charge distribution at E_{ref} are made. A minimum of three energy scans are needed in order to estimate the correlation between the mean kinetic energy and the ionic charge (strongly dependent on the target) and be able to correct the ionic charge distribution of the energy bias. In addition, the time evolution of the fissile material quantity remaining in the target, so called Burn-Up (BU), see Fig. 3, is carefully estimated by the overall ionic charge and kinetic energy measurement of mass 136, in order to normalize each measurement. For each energy scan, Eq. (1) becomes:

$$N(A) = \frac{\sum_{E_k} N(A, E_k, q_{ref})}{BU(t)\, P(q_{ref})} \tag{3}$$

Therefore, at least three evaluations of the same $N(A)$ are made. When these evaluations are compatible, a mean value taking into account the experimental correlations is extracted with a reduced variance [9]:

$$\overline{N(A)} = \left(\sum_{i,j}^{n,n} (C^{-1})_{i,j} \right)^{-1} \left(\sum_{i,j}^{n,n} (C^{-1})_{i,j} \cdot N_j(A) \right)$$

$$var(\overline{N(A)}) = \left(\sum_{i,j}^{n,n} (C^{-1})_{i,j} \right)^{-1} \tag{4}$$

Where C is the experimental covariance matrix between the $N(A)$'s, n being the number of energy scans. For the very low yields in the symmetric and very asymmetric regions a different procedure has to be implemented due to the very low count rates, but will not be detailed in this document.

The mass yields $Y(A)$ are finally obtained by normalizing the heavy peak to 1. This auto-normalization gives access to absolute yields while assuring independence towards other existing data. Figure 3 shows on the right the results obtained with Lohengrin compared to JEFF-3.1.1. While important discrepancies are visible in the very low yield regions, the agreement is within the uncertainties for the rest of the distribution. Yet, a systematic difference seems to be visible for the light region, a narrower peak is obtained for the Lohengrin measurements resulting in higher yields than

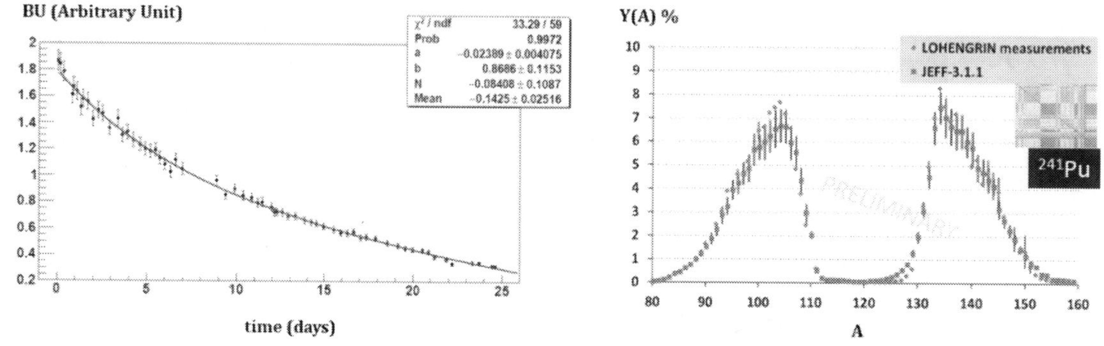

Figure 3. *On the left*, the time evolution of the fissile material quantity in the target, so-called Burn-Up (BU). *On the right*, the ^{241}Pu isobaric fission yields measured at Lohengrin (in blue), and the experimental covariance matrix, compared to JEFF-3.1.1 (in red). Lohengrin measurements are normalized to 1 for the heavy peak. These results are combining two experiments, the first one performed in 2013 [10] and the second one in 2015 [11].

JEFF-3.1.1. Since for ^{241}Pu JEFF-3.1.1 is mainly based on 2E-2v measurements, a possible interpretation of these differences could be a mass resolution effect not perfectly solved.

The uncertainties obtained in this work are taking into account both statistical and systematics uncertainties and are of a few percent for most of the masses and smaller than JEFF-3.1.1 uncertainties.

4 Mean kinetic energy

Two corrections have to be carried out in order to reach the post neutron mean kinetic energy of the fission products. First of all, since the target is evolving through time, so is the mean kinetic energy of the fission products exiting the target. As it can be seen on the left of Fig. 4, the mean of the kinetic energy distribution of mass 136 at ionic charge 23, which is the image of the BU, is decreasing over time. This is mainly because a significant amount of fissile material is implanted on the target support over time, further away from the exit of the target facing Lohengrin. The fission products will then have a bigger mean path through the target, losing more kinetic energy in the process. The Fig. 4, on the right, shows the correction of the measured mean kinetic energies depending on the time at which the measurement has been done The time reference is the beginning of the experiment when the target width is known. This is important since the second correction is to estimate the energy loss in the target. Two layers have to be considered for one target: the Pu layer and the protective Ni foil covering it.

For the energy loss in the Ni foil, the mean energy loss of the mass A is taken to be:

$$\overline{\frac{dE}{dx}}(A) \cdot e \cdot \rho_{mat} = 3.07 \cdot 10^{-4} \cdot \frac{Z_{mat}\, Z_{eff}^2(A)}{A_{mat}\, \beta^2(A)} L(A) \tag{5}$$

Where Z_{eff} is the average equilibrium charge state, Z_{mat} and A_{mat} the nuclear charge and mass of the penetrated material, e its thickness, ρ_{mat} its density, β the reduced fission fragment velocity and L the stopping number. Additional details can be found in [12]. This correction is used for the Ni foil since the needed parameters are given in [12]. The Ni foil thickness of 0.25 μm is given by the constructor with a 20% uncertainty. It is assumed to be by far the major component of the uncertainty for this energy loss.

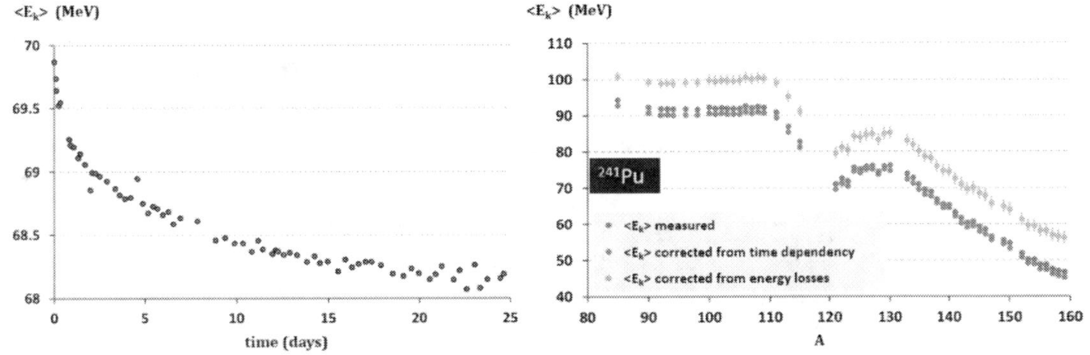

Figure 4. *On the left*, the evolution over time of the measured post-neutron mean kinetic energy for the mass 136 at the ionic charge 23. *On the right*, the measured post-neutron mean kinetic energy as a function of the fission product mass before (in blue) and after correction of the energy dependency to time (in red) and after correction from the energy loss in the Pu and the Ni protective foil (in green).

To estimate the kinetic energy loss in the Pu layer, the stopping range tables from the SRIM software [13] have been used. Some discrepancies have been observed in the kinetic energy loss estimation for the thin Ni layer between SRIM and the correction described in Eq. (5). For this reason, the uncertainty on the loss in the Pu layer is taken equals to be 100%. In fine, the total energy loss (Ni foil + Pu layer) increases from 8 MeV for the light masses up to 11 MeV for the heavy masses, whose around 0.2 MeV are lost in the Pu layer and the rest in the Ni foil, see Fig. 4.

The pre-neutron emission mean kinetic energy can also be accessed using the Monte Carlo deexcitation code Fifrelin [14] sawtooth, $\overline{\nu}_{Fif}(A)$, and using the following formula [15]:

$$\frac{\overline{E}_{k-pre}(A) - \overline{E}_{k-post}(A)}{\overline{E}_{k-pre}(A)} = \frac{\overline{\nu}_{Fif}(A)}{A} \tag{6}$$

The Tab. 1 regroups the results obtained when averaging the mean kinetic energies for the light and heavy masses, respectively $< KE_L >$ and $< KE_H >$, and the total mean kinetic energy $< TKE >$. These three quantities computed for post and pre-neutron emission are compared to existing results. This work presents higher values than previous work but consistent within the uncertainties.

A significant part of the light region has not been measured. As a consequence, the mean kinetic energy of the light fragments, $< KE_L > = \sum_{i \in L} Y(A_i) \cdot < E_k(A_i) >$, can not be properly computed. However, since most of the high yields have been covered, and the mean kinetic energy is not fluctuating much for the light masses, it is still a good estimation.

5 Conclusion

The mass yields for the entire heavy peak and a significant part of the light peak have been measured at Lohengrin and the experimental correlation matrix has been computed. The possibility to normalize the measurements without using nuclear data libraries is an important feature of this work, guaranteeing the independence of the produced fission yields to existing libraries. Further details about the estimations of the correlations between the different Y(A) and the analysis scheme in the particular case of the very low yield regions will be explicated in future publications. The symmetry region

Table 1. Mean kinetic energy for the light fragments, $< KE_L >$, the heavy fragments, $< KE_H >$ and the total mean kinetic energy $< TKE >$ obtained in this work for post and pre neutron emission, compared to previous works and to JEFF-3.1.1.

Post neutron emission	This work	Thierens et al. [16]	JEFF-3.1.1
$< KE_L >$	99.7 ± 0.9	-	-
$< KE_H >$	75.6 ± 1.1	-	-
$< TKE >$	175.8 ± 1.5	176.93 ± 0.06	175.37 ± 0.68
Pre neutron emission	This work	Neiler et al. [17]	
$< KE_L >$	101.3 ± 0.9	103.2 ± 1.0	
$< KE_H >$	76.1 ± 1.2	76.3 ± 0.8	
$< TKE >$	177.5 ± 1.6	179.6 ± 1.8	

is particularly interesting since it is a sensitive region to probe fission mode signature in the kinetic energy distribution, this work is still on going. The mean kinetic energies obtained are similar to previous works, especially before neutron emission correction.

This work was supported by CEA, IN2P3 and "le défi NEEDS". The authors are grateful for the support of the ILL and all the staff involved from CEA Cadarache and LPSC.

References

[1] N. Terranova, PhD thesis, *Covariance Evaluation for Nuclear Data of Interest to the Reactivity Loss Estimation of the Jules Horowitz Material Testing Reactor* (Université di Bologna, 2016) Chapter 6

[2] N. Terranova, O. Serot, P. Archier, C. de Saint Jean and M. Sumini, Nucl. Data Sheets **123**, 225–230 (2015)

[3] N. Terranova, O. Serot, P. Archier, C. de Saint Jean and M. Sumini, accepted for publication in Annals of Nucl. En. (2017)

[4] L. Fiorito, A. Stankovskiy, G. Van den Eynde, C.J. Diez, O. Cabellos, P.E. Labeau, Annals of Nucl. En. **88**, 12–23 (2016)

[5] D. Rochman, P. Leray, A. Vasiliev, H. Ferroukhi, A.J. Koning, M. Fleming, J.C. Sublet, Annals of Nucl. En. **95**, 125–134 (2016)

[6] O. Serot, C. Amouroux, A. Bidaud, N. Capellan, S. Chabod, A. Ebran, H. Faust, G. Kessedjian, U. Köster, A. Letourneau, O. Litaize, F. Martin, T. Materna, L. Mathieu, S. Panebianco, J.-M. Regis, M. Rudigier, C. Sage, and W. Urban, Nucl. Data Sheets **119**, 320-323 (2014)

[7] G. Fioni, H.R. Faust, M. Gross, M. Hesse, P. Armbruster, F. Gönnenwein and G. Müzenberg, Nucl. Instrum. Meth. A **332**, 175-180 (1993)

[8] H.R. Faust, P. Geltenbort, F. Gönnenwein and A. Oed, ILL Internal Scientific Report **81FA45S** (1981)

[9] M. Schmelling, Physica Scripta **51**, 676-679 (1995)

[10] F. Martin, PhD thesis, *Étude des distributions en masse, charge nucléaire et énergie cinétique des produits de fission de l' $^{233}U(n_{th},f)$ et du $^{241}Pu(n_{th},f)$ mesurées auprès du spectromètre de masse Lohengrin (ILL)* (Université de Grenoble, 2013) Chapter 7

[11] S. Julien-Laferrière, ongoing PhD thesis

[12] G. Knyazheva et al., Nucl. Instrum. Meth. B **7**, 248 (2006)

[13] J.F. Ziegler, M.D. Ziegler and J.P. Biersack, Nucl. Instrum. Meth. B **268**, 1818–1823 (2010)
[14] O. Litaize, O. Serot and L. Berge, Eur. Phys. J. A **51**, 117 (2015)
[15] C. Wagemans, *The Nuclear Fission Process* (CRC Press, Boca Raton, Florida, 1991) 319
[16] H. Thierens, E. Jacobs, P. D'Hondt, A. de Clercq, M. Piessens, and D. de Frenne, Phys. Rev. C **29**, 498-507 (1984)
[17] J.N. Neiler, F.J. Walter, and H.W. Schmitt, Phys. Rev. **149**, 894 (1966)

Determination of the fast-neutron-induced fission cross-section of ^{242}Pu at *n*ELBE

Toni Kögler*, *Roland* Beyer, *Arnd R.* Junghans**, *Ronald* Schwengner, and *Andreas* Wagner
Helmholtz-Zentrum Dresden - Rossendorf, Postfach 510 119, 01314 Dresden, Germany

Abstract. The fast-neutron-induced fission cross section of ^{242}Pu was determined in the energy range of 0.5 MeV to 10 MeV at the neutron time-of-flight facility *n*ELBE. Using a parallel-plate fission ionization chamber this quantity was measured relative to ^{235}U(n,f). The number of target nuclei was thereby calculated by means of measuring the spontaneous fission rate of ^{242}Pu. An MCNP 6 neutron transport simulation was used to correct the relative cross section for neutron scattering. The determined results are in good agreement with current experimental and evaluated data sets.

1 Introduction

Neutron-induced fission cross sections of actinides such as the Pu-isotopes are of relevance for the development of nuclear transmutation technologies. Apart from ^{244}Pu, whose abundance in spent nuclear fuel could be neglected [1], ^{242}Pu is the plutonium isotope with the longest half-life ($T_{1/2}$ = 373 300 y). Thus, there is a special interest to investigate the fission of this particular isotope using fast neutrons.

The fast neutron-induced fission cross section of ^{242}Pu was studied first by Butler in 1960 [2]. A brief summary of the available experimental data acquired since then is given in Ref. [3]. Results of measurements at the Los Alamos National Laboratory by Tovesson et al. [4], at the Joint Research Center Geel by Salvador-Castiñeira et al. [3] and at the National Physical Laboratory of the United Kingdom in Teddington by Matei et al. [5] have been recently published and tend to be lower than the current european data evaluation JEFF 3.2 [6].

For ^{242}Pu(n,f) current uncertainties are of about 21% in an energy range of 0.5-2.23 MeV. Sensitivity studies [7, 8] show that the total uncertainty has to be reduced to below 5% to allow for reliable neutron physics simulations of plutonium using reactor concepts.

This challenging task was performed at the neutron time-of-flight facility *n*ELBE of the Center for High-Power Radiation Sources ELBE at Helmholtz-Zentrum Dresden - Rossendorf. Improved experimental conditions (low radiation background) and neutron beam intensity, paired with the right spectral shape of the neutron source provide excellent conditions to achieve this aim.

The present work reports an experiment at *n*ELBE studying the neutron-induced fission cross section of ^{242}Pu relative to ^{235}U(n,f).

*First author e-mail: `t.koegler@hzdr.de`
**Responsible author e-mail: `a.junghans@hzdr.de`

2 Experiments

2.1 Fast neutrons at ELBE

At *n*ELBE fast neutrons with energies between 10 keV and 20 MeV are generated by a photo-neutron source. Impinging electrons of the ELBE beam produce bremsstrahlung while getting decelerated in a liquid lead target [9]. Neutrons are subsequently produced via (γ, n) reactions on the lead nuclei. The compactness of the neutron producing target together with the excellent timing of the ELBE electron beam enables high resolution neutron time-of-flight experiments even at short flight paths of around 6 m. Further details of *n*ELBE can be found in Ref. [10]. For this experiment a repetition rate of 406.25 kHz was chosen as a compromise between avoiding pulse overlap and maximum beam intensity. The electron beam energy was 30 MeV and the average bunch charge 73 pC. The used fission chambers, which will be described in more detail in the following section, were placed at a distance of 5.95–6.35 m to the neutron source. The neutron beam diameter at this point was determined to be 56 mm.

2.2 Fission chambers

Within the TRAKULA project [11], large (Ø 74 mm) and homogeneous deposits of ^{235}U and ^{242}Pu have been produced successfully [12] and two fission ionization chambers were constructed at HZDR [13, 14], one assembled with uranium (UFC) and one with plutonium (PuFC). In the present experiment, the PuFC was used in connection with the transfer device H19 of the PTB Braunschweig [15, 16]. The well characterized ^{235}U fission chamber was used for the determination of the incident neutron flux. Both fission chambers were operated in the forward biasing mode, which means that the five double-sided fission samples of the H19 and the eight single-sided samples of the PuFC are "cathodes" on positive potential. The induced charge on the anodes of the PuFC were read out by in-house developed preamplifiers with nanosecond discharge times.

2.3 Data acquisition

The short signal length of the ns-preamplifier reduces the α-pile-up probability by nearly a factor of 5 compared to conventional µs-shaping time preamplifiers. In addition it enables the use of a charge-to-digital converter (QDC) instead of the combination of a spectroscopic amplifier and an analog-to-digital converter (ADC). A scheme of the VME-based data acquisition electronics is shown in Fig. 1. Further details of the used experimental setup and electronics can be taken from Ref. [17].

2.4 Data analysis

The recorded list-mode data were analyzed using the Qt/ROOT-based software framework Go4 [18]. The produced time-of-flight spectra (cf. Fig. 2) were calibrated with respect to the bremsstrahlung peak. A gate on the QDC values was applied to filter out the α-particle background coming from the radioactive decay of the target nuclei. Time-independent fission events caused by room-return neutron background and spontaneous fission were approximated by a constant determined by averaging the number of events in the ranges in front of the bremsstrahlung peak, between this peak and the begin of the neutron-induced fission events and at the end of the spectrum.

Both, time-of-flight spectrum and background were translated into kinetic energy spectra by using Eq. (1).

$$E_n(t) = m_n c^2 \left(\frac{1}{\sqrt{1 - \left(\frac{L}{tc}\right)^2}} - 1 \right), \tag{1}$$

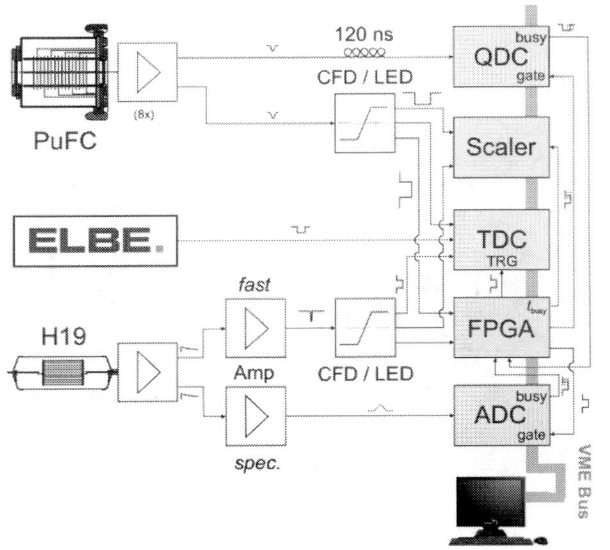

Figure 1. Scheme of the electronic setup and the data acquisition system. The output signals of the charge-sensitive (ns-) preamplifiers connected to the PuFC are split to determine the timing and the collected charge. The pulse heights of the H19 signals are acquired with an ADC after getting shaped by a spectroscopic amplifier, whereas the charge of the eight PuFC channels (only one is shown here) is determined by a QDC. The second output signal is converted to a logical signal by an in-house developed constant fraction discriminator (CFD). The production of a fast trigger makes the use of a timing-filter-amplifier in the fast branch of the H19 necessary. The logical signals are used to determine the timing in a time-to-digital converter (TDC) and to produce a trigger for the data acquisition in an FPGA. The radio-frequency of the ELBE accelerator is used as reference for the time-of-flight measurement.

where $m_n c^2$ stands for the neutron rest energy, L for the flight path and t for the time-of-flight of the neutron to a certain actinide target.

To combine the time-of-flight spectra of different targets, the same energy binning has to be used. For that, a binning was chosen, which corresponds to the flight path of the H19 which was placed in front of the PuFC. The time-of-flight spectra of the PuFC were rebinned to that particular binning using a Monte-Carlo approach. This approach generates "pseudo"-time-of-flights within a given time-of-flight bin, calculates the kinetic energy using Eq. (1) and the flight path to the considered target. The kinetic energy is then filled to the universal binning.

Subtracting the estimated background only after this conversion helps to eliminate fluctuations introduced by the Monte-Carlo during the rebinning process. The result of this procedure is shown in Fig. 2 on the right side.

After background subtraction the relative fission cross section can be calculated by:

$$\frac{\sigma_{(\mathrm{n,f})}^{\mathrm{PuFC}}}{\sigma_{(\mathrm{n,f})}^{\mathrm{H19}}} = K \frac{\sum_i \left(C\varepsilon \dot{N}_{(\mathrm{n,f})}\right)^{\mathrm{PuFC}_i}}{\langle C\rangle^{\mathrm{H19}} \sum_i \left(\varepsilon \dot{N}_{(\mathrm{n,f})}\right)^{\mathrm{H19}_i}}. \tag{2}$$

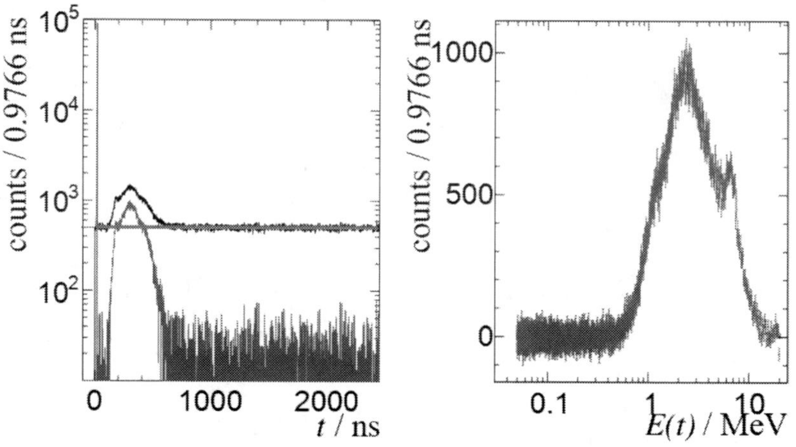

Figure 2. *Left*: neutron time-of-flight spectrum acquired by one channel of the plutonium fission chamber before (in black) and after (in blue) background subtraction. The red line indicates a constant extrapolation of the background induced by room-returned neutrons and spontaneous fission events. *Right*: background subtracted energy spectrum calculated using the time-of-flight spectrum shown on the left side.

Here $\dot{N}_{(n,f)} = \dot{N}_{(n,f)}(E_n(t))$ is the detected neutron-induced fission rate at a certain time-of-flight t in the i-th actinide target of the PuFC or the H19. This count rate has to be corrected for the fission fragment detection efficiency ε and for neutron scattering C, which will be discussed later in Sect. 2.5.

The normalization factor K is the ratio of the total areal densities n_A of the two fission chambers and depends only on constants.

Because ^{242}Pu has a high spontaneous fission (SF) rate $\lambda\gamma$, which is in addition known with an uncertainty better than 2% (see Ref. [19, 20]), n_A^{PuFC} can be determined by measuring the spontaneous fission rate $\dot{N}_{(\mathrm{SF})}$. As the target area F^{PuFC_i} is equal for all eight deposits this simplifies to a sum over the effective areal density of each actinide target:

$$n_A^{\mathrm{PuFC}} = \sum_i n_A^{\mathrm{PuFC}_i} = \frac{1}{\lambda\gamma F}\sum_i \left(\alpha\varepsilon\dot{N}_{(\mathrm{SF})}\right)^{\mathrm{PuFC}_i} \tag{3}$$

Here α accounts for the small dead time correction of the DAQ during this measurement. With Eq. (3) the normalization factor can be written as follows:

$$K = \frac{\lambda\gamma F n_A^{\mathrm{H19}}}{\sum_i \left(\alpha\varepsilon\dot{N}_{(\mathrm{SF})}\right)^{\mathrm{PuFC}_i}}. \tag{4}$$

Using this expression in Eq. (2), the relative cross section gets independent from the fission fragment detection efficiency of the PuFC, which is usually difficult to determine [21]. This technique was already introduced by Weigmann et al. in Ref. [21]. Of course one has to correct for the linear and angular momentum introduced by the incident neutrons. This results in a fission fragment anisotropy, which lowers the detection efficiency[22]. Due to the lack of experimental data for this anisotropy, this inefficiency was not corrected for. An estimation of this measure was done using unpublished data of Salvador-Castiñeira et al. and included to the systematic uncertainties.

2.5 Neutron Scattering Corrections

To correct the time-of-flight spectra for influence of neutron scattering one has to consider two different aspects. The first is the attenuation of the neutron beam by passing any material along its track. The second is the loss of the energy to time-of-flight correlation of scattered neutrons. Especially for inelastically scattered neutrons this is an important issue, because they are losing a large amount of their total kinetic energy in one interaction. Measuring their time-of-flight afterwards will lead to a much higher kinetic energy, if the scattering was close to the fission targets. As a result of that, the cross section at high energy would be overestimated.

Both effects have been corrected for using an MCNP 6.1.1 simulation. The outcome of the simulation is a correlation matrix of the true kinetic energy E_n and the kinetic energy $E_n(t)$ one would calculate from their time-of-flight and the assumed undisturbed flight path. Such a correlation matrix for an arbitrary actinide target is shown in Fig. 3.

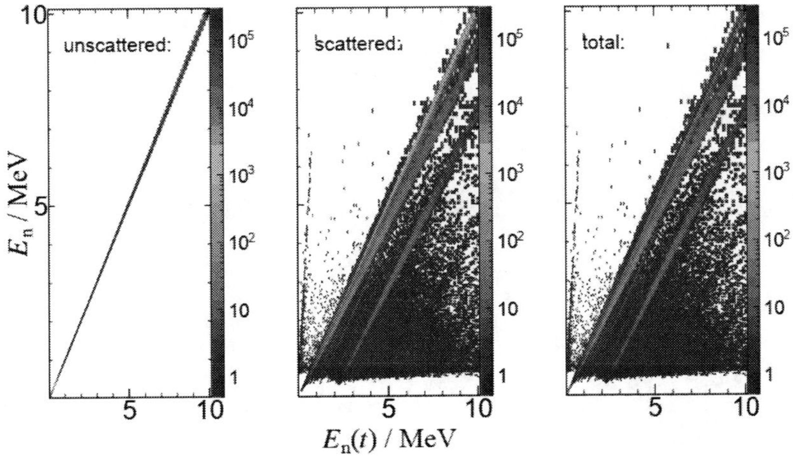

Figure 3. Simulated energy to time-of-flight-correlation of the first PuFC deposit in the neutron beam calculated using MCNP 6.1.1. Shown are from left to right: unscattered neutrons, scattered neutrons and the sum of both. Here the bin content was multiplied columnwise with the ^{242}Pu fission cross section to become proportional to the fission rate. The visible structures in the spectrum arise mainly from neutron elastic and inelastic scattering on target backing (mostly ^{28}Si) and chamber stainless steel windows (mostly ^{56}Fe).

The attenuation of the neutron beam was corrected by the ratio of all impinging neutrons N^i in the i-th actinide target, which have not been scattered on their way to the target (on the left of Fig. 3), to the total number of neutrons N_0 started from the neutron production target. Hence one could define a transmission correction factor T up to the target i as:

$$T^i(E_n(t)) = \frac{N^i(E_\mathrm{n} = E_n(t))}{N_0} \tag{5}$$

The loss of the energy to time-of-flight correlation could be expressed in a similar way. Here the correction factor k^i is the ratio of the detected fission rate of unscattered neutrons to the total detected fission rate. Because the fission rate depends on the cross section, the results of the simulations were multiplied by the JEFF 3.2 evaluated fission cross section of ^{242}Pu:

$$k^i(E_n(t)) = \frac{N^i(E_n = E_n(t), E_n(t))\sigma(E_n(t))}{\int N^i(E_n', E_n(t))\sigma(E_n')\mathrm{d}E_n'} \tag{6}$$

With Eq. (5) and Eq. (6) the neutron scattering correction factor $C^i(E_n(t))$ is defined in the following way:

$$C^i = \frac{k^i}{T^i} \tag{7}$$

As only the sum of all H19 fission targets is available, the arithmetic mean $\langle C\rangle^{\mathrm{H19}}$ was calculated to take care of the neutron scattering within this chamber. Then, the average total correction factor is in the order of 9%.

3 Results

Including the neutron scattering corrections into Eq. (2) one can finally calculate the relative neutron-induced fission cross section of ^{242}Pu. The result of this calculation is shown in Fig. 4.

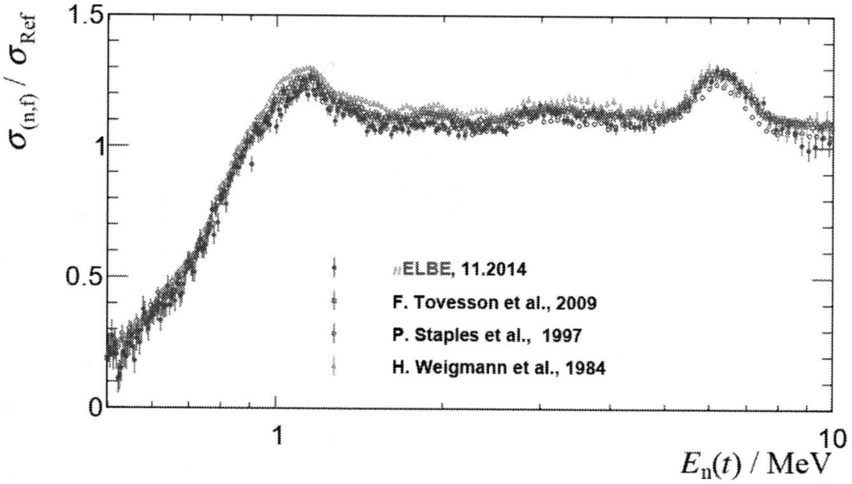

Figure 4. Neutron-induced-fission cross section of ^{242}Pu relative to ^{235}U. The *n*ELBE data are shown in blue together with selected EXFOR-data of Tovesson et al. [4], Staples et al. [23] and Weigmann et al. [21]. Within their uncertainties there is a good agreement of the presented data set with the data of Staples and Tovesson. Small deviations from the re-normalized (ENDF/B-V→ ENDF/B-VII.1) Weigmann data are clearly visible.

The *n*ELBE data are in good agreement in shape and absolute scale to previous experimental data. Compared to the data of Staples et al. [23], the accordance in absolute scale is above 99%. The discrepancy to the Tovesson [4] data set is around 2% in the considered energy range. Larger deviations to the Weigmann data [21] in the order of 5.4% are present especially in the plateau region between 1.2 MeV and 5 MeV. The recently published data of Salvador-Castiñeira et al. [3] are also below the Weigmann cross section ratio. This is of special interest, because the current European evaluation JEFF 3.2 (see Ref. [6]) is mainly based on the Weigmann data.

3.1 Uncertainties

The statistical uncertainty[1] of the presented data ranges from 1.1% in the plateau to ≈ 47% in the threshold and second-chance fission region, where the neutron fluence of *n*ELBE is too low to achieve better statistics within a reasonable measuring time.

The main contribution to the systematic uncertainty is caused by the uncertainty in the normalization factor K ($\sigma_K/K \approx 2.1\%$), where the uncertainty in the determination of the target area F is one of the largest terms. Here a very conservative assumption was used by considering edge effects in the order of 1 mm with respect to the target diameter. This contribution was $\sigma_F/F \approx 1.1\%$. Together with the uncertainty of the remaining quantities in K this gives a 2.1% uncertainty of the normalization factor. If one includes the uncertainty in the reference cross section (≈ 0.7%), the neutron scattering correction (0.21%) and the uncertainty in the fission fragment anisotropy caused by linear and angular momentum transfer (1.6%) one ends up in a total systematic uncertainty of 2.7%.

4 Summary

The fast neutron-induced fission cross section of ^{242}Pu has been measured in the range of 0.5–10 MeV at *n*ELBE. The neutron scattering is a crucial correction which has to be done in neutron time-of-flight experiments. For the described experiment the average scattering correction was around 9%.

The presented data are in good agreement with previous data sets in the threshold region. At the plateau we achieved excellent agreement with the Staples and Tovesson data, but there are deviations from the data of Weigmann et al., which the JEFF 3.2 evaluation is mainly based on. In this energy range we achieved a statistical uncertainty of 1.1%. The average systematic uncertainty is 2.7% caused mainly by the uncertainty in the normalization.

We would like to thank Manfred Sobiella and Klaus Heidel for the support in the construction of the fission chamber and the preamplifiers. We also thank Andreas Hartmann for the continuous support in the preparation of our experiments. Special thanks go to Alessio Vascon and Klaus Eberhardt from the Johannes Gutenberg University of Mainz for their effort in the production of our fission samples. Thanks also go to the ELBE accelerator crew and all of the colleagues, who have participated during the experiments. This work was supported by the German Federal Ministry of Education and Research under contract No. 02NUK13A.

References

[1] A. Schwenk-Ferrero, Science and Technology of Nuclear Installations (2013), Art.ID 293792
[2] A. Butler, Phys. Rev. **117**, 1305 (1960)
[3] P. Salvador-Castiñeira et al., Phys. Rev. C **92**, 044606 (2015)
[4] F. Tovesson et al., Phys. Rev. C **79**, 014613 (2009)
[5] C. Matei et al., Phys. Rev. C **95**, 024606 (2017)
[6] OECD / NEA, *JEFF-3.2 evaluated data library - Neutron data* `https://www.oecd-nea.org/dbforms/data/eva/evatapes/jeff_32/`
[7] Working Party on International Evaluation Co-Operation, OECD / NEA (2008) `http://www.oecd-nea.org/science/wpec/volume26/volume26.pdf`
[8] OECD / NEA, *The Nuclear Data High Priority Request List (HPRL)* `http://www.oecd-nea.org/dbdata/hprl/`
[9] E. Altstadt et al., Ann. Nucl. Energy **34**, 36 (2007)
[10] R. Beyer et al., Nucl. Instr. and Meth. A **723**, 151 (2013)

[1] All uncertainties are given as 1 σ values.

[11] *Das TRAKULA-Projekt (BMBF 02NUK13A)*
`http://www.hzdr.de/db/Cms?pNid=212&pOid=30395`
[12] A. Vascon et al., Appl. Radiat. Isot. **95**, 36 (2015)
[13] T. Kögler et al., Phys. Proc. **47**, 178 (2013)
[14] T. Kögler et al., CERN-Proceedings **2014-002**, 25 (2014)
[15] D.B. Gayther, Metrologia **27**, 221 (1990)
[16] R. Nolte et al., Nucl. Sci. Eng. **156**, 197 (2007)
[17] T. Kögler, Dissertation, Technische Universität Dresden (2016), Dresden, Germany,
`http://nbn-resolving.de/urn:nbn:de:bsz:14-qucosa-223314`
[18] J. Adamczewski-Musch et al., *The Go4 Analysis Framework*, GSI
`http://web-docs.gsi.de/ go4/go4V05/manuals/Go4introV5.pdf`
[19] N.E. Holden et al., Pure Appl. Chem. **72**, 1525 (2000)
[20] M.M. Bé et al., *Table of Radionuclides*, Vol. 5 (2010),
`http://www.bipm.org/en/publications/scientific-output/monographie-ri-5.html`
ISBN 92-822-2234-8
[21] H. Weigmann et al., Nucl. Phys. A **438**, 333 (1985)
[22] G. W. Carlson., Nucl. Instr. and Meth. **119**, 97 (1974)
[23] P. Staples et al., Nucl. Sci. Eng. **129**, 149 (1998)

Measurement of the ROT effect in the neutron induced fission of ^{235}U in the 0.3 eV resonance at a hot source of polarized neutrons

Yuri Kopatch[1,2,⋆], *Vadim* Novitsky[1,2], *Gadir* Ahmadov[1,6], *Alexei* Gagarsky[3], *Daniyar* Berikov[1,7], *Gevorg* Danilyan[1,2], *Vladimir* Hutanu[4], *Jens* Klenke[5], and *Sergey* Masalovich[5]

[1] *Joint Institute for Nuclear Research, 141980 Dubna, Russia*
[2] *Institute for Theoretical and Experimental Physics of National Research Centre "Kurchatov Institute", 117218 Moscow, Russia*
[3] *Petersburg Nuclear Physics Institute of National Research Centre "Kurchatov Institute", 188300 Gatchina, Russia*
[4] *Institut für Kristallographie, RWTH Aachen and Jülich Centre for Neutron Science at Heinz Maier-Leibnitz Zentrum (MLZ), Lichtenbergstr. 1, 85748 Garching, Germany*
[5] *Forschungs-Neutronenquelle Heinz Maier-Leibnitz, 85747 Garching, Germany*
[6] *National Nuclear Research Centre, Baku, Azerbaijan*
[7] *Institute of Nuclear Physics of the National Nuclear Center of Kazakhstan, 050032 Almaty, Kazakhstan*

Abstract. The TRI and ROT asymmetries in fission of heavy nuclei have been extensively studied during more than a decade. The effects were first discovered in the ternary fission in a series of experiments performed at the ILL reactor (Grenoble) by a collaboration of Russian and European institutes, and were carefully measured for a number of fissioning nuclei. Later on, the ROT effect has been observed in the emission of prompt gamma rays and neutrons in fission of ^{235}U and ^{233}U, although its value was an order of magnitude smaller than in the α-particle emission from ternary fission. All experiments performed so far are done with cold polarized neutrons, what assumes a mixture of several spin states, the weights of these states being not well known. The present paper describes the first attempt to get "clean" data by performing the measurement of gamma and neutron asymmetries in an isolated resonance of ^{235}U at the POLI instrument of the FRM2 reactor in Garching.

1 Introduction

The T-odd effects in fission of heavy nuclei have been known since more than a decade. The first effect of this type, the so called TRI-effect was discovered at the ILL reactor (Grenoble) by a collaboration of Russian and European institutes [1,2] in an experiment aimed at the search for violation of time reversal invariance (TRI) following the idea, proposed in [3]. It was found that the probability of emission of an alpha-particle in ternary fission in the direction perpendicular to the plane formed by the neutron spin and the fragment momentum shows a pronounced anisotropy. The magnitude of the effect turned out to be surprisingly large, and the current explanation doesn't imply the existence of such a violation, but is based on the final state interaction of the reaction products. In other words, the

⋆e-mail: `kopatch@nf.jinr.ru`

effect is not connected with the violation of the time reversal invariance, but with the mechanism of the fission process. Nevertheless, the effect is still called the "TRI-effect" in literature.

Furthermore, it was noticed that when reversing the direction of polarization of the neutron beam, the angular distribution of α-particles is shifted by a small angle relative to the axis of fragment emission, the offset direction being determined by the direction of polarization of the neutron beam. The authors called this effect the ROT-effect [4]. Both, TRI and ROT effects are formally T-odd, but have no direct connection with the violation of the time reversal invariance.

From the semiclassical description of the ROT effect, which assumes the rotation of a polarized nucleus before it splits into two (or three) fragments it follows that a similar phenomenon can also be discovered in the angular distribution of some other particles accompanying the fission of a nucleus into two fragments. The effect can be observed only if this distribution is anisotropic with respect to the deformation axis of the fissioning nucleus at the moment of scission, and the asymmetry with respect to the initial direction of the deformation axis is fully or partly conserved after the escape of the fragments to infinity. Indeed, a similar effect has been observed in the emission of prompt gamma rays and neutrons in fission of ^{235}U and ^{233}U, although its value was an order of magnitude smaller than in the α-particle emission from ternary fission [5–7].

At present, there are several theoretical models which can describe both effects [8–13]. According to the model, proposed in [13], both effects depend on the quantum numbers J and K (the total angular momentum and its projection on the deformation axis), which characterize the fission channel. For the thermal (or cold) neutron induced fission (where all previous data are obtained), there is a mixture of several spin states, and the weights of these states are not known. The only way to get "clean" data is to perform measurements in isolated resonances. Such an experiment was performed at the POLI instrument of the FRM2 reactor in Garching, which provides the necessary polarized neutron beam with an energy of 0.27 eV, corresponding to the lowest resonance of ^{235}U. Preliminary results of this experiment are presented in this paper.

2 Experiment

We used the polarized hot neutron beam provided by the POLI instrument [14] at the FRM-II reactor in Garching. The schematic view of the experimental setup is shown in Fig. 1. A monochromator made of a mosaic of Cu crystals was used to select a narrow neutron beam with the mean energy of 270 meV (λ=0.55 Å). This energy exactly coincides with the position of the lowest resonance of ^{235}U [15]. The monochromator also allows simultaneous focusing of the neutron beam on the target position providing the maximum intensity of unpolarized neutrons of about 4×10^6 n/cm^2/sec.

The neutrons were polarized using specially designed ^{3}He gas cells [16]. The same type of cell was also used as analyzer for measuring beam polarization. Since polarized nuclei of ^{3}He possess very high spin-dependent neutron absorption efficiency over a wide range of energies, the ^{3}He cell can be used as a broadband neutron polariser or analyser, with the possibility to optimise its efficiency for nearly all neutron wavelengths. In our experiment, the size of the cells was Ø60 × 130 mm and the gas pressure 2.5 bar (0.25 MPa), which provided the maximal neutron polarization of about 70%. The polarizer and analyzer cells were polarized in an external lab and placed into a special magnetic housing with highly homogeneous constant magnetic field. The polarization of ^{3}He in the cell exponentially decreased with the time constant of about 40 hours, therefore both cells were replaced every 24 hours.

Both, polarizer and analyzer provided vertical polarization of the neutron beam while the searched effect requires horizontal (longitudinal) polarization. For changing the polarization direction from vertical to horizontal, a specially designed spin control system was used, consisting of several μ-metal

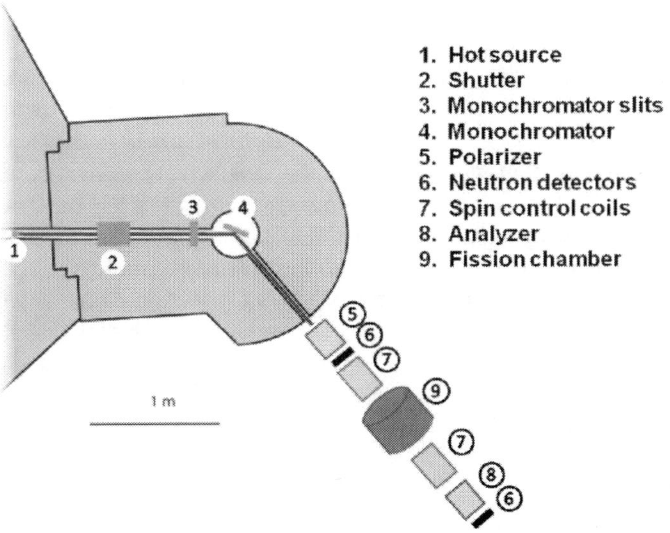

Figure 1. Schematic view of the experimental setup at the POLI facility of FRM-II.

shielded magnetic coils, which allowed also flipping the spin at the target position by 180 degrees every 1.3 seconds.

The schematic view of the fission chamber surrounded by a set of gamma-ray detectors is shown in Fig. 2. The chamber was filled with CF_4 gas at a pressure of about 10 mbar. A uranium target containing about 82 mg of ^{235}U (99.99%) oxide-protoxide deposited on the two sides of a thick 40 $\times$ 100 mm^2 aluminium backing was arranged along the chamber axis. Thin low-pressure multiwire proportional counters (MWPC) were used as fission fragment detectors, being placed on two sides of the target at a distance of ~3 cm (start detector) and ~11 cm (stop detector). Start and stop detectors are supposed to be used for measuring the fragment velocities (momenta). Eight cylindrical plastic scintillators were inserted in a rotatable holder at a distance of about 30 cm from the target center that ensures subsequent measurements of coincidences of prompt fission gamma rays and neutrons with fission fragments at angles of ±22.5, ±67.5, ±112.5 and ±157.5 degrees with respect to the mean axis of the detection of fragments. The detectors of gamma rays and fission fragments were arranged in the plane orthogonal to the neutron beam direction, which also coincides with the axis of the polarization of ^{236}U nuclei.

Prompt neutrons could be rather well separated from the prompt gamma-rays using the time-of-flight method (see Fig. 3). Every event matching coincidence of the signals from the gamma/neutron and fragment detectors is digitized by a multichannel TDC CAEN V775N and stored together with the information about the direction of polarization of the neutron beam. A reversal of the polarization occurs at a frequency of 1.3 Hz, the input of the TDC being inhibited by the time of the neutron spin flip. At the same time, for the on-line control of the installation, the coincidence count rates of neutrons/γ-rays and the fission fragments were recorded by counters, which were read out every 5min for each detector. The values of the asymmetries, calculated by the formula:

$$R = (N^+ - N^-)/(N^+ + N^-) \qquad (1)$$

were constantly monitored. Here N^+ and N^- are the coincidence count rates for opposite directions of the neutron polarization. Simultaneously, the asymmetry of the fragment count rates was measured and controlled.

The total time allocated for the experiment at the POLI facility was 11 days. Four days were spent for the installation of the setup, calibration of the detectors, alignment of the spin control system etc. The statistic was accumulated during 7 days. Because of the rather short measurement time, to increase the coincidence rate only "start" fission fragment detectors were used in the analysis, which made the determination of the fragment mass (separation of light and heavy fragment groups) impossible and increased the angular spread of the fission fragments. For the ROT-effect the light-heavy fragment separation is not necessary, as the effect is mass-symmetric. The use of only the "start" detector increased the solid angle of the fission fragment detection which could lead to a decrease of the potentially observed effect, but significantly increased the accumulated statistics, which was more important considering that the main aim of this pioneering experiment was to observe the effect, but not to measure its absolute value.

3 Results and discussion

Figure 4 shows the anisotropy ratio R determined from the experimental data according to formula (1), for prompt gamma-rays (left) and neutrons (right), detected in coincidence with one of the fission fragments. Each point corresponds to the angular detector position with respect to the average direction of emission of the detected fission fragments. The angular dependence at first approximation can be fitted by the function $F = A \cdot \sin(2\theta)$, which is shown on the plots. The anisotropy parameter A could be determined from the fit and equals to $A_\gamma = (\mathbf{-6.5 \pm 3.9}) \times \mathbf{10^{-5}}$ for the gamma-rays and $A_n = (\mathbf{+3.8 \pm 4.1}) \times \mathbf{10^{-5}}$ for the neutrons, χ^2/N being 0.98 and 1.70, respectively. These results can be compared to the corresponding values for ^{235}U, obtained with cold neutrons: $A_\gamma = (\mathbf{-16.6 \pm 1.6}) \times \mathbf{10^{-5}}$ (at 45 degrees) and $A_n = (\mathbf{-21.2 \pm 2.5}) \times \mathbf{10^{-5}}$ (at 22.5 degrees). Although the statistical accuracy is not sufficient to claim that the ROT effect has been observed in the lowest resonance of ^{235}U, one can conclude that the effect is definitely smaller than that in the cold neutron induced fission.

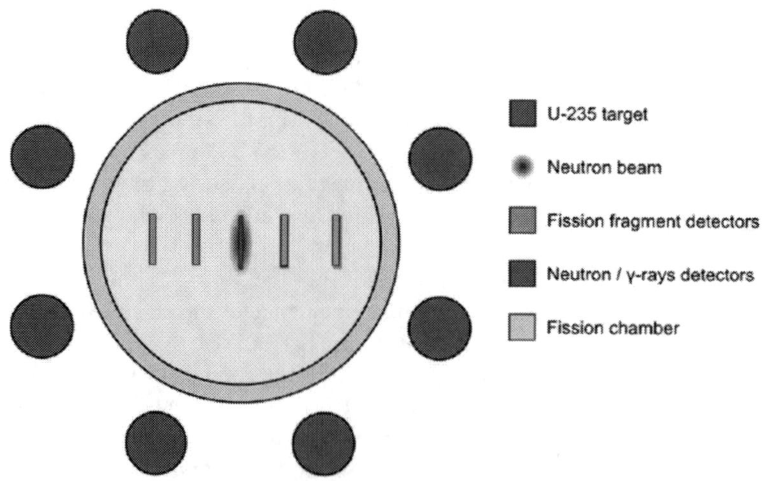

Figure 2. Layout of the experimental facility. View from the beam direction.

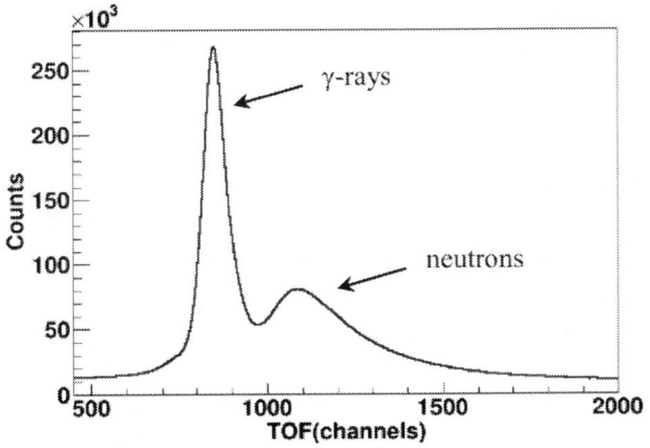

Figure 3. Time-of-flight spectrum from one of the plastic detectors.

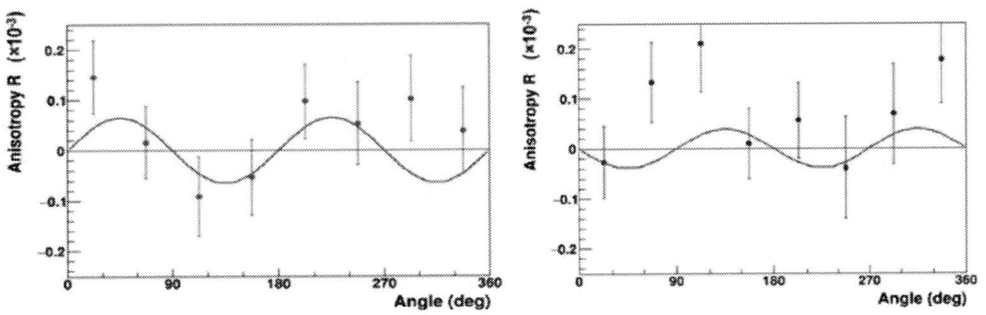

Figure 4. Anisotropy ratio R as a function of angle for the gamma-rays (left) and neutrons (right).

It should be mentioned that the authors of [13], who developed one of the most comprehensive models of the TRI- and ROT-effects, predicted such a decrease of the anisotropy coefficient for the 0.27 eV resonance of ^{235}U, based on the known contributions of the J=3 and J=4 partial cross sections for this nucleus and on the value of the most probable K-channel for these spins, derived from their work. Thus, the results of our experiment are in agreement with the most modern theoretical model prediction.

We believe that it is important to continue this type of experiments in order to gain better statistical accuracy for the 0.27 eV resonance and extend the measurements to higher energies, e.g. to the 1.14 eV resonance where the effect should be larger than for cold neutrons and where practically only the J=4 spin state is present.

This work has been supported by the Russian Ministry for Science and Education, German Ministry for Education and Research BMBF through the project 05K13PA3. The instrument POLI is operated by RWTH Aachen in cooperation with JCNS FZ Jülich (Jülich Aachen Research Alliance JARA). We're grateful to A. Sazonov, W. Luberstetter, and P. Stein for their help in setting up and conducting the experiment.

References

[1] P. Jesinger, G.V. Danilyan, A.M. Gagarski, et al., Phys. At. Nucl. **62**, 1608 (1999)
[2] P. Jesinger, A. Kotzle, A. Gagarski, et al., Nucl. Instrum. Methods Phys. Res. A **440**, 618 (2000)
[3] K. Schreckenbach, J. van Klinken, and J. Last, *Proceedings of the 2nd International Workshop on Time Reversal Invariance and Parity Violation in Neutron Reactions*, Dubna, Russia, May 4–7, 1993, (World Scientific Publishing Company, Singapore, 1994), p. 187
[4] F. Gönnenwein, M. Mutterer, A. Gagarski, et al., Phys. Lett. B **652**, 13 (2007)
[5] G.V. Danilyan, J. Klenke, V.A. Krakhotin, et al., Phys. At. Nucl. **72**, 1812 (2009)
[6] G.V. Danilyan, J. Klenke, V.A. Krakhotin, et al., Phys. At. Nucl. **74**, 671 (2011)
[7] G.V. Danilyan, J. Klenke, Yu.N. Kopach, et al., Phys. At. Nucl. **77**, 677 (2014)
[8] V.E. Bunakov and S.G. Kadmensky, Phys. At. Nucl. **66**, 1846 (2003)
[9] V.E. Bunakov, S.G. Kadmensky, and S.S. Kadmensky, Phys. At. Nucl. **71**, 1887 (2008)
[10] V.E. Bunakov and S.G. Kadmensky, Phys. At. Nucl. **74**, 1655 (2011)
[11] A.L. Barabanov, in *Proceedings of the IX International Seminar on Interaction of Neutrons with Nuclei*, Dubna, May 23–26, 2001 (Dubna, Russia, 2001), p. 93
[12] A.L. Barabanov, Phys. Part. Nucl. Lett. **10**, 336 (2013)
[13] A. Gagarski, F. Gönnenwein, I. Guseva et al., Phys. Rev. C **93**, 054619 (2016)
[14] V. Hutanu, J. Large-Scale Res. Facil. **1**, A16 (2015)
[15] L.C. Leal, G. de Saussure, and R.B. Perez, Nuclear Science and Engineering **109**(1) (1991)
[16] V. Hutanu, M. Meven, S. Masalovich, G. Heger, and G. Roth, J. Phys.: Conf. Ser. **294**, 012012 (2011)

Investigation of the heavy nuclei fission with anomalously high values of the fission fragments total kinetic energy

Vitaly Khryachkov⋆, *Andrei* Goverdovskii, *Vladimir* Ketlerov, *Vecheslav* Mitrofanov, and *Alexei* Sergachev

Institute of Physics and Power Engineering, 1, Bondarenko sq., Obninsk 249030, Russia

Abstract. Binary fission of ^{232}Th and ^{238}U induced by fast neutrons were under intent investigation in the IPPE during recent years. These measurements were performed with a twin ionization chamber with Frisch grids. Signals from the detector were digitized for further processing with a specially developed software. It results in information of kinetic energies, masses, directions and Bragg curves of registered fission fragments. Total statistics of a few million fission events were collected during each experiment. It was discovered that for several combinations of fission fragment masses their total kinetic energy was very close to total free energy of the fissioning system. The probability of such fission events for the fast neutron induced fission was found to be much higher than for spontaneous fission of ^{252}Cf and thermal neutron induced fission of ^{235}U. For experiments with ^{238}U target the energy of incident neutrons were 5 MeV and 6.5 MeV. Close analysis of dependence of fission fragment distribution on compound nucleus excitation energy gave us some explanation of the phenomenon. It could be a process in highly excited compound nucleus which leads the fissioning system from the scission point into the fusion valley with high probability.

1 Introduction

We would like to point out some particular cases of nuclear fission when fission fragments appeared to have no excitation energy, at equilibrium deformation. In such cold-fragmentation events the total energy Q^* transforms into the fragment kinetic energy Ek due to their Coulomb repulsion. Expression for total energy of the reaction is

$$Q^* = Q + Bn + En, \qquad (1)$$

where Q is the energy of nuclear fission, Bn is the neutron binding energy in the fissile nucleus, and En is the neutron kinetic energy.

This phenomenon was studied in several works (see, for example, [1–5]) and it is of great interest because we have a possibility to get information about the evolution of fissile system. Fission fragment kinetic energy can reach the maximum of possible values only if the fragments were produced almost at the saddle point. The probability of such fission events depends directly on the properties of the potential surface corresponding to the fissile system. In particular, the penetrability of the barrier between the fission and fusion valleys is of great importance. The dynamical component of the kinetic

⋆e-mail: hva@ippe.ru

energy depends on the deformation of the fissioning nucleus as it descends from the saddle point to the scission point. In the case of cold fragmentation this stage of the fission process is omitted, because of at the saddle point the nucleus transforms into two fission fragments in the ground state. They have no excitation energy and therefore cannot emit neutrons or gamma rays.

It is very interesting to compare the same fissile system at various values of excitation energies. By estimating the penetrability of the barrier between the fission and fusion valleys at the different excitation energies of the fissile nucleus we can get an idea of the parameters of this barrier.

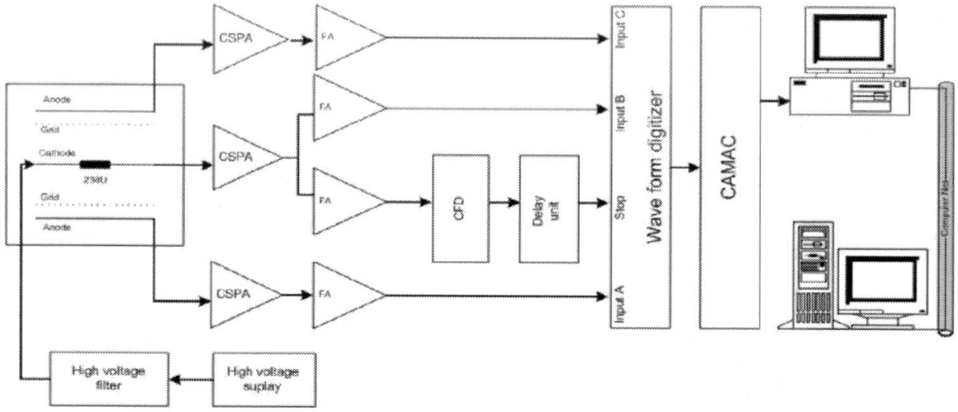

Figure 1. Layout of spectrometer electronics.

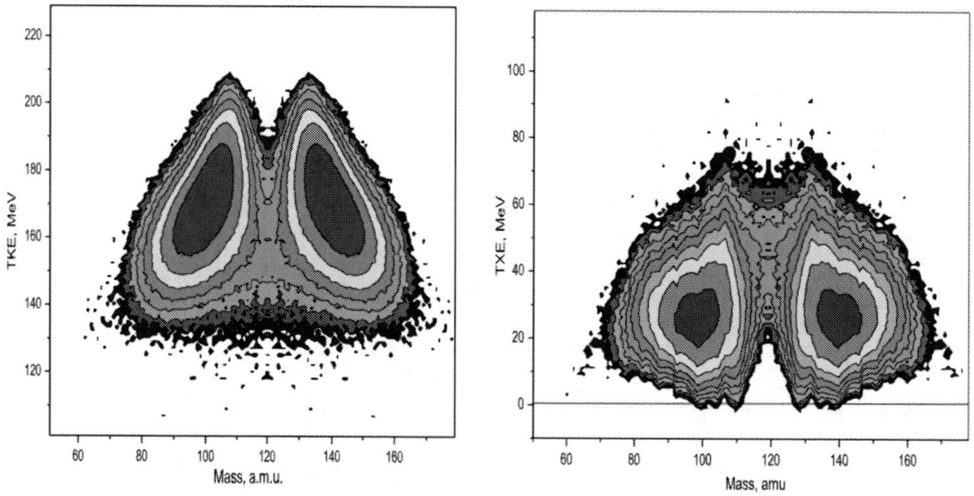

Figure 2. Distribution of ^{239}U fragment mass versus TKE (left) and TXE (right) for the incident neutron energy of 5 MeV.

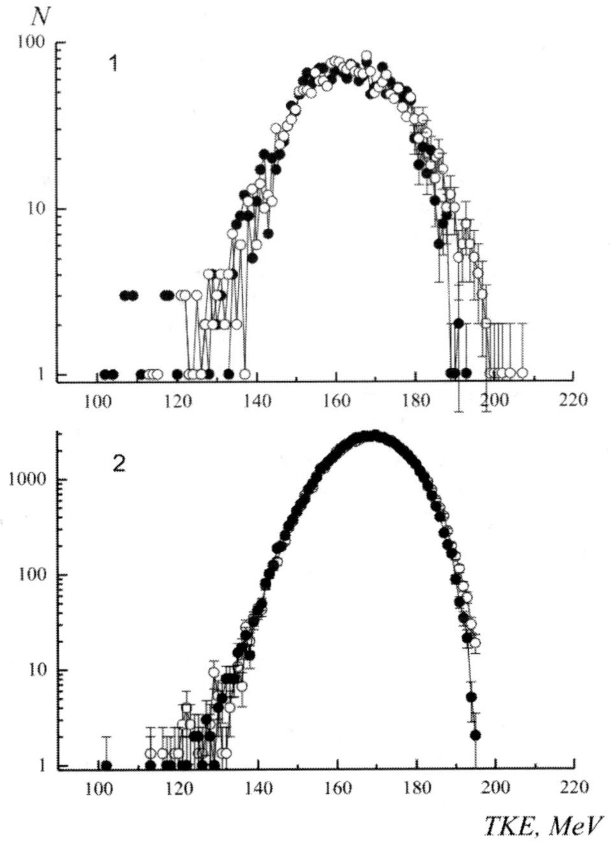

Figure 3. TKE distribution for ^{239}U fragment mass 123 amu (1) 140 amu (2). Full points – neutron energy 5 MeV, open points – 6.5 MeV.

2 Experiment

Fission fragments were detected in a twin ionization chamber with Frisch grids. The ^{238}U target of a diameter of 3 cm and a thickness of 100 μg/cm^2 was mounted on the common cathode. It was prepared by vacuum sputtering of uranium fluoride onto Al_2O_3 backing having a thickness of 50 μg/cm^2. In order to have good electric conductivity over the entire cathode surface the target was additionally coated with 50 μg/cm^{2}of gold. The schematics of the experimental facility is shown in Fig. 1. The signals from the anodes were collected as digital oscillograms of 512 points with a step time of 12.5 ns from the LeCroy 2262 pulse-shape digitizer on the data acquisition computer. Offline digital signal processing and analysis of saved oscillograms allows us to determine the kinetic energy, the mass, and the emission angle [4] of the pair of fission fragments. The energies of fission fragments were obtained with the corrections for the Frisch grid inefficiency [6], for the energy loss in the target [7], and for the pulse height defect [4]. Additional filter was performed for the pile-ups from alpha particles or scattered protons.

The experiments were carried out in a neutron beam of the EG-1 accelerator of the Institute for Physics and Power Engineering (Obninsk, Russia). The reaction D(d, n)^{3}He was used to obtain 5 and 6.5 MeV neutrons. The mean current at the target was 50 μA.

During a one month long measurement, we were able to accumulate and analyze 2.5 million digital oscillograms corresponding to ^{239}U fission events. In order to calibrate the primary distributions, we took data reported in [8], where one can find high precision mass and energy distributions for ^{239}U fission over a wide range of excitation energies. Signals from the detector were processed in two steps. At the first step, the masses and energies were reconstructed in an iterative procedure with corrections for the pulse height defect and for the emission of prompt neutrons from fission fragments. The resulting energy, mass, and angular distributions were found to be in good agreement with data presented in [8]. The mean total kinetic energy value was found to be 169.8 MeV against 170.05 MeV in [8]. But in such approach corrections for neutron emission performed based on the mass of nascent fragments only, without taking into account fission fragment excitation energy. This leads to a distortion in the region of heavy fragments spectrum, where neutron emission is energetically forbidden. Since this region is of special interest to present work, it is necessary to perform an additional processing of data using the energy calibration from the first step. Repeated iterations were done without the correction for neutron emission. This time the information about masses and energies appears to be incorrect for the majority of the fission fragments, but the high values region of the spectrum is correct.

3 Results

Two-dimensional spectrum with axis of fragment mass versus TKE for incident neutron energy of 5 MeV is shown in Fig. 2 (left). Two-dimensional spectrum with axis of fragment mass versus TXE for the same neutron energy is shown in Fig. 2 (right).

The yields of nuclear-fission events where the heavy-fragment masses are 123 and 140 amu versus the total kinetic energy (TKE) of fragments are shown in Fig. 3. Data for ^{238}U fission induced by 5 and 6.5 MeV neutrons presented at the same plots for comparison. The distribution of total kinetic energies changes significantly with increasing of the excitation energy of the fissioning nucleus, especially in the high values region.

We would like to make special notes for several facts.

(i) The highest observed kinetic energy of fragment pairs where the heavy-fragment mass is 140 amu have the same value for both values of the incident neutron energy (6.5 and 5 MeV), even though the production of fragments with greater value of TKE is energetically possible in the former case. This means that the entire energy introduced in the system goes to the excitation of fragments.

(ii) In the region of the highest values of *Ek* the yield of fragment pairs where the heavy-fragment mass is 140 amu is one order of magnitude greater in the fission induced by 6.5 MeV neutrons than for 5 MeV neutrons.

(iii) The situation with fission fragments pairs where the heavy fragment mass is 123 amu is different. For the incident neutron energy of 5 MeV the measured maximum of TKE of such fragment pairs is 190 MeV, while for E_n = 6.5 MeV, it is by 10 MeV higher, close to the value of the total implementable energy of Q^* = 202.7 MeV.

Figure 4 represents the difference between Q^* and the maximum measured total kinetic energy of fragments as a function of the heavy-fragment mass for ^{238}U(n,f) with En = 5 MeV and En = 6.5 MeV. In the case of smaller incident neutron energy the maximum measured TKE reaches the value of Q^*

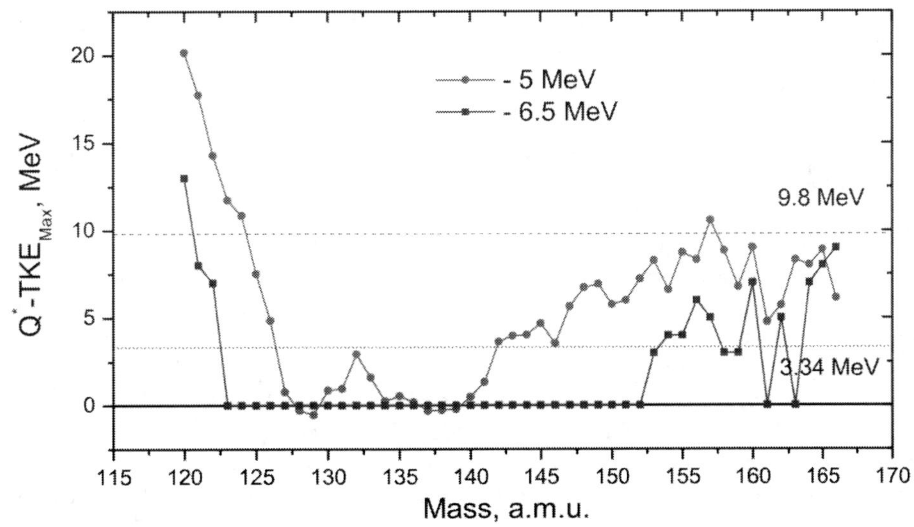

Figure 4. Mass dependence of available TXE value for 5 MeV (red points) and 6.5 MeV (blue points) incident neutron energy (data for $^{238}U(n,f)$).

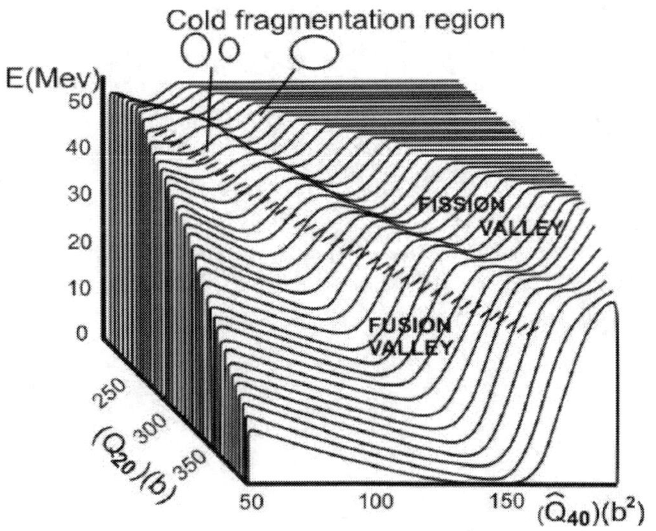

Figure 5. Potential-energy surface of ^{240}Pu versus the elongation of the nucleus, Q_{20}, and the decrease in the thickness of its neck, Q_{40}.

only for a small number of events where the heavy-fragment masses are 128, 129, and 134–139 amu. For En = 6.5 MeV, this is true for all events where the heavy-fragment mass falls within the range of 123–152 amu. Despite of the fact that the increase in the excitation energy of the fissioning nucleus is only 1.5 MeV, the increase of maximum value of the measured total kinetic energy is 5–10 MeV.

https://doi.org/10.1051/epjconf/201816900011

Figure 5 shows the surface of potential energy for the ^{240}Pu fissile nucleus [9]. Two valleys are clearly seen: the fission valley and the fusion valley. There is a potential barrier between them. Its height decreases with increasing of the fissioning nucleus deformation. At some deformation, the fissile system moves from the fission valley to the fusion valley, from one nucleus to two separate nuclei.

By this figure we are trying to understand qualitatively the results of our measurements in Figs. 3 and 4. The increase of 1.5 MeV in the excitation energy of the fissile nucleus leads to a substantial increase in the yield of events in which Ek = Q^*. This is because the system then overcomes much more readily the potential barrier between the fission valley and the fusion valley, where the fragments already exist as individual nuclei. For those ways of nuclear mass separation for which, at the neutron energy of 5 MeV, the measured highest total kinetic energy of fragments already reached Q^*, a further increase in the neutron energy only leads to an increase in the yield of such events. For a number of those ways of the separation of the nucleus where, at the neutron energy of 5 MeV, the highest measured values of Ek are smaller than Q^* by 5 to 10 MeV, the increase of 1.5 MeV in the neutron energy leads to a substantial increase in the probability for overcoming the barrier in the case of events in which Ek = Q^*; at the total number of recorded fission events that is equal to about two millions, it also becomes possible to discover experimentally cold fragmentation for such ways of nuclear-mass separation.

References

[1] V.A. Khryachkov et al., Phys. At. Nucl. **57**, 1154 (1994)

[2] A.A. Goverdovski et al., Phys. At. Nucl. **62**, 901 (1999)

[3] A.A. Goverdovski et al., JETP Lett. **67**, 793 (1998)

[4] C. Budtz-Jorgensen et al., Nucl. Instrum. Methods Phys. Res. A **258**, 209 (1987)

[5] V.A. Khryachkov et al., Preprint No. 2740, Inst. Phys. Power Eng. (1998)

[6] V.A. Khriatchkov et al., Nucl. Instrum. Methods Phys. Res. A **394**, 261 (1997)

[7] M.V. Dunaev et al., *Proceedings of VIII Int. Seminar on Interaction of Neutrons with Nuclei*, Dubna (2000), p. 363

[8] F. Vives et al., *Proceedings of International Conference "Nuclear Fission and Fission-Product Spectroscopy"*, Seyssins, France (1998), p. 435

[9] J.F. Berger, M. Girod, and Gogny. Nucl. Phys. A **502**, 85 (1989)

Influence of primary fragment excitation energy and spin distributions on fission observables

Olivier Litaize[1,⋆] , *Loïc* Thulliez[1,2], *Olivier* Serot[1], *Abdelaziz* Chebboubi[1], and *Pierre* Tamagno[1]

[1]*CEA, DEN, DER, SPRC, Cadarache, 13108 Saint-Paul-lez-Durance, France*
[2]*CEA, Irfu, Université Paris-Saclay, 91191 Gif-sur-Yvette, France*

Abstract. Fission observables in the case of ^{252}Cf(sf) are investigated by exploring several models involved in the excitation energy sharing and spin-parity assignment between primary fission fragments. In a first step the parameters used in the FIFRELIN Monte Carlo code "reference route" are presented: two parameters for the mass dependent temperature ratio law and two constant spin cut-off parameters for light and heavy fragment groups respectively. These parameters determine the initial fragment entry zone in excitation energy and spin-parity (E*, J^π). They are chosen to reproduce the light and heavy average prompt neutron multiplicities. When these target observables are achieved all other fission observables can be predicted. We show here the influence of input parameters on the saw-tooth curve and we discuss the influence of a mass and energy-dependent spin cut-off model on gamma-rays related fission observables. The part of the model involving level densities, neutron transmission coefficients or photon strength functions remains unchanged.

1 Introduction

The primary fission fragment (FF) entry zone in excitation energy and spin-parity (E*, J^π) is one of the key points in the simulation of the fission process leading to prompt particle emission. Theory as well as experiments are not yet able to clearly identify these initial primary FF nuclear states. Models that are used in Monte Carlo codes allow to reproduce the neutron multiplicity or the gamma multiplicity but it is quite difficult to achieve both of them in one shot. Neutron and gamma multiplicities are the end point of the whole de-excitation process. Several models have to be used in order to calculate level densities, photon strength functions, optical model potential for neutron transmission coefficients. At low energies the nuclear level scheme, that is to say the knowledge of level energies, spins, parities, half-lives, intensities of the transitions from level to level have to be accounted for through experimental databases. In the FIFRELIN Monte Carlo code, the reference calculation route usually involves five input parameters (here one of them, the fraction of rigid body moment of inertia is fixed to 1.). They are chosen to calculate the initial states of primary fragments and reproduce a fission target observable such as the average prompt neutron multiplicity $\overline{\nu}$. We will show in this work how the main four parameters influence the average prompt neutron multiplicity as a function of pre-neutron fragment mass (the so-called saw-tooth). On the other hand we will discuss the influence of the spin cut-off parameter model used to assign the initial FF spin.

⋆e-mail: olivier.litaize@cea.fr

2 Quick reminder of the FIFRELIN model

Pre-neutron fission fragment mass A and kinetic energy KE distributions are experimental data used as input data in the code. Data from [1] are used in this work for the study of ^{252}Cf(sf). Recent available data have been published [2] and will be used in a near future for comparison but this is out of the scope of this paper. Details related to the FIFRELIN "calculation route" can be found in [3] and [4]. The main steps are reminded hereafter.

Nuclear charge Z is sampled from the so-called Wahl Z_p model. When the light fragment A_L, Z_L, KE_L characteristics are sampled, the complementary heavy fragment of a binary fission is selected by mass, charge and linear momentum conservation. Before starting the de-excitation process, the excitation and spin-parity of both fragments have to be known. The total excitation energy TXE is calculated from the kinetic energy of both FF and the Q-value of the fission process knowing the binding energies of the fissioning nucleus and the light and heavy primary fragments. This intrinsic part of TXE is shared between the two fragments E_L^* and E_H^* through a $E^* = aT^2$ relation where a stands for the level density parameter (Ignatyuk's prescription) and T the temperature. Light and heavy FF temperatures are linked together through a mass-dependent temperature ratio law [3]. Before that, the rotational energy, considering a simple rotating liquid drop, has been subtracted from TXE. The calculation of the rotational energy requires a moment of inertia and a total angular momentum which is spin cut-off model-dependent.

$$TXE = E_L^* + E_H^* + E_L^{rot} + E_H^{rot} \tag{1}$$

where E_L^{rot}, E_H^{rot} are the rotational energies of the light and heavy fragments and E_L^*, E_H^* are their intrinsic excitation energies. The rotational energy E^{rot} is simply expressed considering a rotating liquid drop:

$$E^{rot} = \frac{\hbar^2 J(J+1)}{2\mathcal{I}} \tag{2}$$

where J stands for the total angular momentum sampled from Eq. (3) following Bethe's work.

$$P(J) = \frac{J+1/2}{\sigma^2} \exp\left(-\frac{(J+1/2)^2}{2\sigma^2}\right) \tag{3}$$

The factor σ^2 is the spin cut-off parameter which can be constant or energy-dependent and will be studied in section 4. The nuclear moment of inertia $\mathcal{I}$ is defined as a proportion k_{rig} of a rigid spheroid $\mathcal{I}_{rig}$ defined by:

$$\mathcal{I}_{rig} = \frac{2}{5} AMR^2 (1 + 0.31\beta_2 + 0.44\beta_2^2 + \cdots) \tag{4}$$

where A, M, R and β_2 respectively stand for the mass number, the nucleon mass, the radius ($R = 1.2A^{1/3} fm$) and the quadrupole deformation parameter of the nucleus in its ground state taken from a Myers-Swiatecki parameterization of the mass formula. The total amount of excitation energy is deduced from the energy release during the fission process Q based on binding energies from [5] and the sampled total kinetic energy:

$$TXE = Q - TKE \tag{5}$$

Here we consider the spontaneous fission of ^{252}Cf. Only the intrinsic excitation energy E^* corresponding to $TXE - E^{rot}$ is treated within a Fermi gas approximation in aT^2 where a and T stand for the level density parameter and the nuclear temperature. An iterative procedure is used to determine

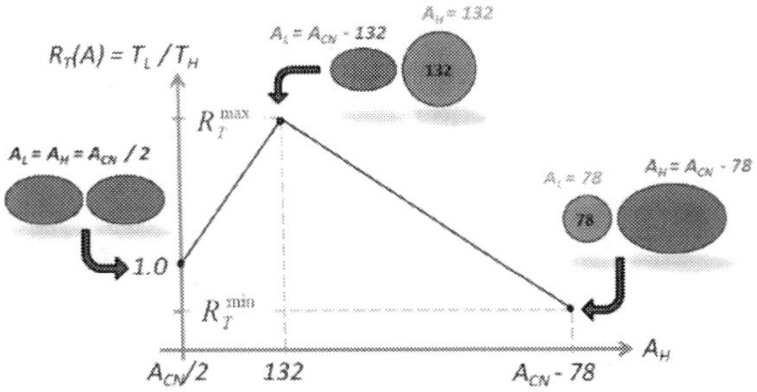

Figure 1. Temperature ratio law $R_T(A)$: three anchor points are considered from macroscopic nucleus shapes at scission to determine the temperature ratio of two complementary fission fragments after full acceleration when they have recovered their ground state deformation (see text for details).

the fragment excitation energy because the level density parameter is energy/temperature dependent following the Ignatyuk prescription:

$$a = \overline{a}\left\{1 + \frac{\delta W}{U^*}\left(1 - e^{-\gamma U^*}\right)\right\} \tag{6}$$

where $\overline{a}(A)$ is the asymptotic level density parameter, δW accounts for the shell corrections that can be estimated by the difference between measured and calculated mass excesses (with a liquid drop model for example), and $U^* = E^* - \Delta$ is an effective excitation energy function corrected by pairing Δ while γ is a damping factor. Several implementations of these parameters are driven through the RIPL-3 documentation [6].

For a given fragmentation, the temperature of the light fragment $T_L = T(A_L)$ is linked to the temperature of its heavy partner $T_H = T(A_H)$ by a mass dependent temperature ratio law $R_T(A) = T_L/T_H$ (Fig. 1). This law has been already described elsewhere [3]. Briefly the maximum of this law (R_T^{max}) is supposed to correspond to a minimum temperature of the heavy fragment which is linked to a minimum in the deformation energy (minimum in the excitation energy). This minimum occurs for the doubly closed spherical shell nuclei (Z=50, N=82 and A=132). The situation is completely reversed with a minimum in the temperature ratio (R_T^{min}) when the light fragment has a proton and a neutron closed spherical shell (Z=28, N=50 and A=78). Finally for symmetric fission, the temperature of the two identical complementary fragments is the same, the ratio is equal to one. In between these anchor points, the evolution is supposed to be linear. Once this law is established, all the ingredients are known and can be used to calculate the excitation energies of the two complementary fragments through the iterative scheme:

$$\begin{cases} E_L^* & = \frac{TXE - E^{rot}}{1 + \frac{a_H}{a_L R_T^2}} \\ E_H^* & = \frac{TXE - E^{rot}}{1 + \frac{a_L R_T^2}{a_H}} \end{cases} \tag{7}$$

At this stage, the characteristics (A, Z, KE, E^*, J, π) of the fission fragments are completely known and the deexcitation process can start by neutron, gamma and conversion electron emission from primary excited fragments (Fig. 2).

The statistical de-excitation process in an energy, spin, parity ensemble, often called Hauser-Feshbach formalism, has been implemented in FIFRELIN code using the notion of nuclear realization first introduced for gamma emission [7] and extended to coupled neutron/gamma emission [8].

Figure 3 shows an example of two possible cascades with or without neutron emission. Neutron emission is taken into account by pre-tabulated neutron-transmission coefficients at the very beginning stage of the de-excitation. Gamma emission requires the calculation of photon strength functions and level densities from unknown levels to other accessible levels. Experimental intensities are used at low energies when nuclear structure is known. Conversion electron are also accounted for through conversion coefficients.

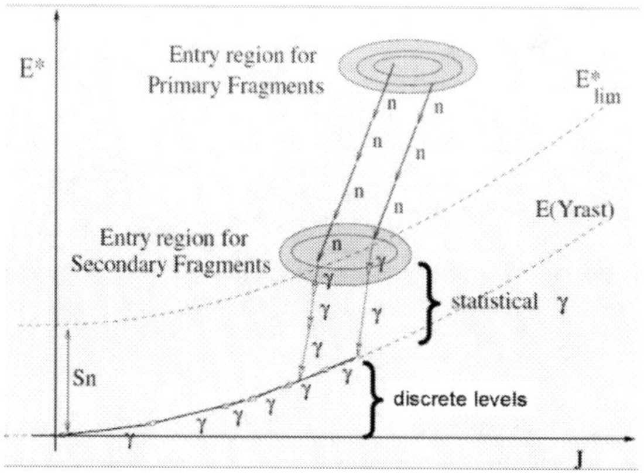

Figure 2. Schematic representation of FF de-excitation in a (E^*, J) plan.

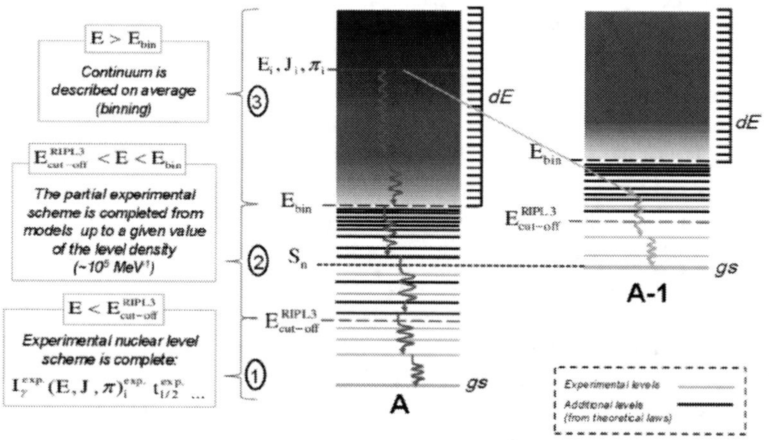

Figure 3. Nuclear de-excitation scheme in FIFRELIN.

3 Influence of the free input parameters

The four free input parameters of the reference calculation route are listed in the first line of Table 1 and reported hereafter:

$$\{R_T^{\min}/R_T^{\max}/\sigma_L/\sigma_H\} = \{0.35/1.40/11.0/10.5\}$$

Table 1. Influence of the free parameters on the neutron multiplicities and gamma multiplicities (without threshold energy and with a $3ns$ coincidence time window). Statistical uncertainties are about 0.002 for neutron and 0.01 for gamma multiplicities.

Input parameters	Neutron multiplicities			Gamma multiplicities		
R_T^{min} ; R_T^{max} ; $\overline{\sigma}_L$; $\overline{\sigma}_H$	$\overline{\nu}_L$	$\overline{\nu}_H$	$\overline{\nu}$	$\overline{M}_{\gamma,L}$	$\overline{M}_{\gamma,H}$	$\overline{M}_\gamma$
0.35 ; 1.4 ; 11.0 ; 10.5	2.07	1.68	3.75	6.45	5.13	11.58
0.65 ; 1.4 ; 11.0 ; 10.5	2.19	1.55	3.74	6.40	5.17	11.57
0.35 ; **1.1** ; 11.0 ; 10.5	1.71	2.08	3.79	6.55	5.00	11.55
0.35 ; 1.4 ; **08.0** ; 10.5	2.08	1.79	3.87	4.89	5.09	9.98
0.35 ; 1.4 ; 11.0 ; **08.5**	2.09	1.73	3.82	6.46	4.15	10.6

This set of parameters allows to reproduce the total average prompt neutron multiplicities of the ^{252}Cf(sf) reaction $\overline{\nu}_L = 2.07$, $\overline{\nu}_H = 1.68$ and $\overline{\nu} = 3.75$ compared to the experimental reference values (target observables) taken from Vorobyev [9]: $\overline{\nu}_L = 2.05$, $\overline{\nu}_H = 1.70$ and $\overline{\nu} = 3.76$. The average multiplicity as a function of mass can also be observed in Fig. 4 compared with experimental data.

The influence of the free input parameters is studied from different variations of these parameters. The rest of the model remain unchanged:

- Composite Gilbert-Cameron model for level densities [10],
- AME-2012 for binding energies [5],
- RIPL3-2015 for nuclear level schemes [11],
- Koning-Delaroche optical model potential for neutron transmission coefficients [12],
- Enhanced Generalized Lorentzian model for E1 photon strength functions while additional XL-types are accounted for using RIPL-3 recommendations [6].

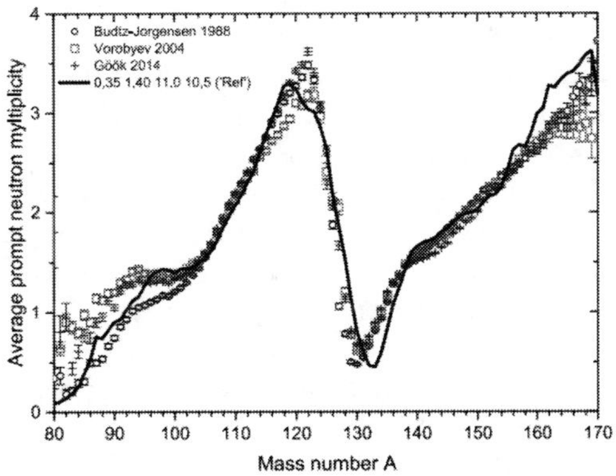

Figure 4. Average prompt neutron multiplicity as a function of pre-neutron fragment mass. Symbols are experimental data from [2, 13, 14]. Black line corresponds to the reference calculation.

A change in the R_T law greatly affects the saw-tooth shape of the average prompt neutron multiplicity as a function of pre-neutron fission fragment mass (Fig. 5) and then the values for light and heavy

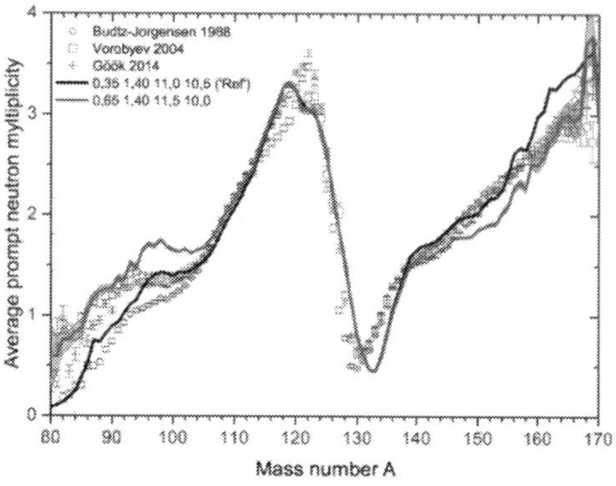

(a) $R_T^{\min}$ increases from 0.35 to 0.65

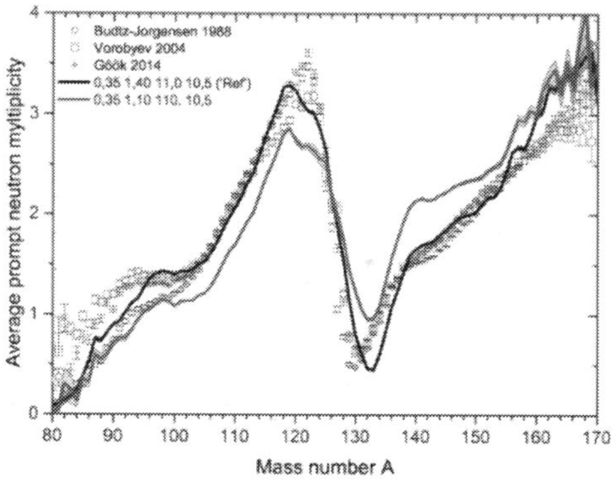

(b) $R_T^{\max}$ decreases from 1.40 to 1.10

Figure 5. Average prompt neutron multiplicity as a function of pre-neutron fragment mass. Symbols are experimental data. The influence of a modification of the $R_T(A)$ law is shown as red line compared to the reference calculation as black line.

groups $\overline{\nu}_L$ and $\overline{\nu}_H$. Due to a compensation between light and heavy groups, this does not influence so much the total average prompt neutron multiplicity $\overline{\nu}$ as it can be seen in Table 1 (second and third lines compared to first one).

Because $R_T = T_L/T_H$, an increase of R_T (for instance $R_T^{\min}$ goes from 0.35 to 0.65) leads to a higher excitation energy for light fragments and a lower excitation energy for heavy fragments. Consequently the average prompt neutron multiplicity increases for light fragments and decreases for heavy fragments (second line of Table 1 and Fig. 5a). In the same way a decrease of R_T (for instance $R_T^{\max}$ goes from 1.40 to 1.10) leads to a lower excitation energy for light fragments and a higher excitation energy for heavy fragments. Consequently the average prompt neutron multiplicity decreases for light fragments and increases for heavy fragments (third line of Table 1 and Fig. 5b).

The corresponding average gamma multiplicity is roughly $\overline{M_\gamma} \sim 11.6$ without gamma energy threshold and with a time coincidence window of about 3 ns. With a 100 keV threshold, this multiplicity is reduced to about $\overline{M_\gamma} \sim 10.6$. Compared to the neutron multiplicity, the gamma multiplicity is less sensitive to the variations of $R_T(A)$ because the very most part of gammas are emitted after neutrons cooling the fragment down to more or less the same nuclear state.

A change of initial spin cut-off parameters affects in a different way the saw-tooth shape of $\overline{\nu}(A)$. The two constant values for the spin cut-off modify the initial spin of primary fission fragments. This does not only affect the position of entry states along the J axis of Fig. 2 but also along the E^* axis because J is used to calculate the total rotational energy (see Eq. (2)). This is why a lower spin cut-off parameter for light fragments as well as for heavy fragments leads to an increase of the average total prompt neutron multiplicity. It is noticeable that it is specifically the neutron multiplicity for heavy fragment group that increases significantly compared to the light fragment group (fourth and fifth lines compared to first line in Table 1). The average gamma multiplicity is closer linked to this spin cut-off modification: if σ_L (σ_H) decreases then the gamma multiplicity for light (heavy) group decreases.

4 Influence of the spin cut-off model

We have seen in the previous section that two different constant values can be used for the spin cut-off parameters of primary light and heavy fragment groups in order to reproduce $\overline{\nu}_L$ and $\overline{\nu}_H$ observables.

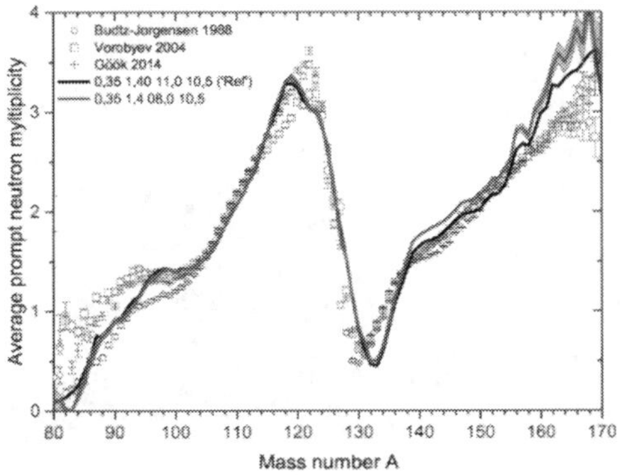

(a) $\bar{\sigma}_L$ decreases from 11 $\hbar$ to 8 $\hbar$

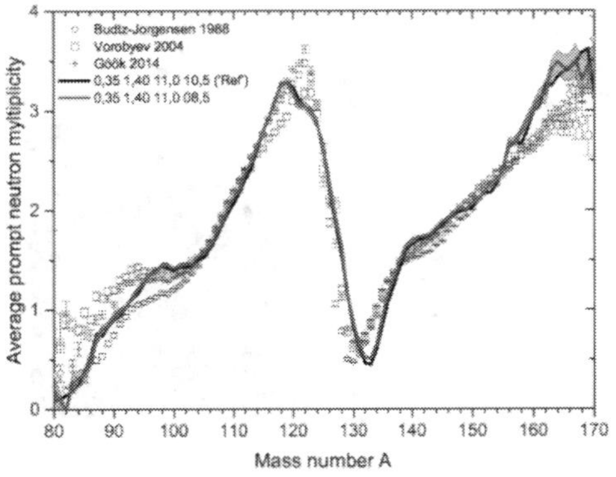

(b) $\bar{\sigma}_H$ decreases from 10.5 $\hbar$ to 8.5 $\hbar$

Figure 6. Average prompt neutron multiplicity as a function of pre-neutron fragment mass. Symbols are experimental data. The influence of a modification of the spin cut-off parameters is shown as red line compared to the reference calculation as black line.

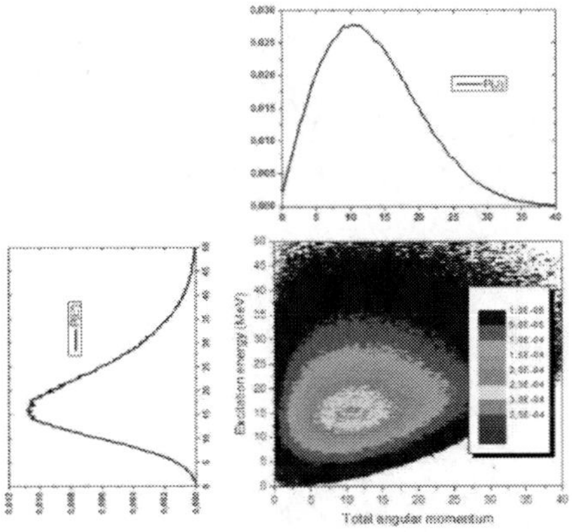

(a) Initial (E^*, J) distribution for light fragments

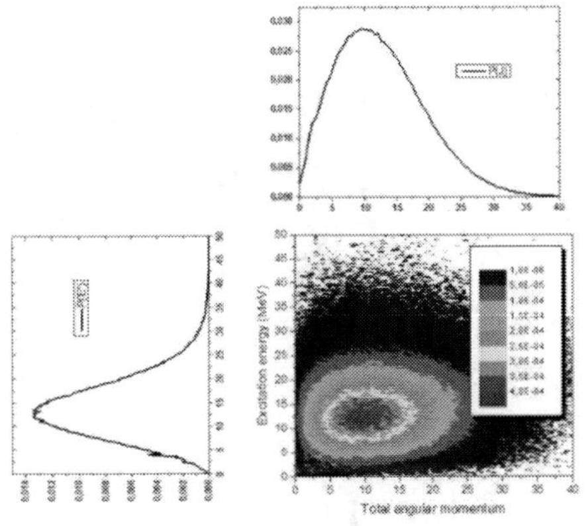

(b) Initial (E^*, J) distribution for heavy fragments

Figure 7. Density of initial fission fragment states with constant spin cut-off parameter.

We can also consider an energy dependent spin cut-off model for sampling the initial total angular momenta (Inertia+ Shell model following preliminary calculations reported in [15]):

$$\sigma^2 = \mathcal{I}\frac{a}{\bar{a}}\sqrt{\frac{U}{a}} \tag{8}$$

where $\mathcal{I}$ stands for the nuclear moment of inertia. This formula accounts for the energy dependence of the spin cut-off parameter as well as the shell effects through the additional term $a/\bar{a}$ where a and $\bar{a}$ stand for the level density parameters of Eq. (6). The spin cut-off parameter is used to sample J (Eq. (3)) and calculate E^{rot} in order to subtract from the total excitation energy. Here, the parameter σ^2 depends also on the excitation energy (which is not known yet at the beginning of the iterative

https://doi.org/10.1051/epjconf/201816900012

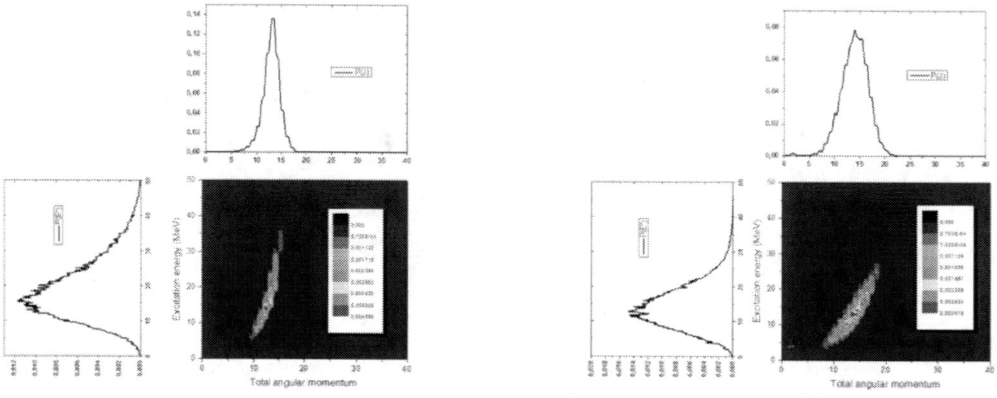

(a) Initial (E^*, J) distribution for light fragments (b) Initial (E^*, J) distribution for heavy fragments

Figure 8. Density of initial fission fragment states (pre-neutron emission) with an energy-dependent spin cut-off parameter.

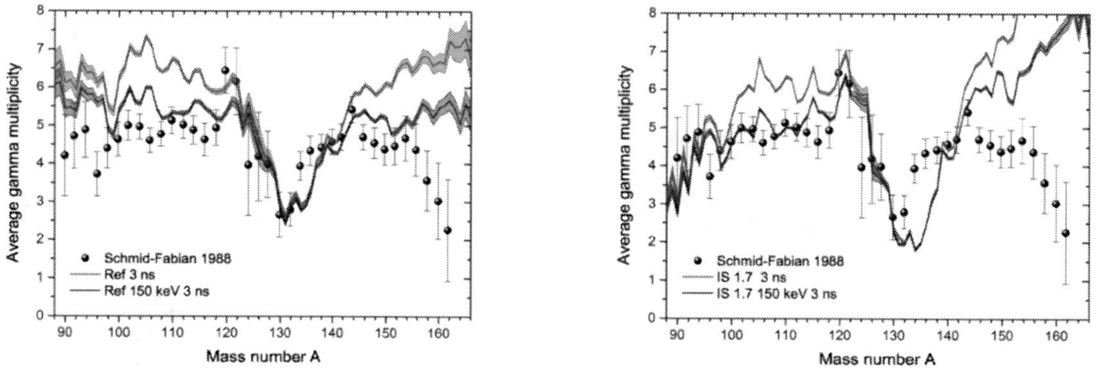

(a) Reference route (b) Inertia+shell model

Figure 9. Average prompt fission gamma multiplicity as a function of mass for the ^{252}Cf(sf) reaction.

procedure used to solve Eq. (7)). We have considered in this work the average value of J which is proportional to σ^2. The (E^*, J) entry zone for primary fragments is shown in Fig. 7 using two constant spin cut-off values for light and heavy groups compared to the entry zone using an energy dependent spin cut-off model in Fig. 8.

A saw-tooth shape is more visible in the average gamma multiplicity as a function of pre-neutron mass when the mass- and energy-dependent spin cut-off model is used instead of constant values for light and heavy groups (Fig. 9). The gamma multiplicity is better reproduced for light fragments but the disagreement still persists for heavy fragments.

The gamma multiplicity distribution is directly linked to the primary spin distribution as we can see in Figs. 7 and 8 (the top part of figures corresponds to a X-projection in the (E^*, J) distribution) compared to Fig. 10.

 https://doi.org/10.1051/epjconf/201816900012

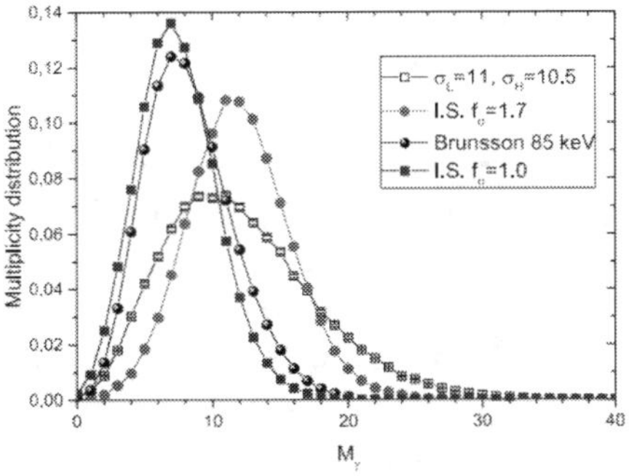

Figure 10. Prompt gamma multiplicity distribution. I.S. stands for Inertia+Shell spin cut-off model.

By using two constant spin cut-off values for primary FF (leading to good neutron multiplicity values) the average gamma multiplicity distribution is far from the double Poisson model used by Brunsson [16] to interpret experimental data from Pleasonton (the threshold of 85 keV explains a part but not the whole effect). If we consider a inertia + shell model (I.S.) without re-scaling (f_σ) the spin cut-off (and the level density) then the calculated gamma multiplicity distribution is clearly improved but the neutron average multiplicity is not reproduced. To reproduce the average neutron multiplicity (even as a function of mass) it is necessary to re-scale the spin cut-off by a factor $f_\sigma = 1.7$ but then the gamma multiplicity distribution is degraded (Fig. 10). The corresponding average value is quite similar to the one obtained with constant spin cut-off values but the distribution is modified as expected.

5 Conclusion

The influence of the free input parameters used in the FIFRELIN code to estimate the primary excitation energy and spin distributions for the calculation of fission observables was studied as well as two different spin cut-off models used for sampling primary fission fragment spins. The present calculation routes are not able to reproduce both neutron and gamma multiplicities in a single run for ^{252}Cf(sf) suggesting that other ingredients of the model have to be improved such as level densities, neutron and photon transmission coefficients through optical models and strength functions respectively.

References

[1] N. Varapai, F.-J. Hambsch, S. Oberstedt, O. Serot, G. Barreau, N. Kornilov, S. Zeinalov, AIP Conf. Proc. **798**, 369 (2005)
[2] A. Göök, F.-J. Hambsch and M. Vidali, Phys. Rev. C **90**, 064611 (2014)
[3] O. Litaize, O. Serot, Phys. Rev. C **82**, 054616 (2010)
[4] O. Litaize, O. Serot and L. Berge, The European Physical Journal A **51**, 177 (2015)
[5] G. Audi, M. Wang, A.H. Wapstra, F.G. Kondev, M. MacCormick and X. Xu, Nuclear Data Sheets **120**, 1 (2014)
[6] R. Capote et al., Nuclear Data Sheets **110**, 3107 (2009)

[7] F. Becvar, Nuclear Instruments and Methods in Physics Research Section A: Accelerators, Spectrometers, Detectors and Associated Equipment, 417 (1988)
[8] D. Regnier, O. Litaize and O. Serot, Computer Physics Communications **201**, 19 (2016)
[9] A.S. Vorobyev, O.A. Shcherbakov, A.M. Gagarski, G.V. Val'ski and G.A. Petrov, EPJ Web of Conferences **118**, 03004 (2010)
[10] A. Gilbert and A.G.W. Cameron, Canadian Journal of Physics **43**, 1446 (1965)
[11] M. Verpelli, R. Capote, Report-INDC(NDS)-0702 (2015)
[12] A.J. Koning and J.P. Delaroche, Nuclear Physics A **713**, 231 (2003)
[13] C. Budtz-Jorgensen and H.-H. Knitter, Nuclear Physics A **490**, 307 (1988)
[14] A. Vorobyev, V. Dushin, F.-J. Hambsch, V. Jakovlev, V. Kalinin, A. Laptev, B. Petrov, O. Shcherbakov, AIP Conf. Proc. ND 2004 **769**, 613–616 (2005)
[15] L. Thulliez, O. Litaize, O. Serot, EPJ Web of Conferences **111**, 10003 (2016)
[16] G.S. Brunson, Los Alamos Report LA-9408-T (1982)

Study of fission using multi-nucleon transfer reactions

Katsuhisa Nishio[1,*], *Kentaro* Hirose[1], *Mark* Vermeulen[1], *Hiroyuki* Makii[1], *Riccardo* Orlandi[1], *Kazuaki* Tsukada[1], *Masato* Asai[1], *Atsushi* Toyoshima[1], *Tetsuya K.* Sato[1], *Yuichiro* Nagame[1], *Satoshi* Chiba[2], *Yoshihiro* Aritomo[3], *Shouya* Tanaka[3], *Tsutomu* Ohtsuki[4], *Igor* Tsekhanovich[5], *Costel M.* Petrache[6], and *Andrei* Andreyev[7,1]

[1]*Advanced Science Research Center, Japan Atomic Energy Agency, 2-4 Shirakata, Tokai, Naka-gun, Ibaraki 319-1195, Japan*
[2]*Laboratory for Advanced Nuclear Energy, Tokyo Institute of Technology, 2-12-1, Ookayama, Meguro-ku, Tokyo 152-8550, Japan*
[3]*Faculty of Science and Engineering, Kindai University, Higashi-Osaka 577-8502, Japan*
[4]*Research Reactor Institute, Kyoto University, Kumatori-cho, Sennangun, Osaka 590-0494, Japan*
[5]*University of Bordeaux, 351 Cours de la Libération, 33405 Talence Cedex, France*
[6]*Centre des Sciences Nucléaire et des Sciences de la Matière, Université Paris-Saclay, CNRS/IN2P3, 91406 Orsay, France*
[7]*Department of Physics, University of York, Heslington, York, YO10 5DD, UK*

Abstract. Multi-nucleon transfer channels of the reactions of $^{18}O+^{232}Th$, $^{18}O+^{238}U$, $^{18}O+^{248}Cm$ were used to measure fission-fragment mass distribution for various nuclides and their excitation energy dependence. Predominantly asymmetric fission is observed at low excitation energies for all the studied cases, with an increase of the symmetric fission towards high excitation energies. Experimental data are compared with predictions of the fluctuation-dissipation model, where effects of multi-chance fission (neutron evaporation prior to fission) was introduced. It was shown that a reliable understanding of the observed fission fragment mass distributions can be obtained only invoking multi-chance fissions.

1 Introduction

Nuclear fission is usually described as an evolution of a nuclear shape on a potential-energy surface which results from the subtle interplay of macroscopic nuclear properties and microscopic shell effects. Also dynamical effects should have an important role to explain various aspects in fission. New experimental techniques and associated new data are indispensable to further understand fission mechanism. Fission-fragment mass distribution (FFMD) is one of the most fundamental data, which exhibits asymmetric shape at low excitation energy due to shell structures. Traditionally, neutron- and charged particle capture reactions as well as spontaneous fission have been used to study low-energy fissions.

Around 2000, GSI in Darmstadt developed a Coulex-induced fission of relativistic RIBs in inverse kinematics, where comprehensive fission studies were performed for several tens of nuclei in the

*e-mail: nishio.katsuhisa@jaea.go.jp

neutron-deficient Ac-U region [1]. The recent SOFIA experiment at GSI also followed the same approach but with a much improved technique [2]. Recently, β/EC delayed fission was investigated for the very proton-rich nucleus using radioactive beams, and ^{180}Hg was found to show an asymmetric fission as a new region of mass-asymmetric fission [3]. Advancement of new technique for fission studies is reviewed in [4].

Multi-nucleon transfer (MNT) reactions is another unique reaction which allow us to populate neutron-rich nuclei which cannot be accessed by other reactions such as particle capture and/or heavy-ion fusion reactions. Furthermore, excited states of CN range widely from under the fission barrier to higher energies, allowing us to measure the excitation energy dependence of FFMDs. Recently, an inverse kinematics technique was applied in the MNT channels of the ^{238}U+^{12}C reaction to study fission using the large-acceptance magnetic spectrometer VAMOS at GANIL [5, 6]. In these experiments, sufficiently-high A and Z resolution for FFs was achieved due to their kinematic boost, allowing the simultaneous measurement of the complete mass- and atomic-number distributions of fission fragments (FFs).

At the tandem accelerator facility of the Japan Atomic Energy Agency (JAEA), we studied the MNT channels of the reactions ^{18}O + ^{232}Th,^{238}U,^{248}Cm in normal kinematics to obtain FFMDs and their excitation-energy dependence for various isotopes (data for ^{18}O + ^{232}Th were published in [7]). An obvious advantage of this method is a relatively easy possibility to change the projectile and/or the target nuclei. In particular by using targets of the rarest highly-radioactive neutron-rich isotopes heavier than ^{238}U (e.g. Cm and Cf), nuclei to be studied can be extended to isotopes far heavier than uranium, which cannot be used at the accelerator facilities for the inverse kinematics experiments similar to VAMOS or SOFIA.

2 Experimental methods

An ^{18}O beam was supplied by the JAEA-tandem accelerator at a typical beam intensity of about 0.5 pnA. Beam energies were 157–162 MeV, depending on the different run. Targets were prepared by electrodeposition of oxide-target material on a thin Ni backing. Thickness of the target-material layer was around 35 – 148 μg/cm^2.

For the event-by-event identification of the transfer channel (thus, of the fissioning nucleus) and of respective coincident FFs, a detection system consisting of a ΔE - E silicon detector telescope and four multiwire proportional chambers (MWPC) were used, see Fig. 1. Specific transfer channels were identified by detecting projectile-like (ejectile) nuclei in twelve 75 μm-thick trapezoidal ΔE silicon detectors which were mounted in a cone around the beam axis, each with the azimuthal angle acceptance of $\Delta\phi = 22.5°$. After passing through the ΔE detector, the ejectiles impinged on the 300 μm-thick annular silicon strip detector (E-detector), divided in 16 annular strips, which allows determination of the scattering angle θ. The inner and outer radius of the detector are 24.0 mm and 48.0 mm, respectively, corresponding to the acceptance angle θ between 16.7° and 31.0° relative to the beam direction.

Figure 2 shows the ΔE - E_{tot} spectrum for ejectiles obtained in the ^{18}O+^{232}Th reaction, where the parabolic lines correspond to different transfer channels, including a clear separation of specific isotopes. Isotopic assignment was done in respect of the elastically-scattered peak of ^{18}O and the missing line of ^{8}Be. It was further confirmed with the energy-loss calculation. The data from ΔE - E_{tot} spectra were also used to deduce the excitation energy of the respective fissioning nuclei, which were determined from reaction Q-value and the measured (angle-dependent) ejectile energy E_{tot}. In this procedure we assumed that no excitation energy is given to the ejectile, thus the quoted excitation energies should be considered as upper limits only.

The coincident FFs resulting from the fission of excited nuclei after the MNT reaction are detected by four 200 × 200 mm^2 position-sensitive MWPCs (see in Fig. 1). The MWPCs were operated with

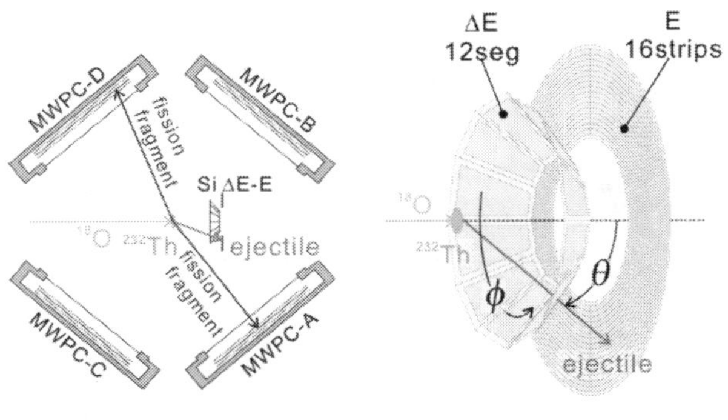

Figure 1. Schematic detection set-up (left) and expanded view of the silicon ΔE-E detector telescope (right). See text for details.

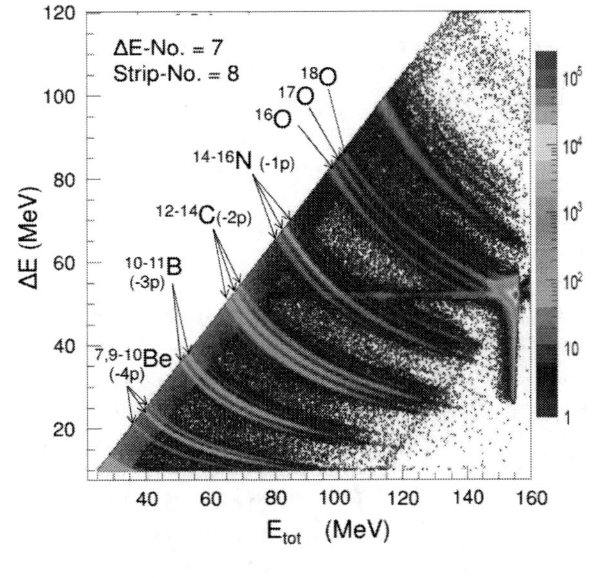

Figure 2. ΔE-$E_{\rm tot}$ spectrum for ejectiles measured by one pair of the ΔE-E detectors obtained in the $^{18}O+^{232}Th$ reaction [7]. The curves corresponding to different ejectiles are marked with the respective isotopes. The scattered ^{18}O is also seen in the plot.

an isobutane gas of about 1.5 Torr [8]. The distance between the target and the center of the cathode was 224 mm, and each MWPC covers a solid angle of 0.67 sr. The positions of FFs's incidence on the MWPC were determined with a position resolution of 4.0 mm. Charge induced in the cathode of the MWPC was recorded to separate FFs from other reaction products. Typical rise time of the MWPC is 5 ns. Both FFs were detected in coincidence with a pair of MWPC facing both sides of the

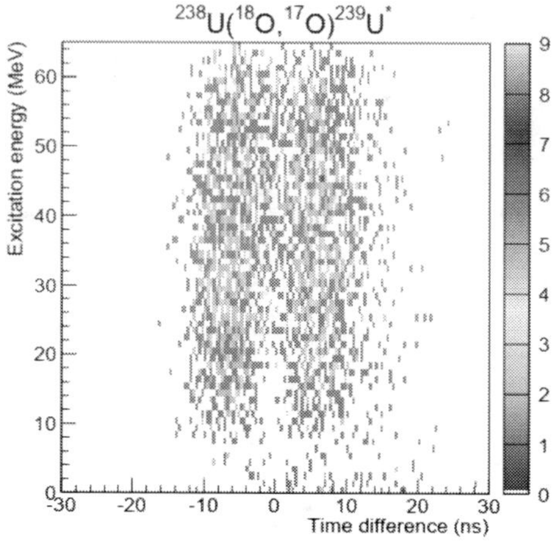

Figure 3. Fission events recorded on the time difference between the signals from coincided MWPCs and excitation energy obtained in one neutron transfer reaction $^{238}U(^{18}O,^{17}O)^{239}U^*$.

target, (+50.1°, −129.9°) or (−50.1°, +129.9°) relative to the beam direction. Fission-fragments time difference, ΔT, between two coincident MWPCs were measured to determine the masses of both FFs. Figure 3 shows an example of recorded FFs on the time difference and excitation energy in the transfer channel of $^{238}U(^{18}O, ^{17}O)^{239}U^*$. Two regions are clearly observed in the low-excitation fissions, corresponding to the light- and heavy-fragment groups, which smear at high-excitation energies.

3 Results

FFs masses were determined event-by-event from the kinematic analysis, where the measured ΔT values and incident positions of both FFs were used. The momentum of the target-like fissioning recoil nucleus is determined by the measured momentum of ejectile under the assumption of a binary reaction process. Figure 4 shows the comparison of FFMDs for $^{239}U^*$, populated in the $^{238}U(^{18}O,^{17}O)^{239}U^*$ reaction [9], with $n + {}^{238}U$ [10]. The obtained FFMDs from MNT reactions agree well with the neutron-induced data, particularly the mass asymmetry at the peak positions at the lowest energy data and the increase of the symmetric fission with excitation energy are noteworthy. The result demonstrates that ^{18}O-induced neutron-transfer reaction can be a surrogate of neutron-induced fission to give FFMDs. In Fig. 4, FFMDs of $^{237-238}U^*$, populated both by the $^{238}U(^{18}O,^{19-18}O)^{237-238}U^*$ reactions and by the $^{232}Th(^{18}O,^{13-12}C)^{237-238}U^*$ reactions are shown. Fairly good agreement are found for each nuclide and its excitation energy dependence, indicating the insensitivity of the FFMDs to the number of transferred nucleons. It is also noted that the FFMD data for $^{233}Pa^*$, $^{233}Th^*$ and $^{236}U^*$ from the MNT reactions of ^{18}O+^{232}Th [7] agree with literature data obtained in proton- and neutron-induced fissions [11–14].

Figure 5 shows the FFMDs for nuclei of $^{238-240}U^*$, $^{239-241}Np^*$, $^{241-243}Pu^*$; selection from the MNT-channels of the $^{18}O + {}^{238}U$ reaction [9]. The FFMDs of the $^{240}U^*$, $^{240,241}Np^*$ were obtained for the first time in this experiment. For the other nuclei, the known FFMD data were systematically extended to excitation energies as high as 60 MeV. It follows from Fig. 5 that mass-asymmetric fission dominates at low excitation energies for all the measured nuclei. The yield in the mass-symmetric

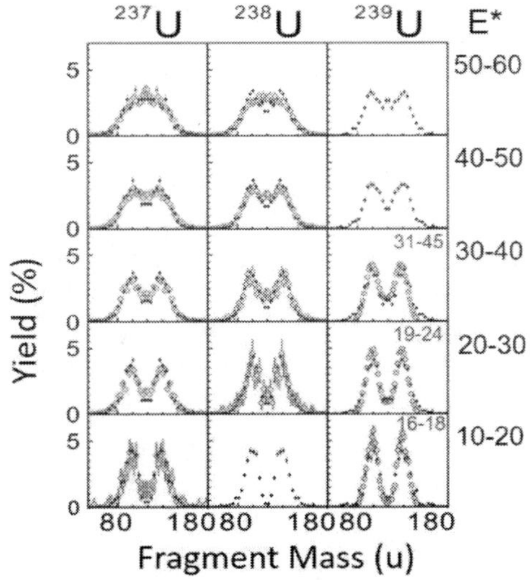

Figure 4. Experimental FFMDs (black dots with error bars) obtained in the ^{238}U(^{18}O,$^{19-17}$O)$^{237-239}$U* reactions. Excitation energy ranges are indicated on the right side. Data are compared with those from the n+^{238}U [10] (open red circles) from the similar excitation energies (shown by the red character). Data for fissions of ^{237,238}U* are compared with the MNT fissions of ^{232}Th(^{18}O,$^{13-12}$C)$^{237-238}$U* [7] (open magenta circles).

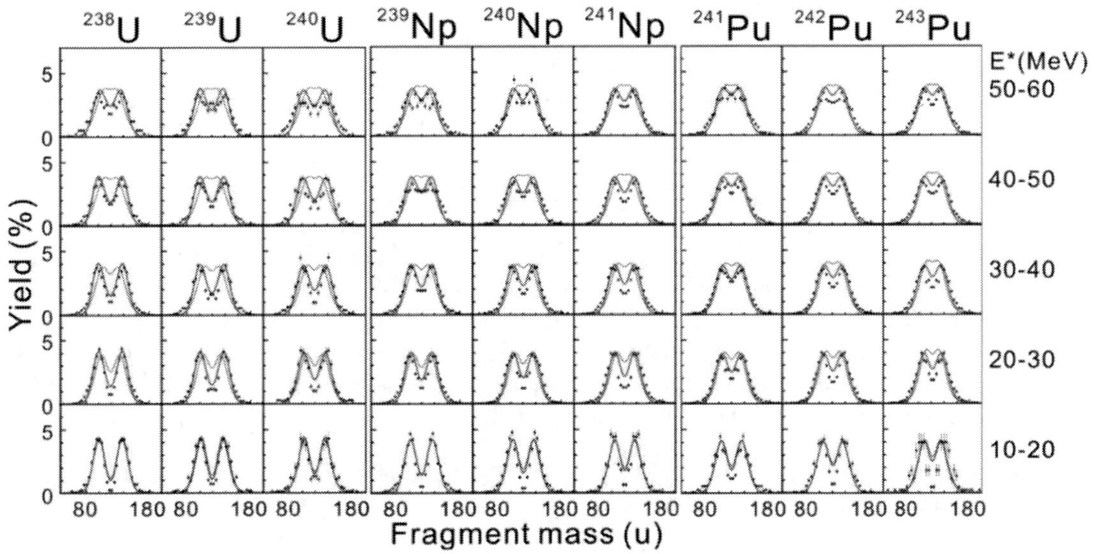

Figure 5. Experimental FFMDs (points with error bars) of the U, Np and Pu isotopes and their dependence on the excitation energy in the range of $E^* = 10$–60 MeV. The experimental FFMDs are compared with the Langevin calculations without (blue curves) and with (red curves) the inclusion of the multi-chance fission (see text).

fission region increases with excitation energy (see also Fig. 4) and the double-peaked shapes tend to be washed out. However, even at the highest energies, $E^* = 50$–60 MeV, double-peak structure is preserved for all the studied nuclides in Fig. 5. It is also interesting to note that the measured spectra reveal smaller peak-to-valley ratio in the FFMDs for heavier elements as can be seen, for example, in the spectra of $E^* = 30$–40 MeV. All these features will be discussed in the following subsection in comparison with a fluctuation-dissipation fission model.

In the recent measurement of ^{18}O+^{248}Cm, new FFMDs of eleven nuclei are further generated, ^{247,249}Cm, ^{249,250,251,252}Bk, ^{251,253}Cf, ^{254,256}Es, and ^{255}Fm [15].

The evolution of the center of the light- and heavy-fragment groups ($\bar{A}_L$ and $\bar{A}_H$) with the mass of the CN in low energy fissions of $10 < E^* < 20$ MeV is shown in Fig. 6, where data obtained from the three MNT reactions, ^{18}O+^{232}Th, ^{18}O+^{238}U, and ^{18}O+^{248}Cm are used. It is found that the $\bar{A}_H$ values are kept constant around 141, whereas $\bar{A}_L$ increases linearly with mass of fissioning nucleus. The trend shows the dominant influence of the shell structure in heavy fragments, well-known in fission studies.

4 Discussions

The measured FFMDs from the MNT reactions are compared with calculations based on the fluctuation-dissipation model developed in [16], where description of fission in Langevin equations from the low-excited state were attempted, and a good reproduction of the measured FFMDs for ^{234,236}U* and ^{240}Pu* from $E^* = 20$ MeV was obtained. As described in [16], the nuclear shape and the corresponding energy is calculated by a two-center shell model [17]. The nuclear shape is defined by three parameters (distance between two potential centers, deformation of fragments, and mass-asymmetry), and the corresponding energy is given by a sum of the liquid-drop energy V_{LD} and the shell correction energy V_{shell}. The latter term is represented as $V_{shell}(0)\exp(-E^*/E_d)$ using the shell correction energy at the zero temperature $V_{shell}(0)$ and shell damping parameter E_d, where $E_d = 20$ MeV was chosen as in [16]. For simplicity, we assumed that the total excitation energy of the system after the MNT reactions is given to the initial excitation energy (E^*) of the fissioning nucleus.

The results of Langevin calculation are shown in Fig. 5 by thin blue curves. Calculated FFMDs are broadened with the experimental mass resolution ($\sigma = \sim 6.5$ u). Under this assumption, the mass asymmetry, i.e. the peak positions of the double-humped FFMD, for all isotopes are reproduced below $E^* = 20$ MeV with clear deviations seen for higher energies. At the highest energy, the calculation shows structure-less symmetric fission in contrast to the measurement. It is seen that the peak-to-valley ratio is reproduced only for the uranium isotopes, $^{238-240}$U of $E^* = 10$–20 MeV, as well as nuclei $^{231-234}$Th, $^{232-236}$Pa and $^{234-237}$U [7] studied in the ^{18}O+^{232}Th reaction. For the neptunium and plutonium isotopes of $E^* = 10$–20 MeV, the calculated peak-to-valley ratio is smaller than the experiments. One of the possible reasons for the deviation could be in the treatment of the neck parameter ϵ ($0 < \epsilon < 1$) [17] to define the shape of nucleus, where $\epsilon = 0.35$ was adopted in the present calculation.

In the above calculation we assumed that all the fission events originate from the initial excitation energy populated by MNT channel. As a next step we attempted to take into account the multi-chance fission (MCF). It is defined as a fission occurring after neutron emission from CN, thus FFs from low-excited and neutron-less excited residual nucleus can contribute (second chance fission). When the residual excited nucleus has enough high excitation energy, further competition between neutron-evaporation and fission (third chance fission) can take place. The higher chance fission successively occurs until the competition terminates. The experimentally observed FFMD is represented by a superposition of all the possible fission chances. These features are demonstrated by Fig. 7, which compares the experimental data for fission of ^{240}U* at the initial excitation energy $E^* = 40$–50 MeV

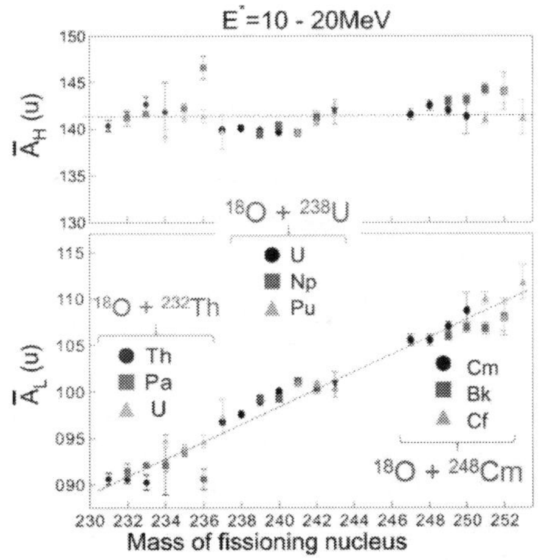

Figure 6. Center of the light and heavy fragment groups ($\bar{A}_L$ and $\bar{A}_H$) as a function of mass of the fissioning nuclei in low-excitation fission of $10 < E^* < 20$ MeV. Data are obtained from the reactions of $^{18}O+^{232}Th$, ^{238}U and ^{248}Cm using the same setup shown in Fig. 1.

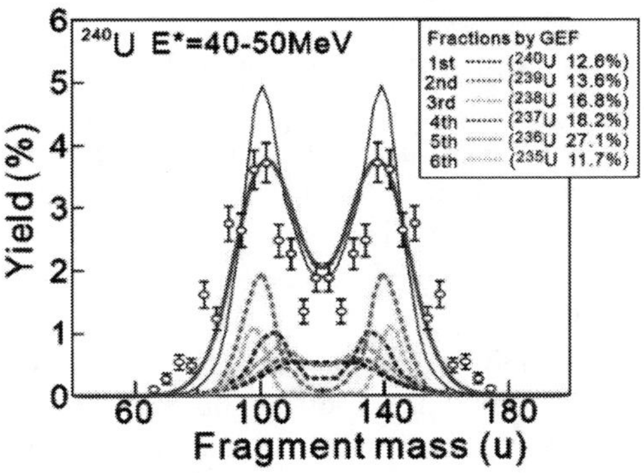

Figure 7. Experimental FFMD of $^{240}U^*$ (open circles) measured at the initial excitation energy of 40–50 MeV obtained from the two neutron-transfer channel $^{238}U(^{18}O,^{16}O)^{240}U^*$, is compared with the Langevin calculation [16] taking into account multi-chance fission. Thin gray curve is obtained by summing all the fission-chances, which is then broadened with the experimental mass resolution (red curve).

with the Langevin calculation taking into account the MCF. As the excitation energy for the calculation, the middle value 45 MeV for the bin-width (40–50 MeV) was used. Probabilities for each fission chances were calculated by the GEF code (Version 2015v2.2) [18], where spins of the compound nucleus were set at zero for simplicity. The reduction of the excitation energy of the compound nucleus due to the neutron emission was determined from neutron binding energies [19] and a mean energy of the emitted neutron, ~1.9 MeV, obtained by the PACE2 code [20]. At each step of MCF, the potential energy surface for the respective compound nucleus was adopted. The finally calculated FFMD shown by the thin gray line is the sum of the FFMDs over the possible chance fissions. It reproduces the observed peak positions of the experimental FFDM, but has narrower peaks than the measured ones. However, after introducing the experimental mass resolution ($\sigma = 6.5$ u) as shown by the thick solid curve, the calculation well reproduces also the peak-to-valley ratio and the total width of the FFMD. It is seen that at this initial energy, the 1st- and 2nd-chance fission occur with somewhat lower probabilities, which exhibits more symmetric-like fission. On the contrary, the higher-fission chances, after emission of several neutrons (2–5, in this case), lead predominantly to an asymmetric mass split. It is evident that the mass-asymmetric fission observed in the data even at the high excitation energy originates from the lower-energy 4th-, 5th-, and 6th-chance fissions (^{235,236,237}U) [9].

The same calculation procedure was applied for all the cases as displayed in Fig. 5, and the results are shown by the red thick curves. In contrast to the calculation without MCF (thin blue curves), the calculation with MCF well explains the variation of FFMDs with the excitation energies. Also mass-asymmetry and peak-to-valley ratio observed at the higher excitation energies are well reproduced. The calculation also demonstrates the decreasing peak-to-valley ratio of FFMDs for heavier elements (from uranium to plutonium), observed for example in the $E^* = 30$–40 MeV range, whereas the analysis without MCF predicts almost the same distributions through the isotopes. It should also be noted that the consideration of MCF validates that the shell effect responsible for mass-asymmetric fission disappears around E^*=30–40 MeV (blue curves in Fig. 5), resulting from the shell-damping energy $E_{\mathrm{d}} = 20$ MeV entering in the excitation-energy dependence of the shell correction energy V_{shell}.

5 Summary and outlook

It is shown that the multi-nucleon transfer reaction is a powerful tool to study fission for nuclei which cannot be accessed by particle-capture and/or heavy-ion fusion reactions. An advantage in the normal kinematics is that the nuclei to be studied can be significantly expanded by using available high-purity radioactive targets. Fission studies using the MNT reactions with other targets, such as ^{243}Am, ^{231}Pa, and ^{249}Cf, are planned at the JAEA tandem facility. Furthermore, a reaction using the ^{254}Es target will allow us to study low-energy fissions of fermium isotopes, where sharp transition from the mass-asymmetric fission (e.g. ^{256}Fm) to the sharp symmetric fission (e.g. ^{258}Fm) was observed in the spontaneous fission studies [21].

In addition to investigate the fission-fragment properties, a measurement of prompt neutrons in coincidence with FFs has stated to obtain neutron multiplicity $\bar{\nu}(A)$ from individual fragments with mass A and their excitation energy dependence, by mounting a neutron detector array around the present fission setup.

Special thanks are due to the crew of the JAEA tandem facility for their beam operation. Present study is supported by "Comprehensive study of delayed-neutron yields for accurate evaluation of kinetics of high-burn up reactors" and "Development of prompt-neutron measurement in fission by surrogate reaction method and evaluation of neutron-energy spectra" by the Ministry of Education, Culture, Sports, Science and Technology of Japan (MEXT).

References

[1] K.-H. Schmidt et al., Nucl. Phys. A **665**, 221 (2000)
[2] E. Pellereau et al., Phys. Rev. C **95**, 054603 (2017)
[3] A.N. Andreyev et al., Phys. Rev. Lett. **105**, 252502 (2010)
[4] A.N. Andreyev et al., Rep. on Prog. Phys., in print
[5] M. Caamaño et al., Phys. Rev. C **92**, 034606 (2015)

[6] C. Rodríguez-Tajes et al., Phys. Rev. C **89**, 024614 (2014)
[7] R. Léguillon et al., Phys. Lett. B, **761**, 125 (2016)
Ibid., M. Wang et al., 1603
[8] K. Nishio et al., Phys. Rev. C **77**, 064607 (2008)
[9] K. Hirose et al., submitted
[10] V.D. Simutkin et al., Nucl. Data Tables **119**, 331 (2014)
[11] S.I. Mulgin et al., Nucl. Phys. A **824**, 1 (2009)
[12] I. Nishinaka, et al., Phys. Rev. C, **70**, 014609 (2004), and Proceedings of the Fourth International Conference on Fission and Properties of Neutron-rich Nuclei, Sanibel Island, Florida, USA, 11–17 Nov. 2007, pp. 206-211, World Scientific
[13] I.V. Ryzhov et al., Phys. Rev. C **83**, 054603 (2011)
[14] Ch. Straede, B. Jørgensen and H.-H. Knitter, Nucl. Phys. A **462**, 85 (1987)
[15] K. Nishio et al., International Conference on Nuclear Data for Science and Technology (ND2016), 11–16 Sep. 2016, Bruges, Belgium
[16] Y. Aritomo and S. Chiba, Phys. Rev. C **88**, 044614 (2013)
[17] K. Sato et al., Z. Phys. A **288**, 383 (1978)
[18] K.-H. Schmidt and B. Jurado and C. Amouroux and C. Schmitt, Nuclear Data Sheets **131**, 107 (2016)
[19] P. Möller et al., Atomic Data and Nuclear Data Tables. **109-110**, 1 (2016)
[20] A. Gavron, Phys. Rev. C **21**, 230 (1980)
[21] D.C. Hoffman, Nucl. Phys. A **502**, 21c (1989)

Prompt gamma rays from ^{252}Cf(sf) and their angular distributions

Andreas Oberstedt[1,⋆], *Robert* Billnert[2,3], *Angélique* Gatera[2], *Alf* Göök[2], and *Stephan* Oberstedt[2]

[1] *Extreme Light Infrastructure - Nuclear Physics (ELI-NP) / Horia Hulubei National Institute for Physics and Nuclear Engineering (IFIN-HH), 077125 Bucharest-Magurele, Romania*
[2] *European Commission, DG Joint Research Centre, Directorate G - Nuclear Safety and Security, Unit G.2 Standards for Nuclear Safety, Security and Safeguards, 2440 Geel, Belgium*
[3] *Present address: Studsvik Nuclear Environmental AB, 61182 Nyköping, Sweden*

Abstract. In a recent measurement of prompt γ rays from the spontaneous fission of ^{252}Cf an experimental setup was chosen that allowed the study of angular correlations between γ rays and the nuclei from which they were emitted. The observed angular distribution of prompt fission γ rays was then analyzed to estimate the relative contributions from radiation of different multipolarity. First preliminary results are presented and compared to previous angular correlation measurements from different fission processes and recent results from calculations with the Monte Carlo Hauser-Feshbach code FIFRELIN.

1 Introduction

The energy release in nuclear fission is distributed in kinetic and excitation energy of the two fragments. The latter manifests itself in fragment deformation and intrinsic excitation energy. In an early and quite simplified picture, it was assumed that the fragments are first de-excited by the emission of neutrons until the fragments' remaining excitation energy is lower than the neutron binding energy, i.e. on average at about half of it. Only then de-excitation would continue by the emission of γ rays, eventually reaching the ground state [1]. These γ rays may be divided into two categories, commonly called statistical and discrete γ rays, respectively. The first ones originate mainly from electric dipole transitions and carry away most of the remaining excitation energy of the fragments, while the latter stem mainly from electric quadrupole transitions along the Yrast line and carry away most of the angular momentum of the fragments [2]. This so-called prompt γ-ray emission is a fast process, which basically takes place within the first few nanoseconds after scission [3] and is followed by β decay(s) and thereafter by delayed emission of neutrons and γ rays towards the valley of β stability. The average prompt fission γ-ray multiplicity, i.e. the average number of photons per fission, amounts typically to 8 [4], as corroborated by our previous measurements [5–10].

In this work we have chosen an experimental setup that allowed measuring angular correlations between fission fragments and prompt γ rays, in order to study the de-excitation of the fragments by prompt γ-ray emission and to deduce information about the relative contributions from radiation of different multipolarity. Below we compare our observations to those from previous angular correlation measurements and give first preliminary results, which are discussed in terms of results from recent calculations with the Monte Carlo Hauser-Feshbach code FIFRELIN [11].

⋆e-mail: andreas.oberstedt@eli-np.ro

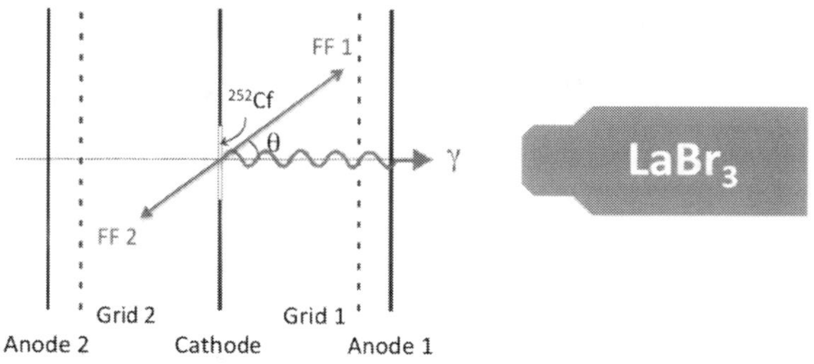

Figure 1. Sketch of the experimental setup used to measure angular correlations between fission fragments (FF 1 or FF 2) and prompt fission γ rays. The fragments were detected with a cylindrical twin Frisch-grid ionization chamber (FGIC) on whose central cathode a ^{252}Cf sample was mounted. For the detection of the γ rays a $LaBr_3$:Ce detector of size 76 mm × 76 mm was employed that was placed along the symmetry axis perpendicular to the sample. The polar angle of the emitted fragments equivalent to the angle relative to the detected photons is denoted by θ.

2 Experimental setup and data treatment

The recent measurement of PFGS from ^{252}Cf(sf) was performed with a 76 mm × 76 mm (diameter × length) $LaBr_3$:Ce scintillation detector that was placed perpendicular to the plane of the sample along its symmetry axis. The ^{252}Cf sample was mounted inside a cylindrical twin Frisch-grid ionization chamber (FGIC) [12] in the center of a common cathode. Fission fragments from a thin ^{252}Cf sample are emitted to both sides of the cathode, ionizing the P-10 counting gas, and being detected by means of the electrons moving towards the respective anodes. A Frisch grid is placed in front of each anode in order to shield the latter from the influence of the ions, but it also allows determining the (polar) emission angle of the fragments (see Fig. 1 for a schematic view of the experimental setup). The FGIC provides the fission trigger for the coincident measurement of γ rays, for which energy and time-of-flight (TOF) between sample and detector are recorded. The latter information is used to distinguish prompt γ rays from other photons, emitted e.g. in prompt fission neutron induced reactions, i.e. mainly inelastic neutron scattering. The experimental techniques applied here are in principle the same as described in Refs. [5–10], where more details may be found.

The collected prompt fission γ rays were sorted into energy spectra for $\cos\theta$ bins between 0 and 1 in steps of 0.05. In order to obtain emission spectra, the response function of the detector has to be determined, which is usually done by means of Monte Carlo simulations of the actual setup with computer codes like e.g. Penelope [13] or Geant4 [14]. Thereafter the response function must be unfolded from the raw spectra. However, as a first step, we have chosen here a different, less time-consuming approach. Adding up all spectra gives an angular integrated raw spectrum that may be compared with others, obtained previously for the same fissioning system. Since they all turned out to exhibit a very similar appearance, we may benefit from the fact that similar raw spectra lead to similar emission spectra, provided that detector and setup are comparable [7]. This is the case here. From a previous measurement [5], for which the proper unfolding was carried out, both measured and emission spectra are known, whose ratio gives a transformation function (see Ref. [7] for details). Multiplying this function with the recently measured PFGS gives then the emission spectrum. In the same way

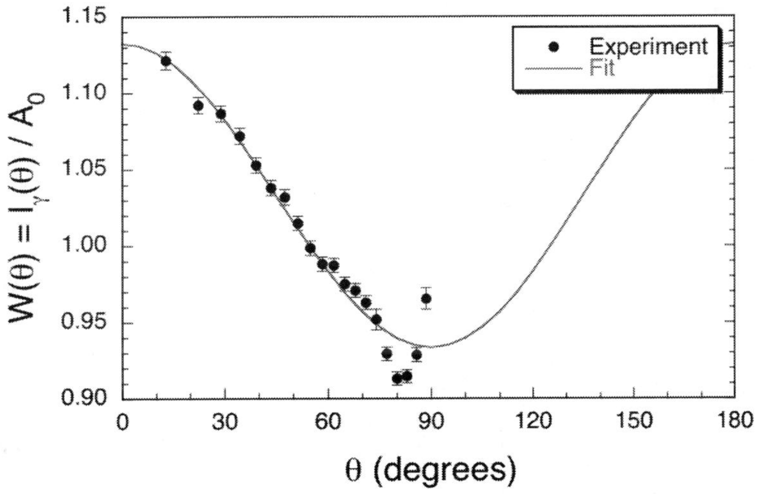

Figure 2. Measured angular distribution of prompt γ rays from ^{252}Cf(sf), depicted as black dots. The error bars correspond to statistical uncertainties only. The line represents the result of a fit according to Eq. (1).

angular-dependent emission spectra were created for each bin, from which PFGS characteristics were determined. A detailed inspection of the obtained angular distribution is performed below.

3 Angular distribution of prompt fission γ rays

As described in the previous chapter, emission spectra of prompt fission γ rays were created for cosθ bins, covering the polar angle range $0° < \theta < 90°$. The integrated total multiplicity, i.e. the average number of emitted photons per fission, was determined to $\overline{M}_\gamma = 8.28 \pm 0.51$, where the statistical uncertainty is 0.07 and the systematic one, mainly due to the transformation function, amounts to 0.44. This result is in good agreement with our last published value $\overline{M}_\gamma = 8.29 \pm 0.13$ [6]. The angular distribution of radiation may be expressed according to

$$W(\theta) = A_0[1 + \{A_2/A_0\}P_2(cos\theta) + \{A_4/A_0\}P_4(cos\theta)], \tag{1}$$

where $P_k(cos\theta)$ denote Legendre polynomials for k = 2, 4 and θ is the emission angle of the fission fragment relative to the γ ray direction, i.e. perpendicular to the cathode plane of the FGIC. Any possible attenuation coefficients are assumed to be 1. The coefficients $\{A_k/A_0\}$ are determined experimentally and may be compared to theory for different types of radiation. For instance, $\{A_2/A_0\} \approx -0.3$ for pure dipole radiation and $\{A_2/A_0\} \approx +0.3$ for pure quadrupole radiation, while $\{A_4/A_0\}$ is close to zero [15]. Figure 2 shows the measured angular distribution as black dots, together with the result of a fit according to the expression given in Eq. (1) shown as line. It should be noted that the data close to both $\theta = 0°$ and 90° suffers from considerable instrumental errors due to absorption and scattering, but these points were taken into account anyway, since they did not influence much the fit result. The obtained coefficient is $\{A_2/A_0\} = 0.13 \pm 0.03$, which according to $(+0.3) \times p + (-0.3) \times (1 - p) = 0.13$ leads to $p = 0.72$, corresponding to 72% quadrupole (L=2) and 28% dipole (L=1) radiation (higher multipolarities are less probable and neglected here).

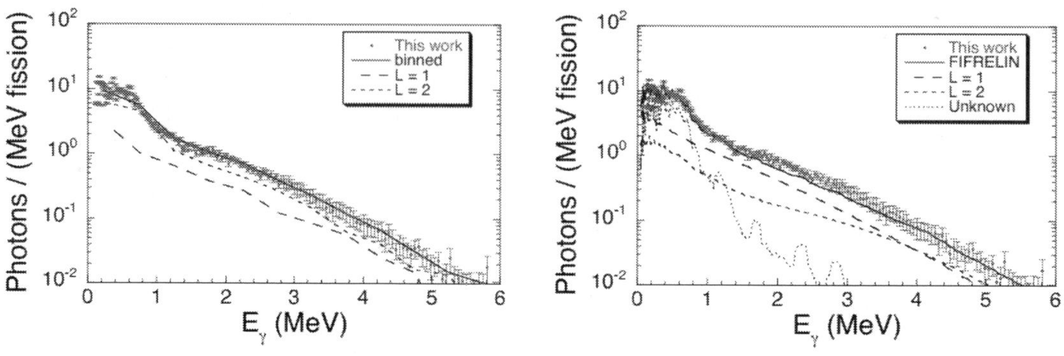

Figure 3. Measured PFGS for ^{252}Cf(sf) from this work, depicted by red dots. In the left part it is compared to the rebinned spectrum of 500 keV intervals (full black line), from which the contributions for dipole (L=1) and quadrupole (L=2) radiation were determined (long and short dashed lines, respectively). The right part shows in addition to the measured spectrum a spectrum and its contributions for L = 1 and L = 2 as well as a contribution for which the multipolarity is unknown (all three as black lines), calculated recently with FIFRELIN (see text for details).

In the next step we tried to assess how the ratio between quadrupole and dipole radiation depends on the energy of the prompt fission γ rays. For that reason the PFGS was rebinned into 500 keV intervals, for which the corresponding angular distributions were fitted according to Eq. (1) as described above for the entire spectrum. As a consequence, the coefficient $\{A_2/A_0\}$ was determined for each energy bin and the contributions for L=1 and L=2 deduced. The resulting multipolarity-dependent spectra are shown on the left-hand side of Fig. 3. These results are discussed in the next section.

4 Discussion and outlook

Prior to this work, angular distributions of prompt γ rays were measured in thermal-neutron induced fission of the target nuclei ^{233}U, ^{235}U and ^{239}Pu [16], as well as the spontaneous fission of ^{252}Cf [17]. A qualitative comparison of all these measurements exhibits a good agreement, indicating a dominating contribution of quadrupole radiation, of which E2 is much more likely than M2. A direct comparison of the coefficients in Eq. (1) from this work and from Ref. [17] has not been made because of the different energy ranges, i.e. only 0.1 to 1.5 MeV are considered in Ref. [17]. However, for the region $E_\gamma = 1.0 - 1.5$ MeV we obtain $\{A_2/A_0\} = 0.077$ and $\{A_4/A_0\} = -0.022$, which is not too far from the corresponding values 0.0683(1) and −0.0308(2) reported in Ref. [17]. If this E2 radiation is assigned to γ decays along the Yrast line (cf. Sect. 1), the corresponding multipolarity-dependent multiplicity would be $\overline{M}_{\gamma,L=2} \approx 6.0$, with a total average multiplicity of $\overline{M}_\gamma = 8.29$ [6]. Each quadrupole photon will then carry away $2\hbar$, which implies an average angular momentum of J $\approx 12\hbar$, which is in rather good agreement with J $\approx 10\hbar$ reported earlier [18] and results from GEF calculations for different fission fragments [19]. It has to be mentioned though that also lower fission fragment spins of $6\hbar$ to $8\hbar$ have been reported (see [17] and references therein). Hence, further experimental studies are necessary. The remaining dipole multiplicity would then be $\overline{M}_{\gamma,L=1} \approx 2.3$ for statistical γ rays.

Calculations based on the Monte Carlo Hauser-Feshbach code FIFRELIN [11] have previously been rather successful in reproducing experimental results for PFGS characteristics as well as the distinct peak structure of the low-energy part of the energy spectra for ^{235}U(n_{th}, f) [8] and ^{252}Cf(sf) [6]. For comparison, the calculated average PFGS multiplicity has been adjusted to the measured one.

Table 1. Prompt γ-ray characteristics for the spontaneous fission of ^{252}Cf. Experimental results from this work for the average γ-ray multiplicity $\overline{M}_\gamma$, the average energy per photon ϵ_γ, and the total energy $E_{\gamma,tot}$ released in fission, are given for the integral spectrum as well as for the multipolarity-dependent spectra and compared to corresponding values recently calculated with FIFRELIN [22] (see text for details). Relative multiplicities are also given, under the assumption that higher multipolarities may be neglected.

	Multipolarity L	Experiment (this work)		Calculations (FIFRELIN)	
$\overline{M}_\gamma$	all	8.28 ± 0.51		8.28	(adjusted)
$\overline{M}_\gamma$	1	2.31	(28%)	3.20	(39%)
$\overline{M}_\gamma$	2	5.97	(72%)	1.45	(17%)
$\overline{M}_\gamma$	unknown			3.63	(44%)
ϵ_γ	all	0.79 ± 0.10 MeV		0.76 MeV	
ϵ_γ	1	0.86 MeV		0.94 MeV	
ϵ_γ	2	0.76 MeV		1.03 MeV	
ϵ_γ	unknown			0.50 MeV	
$E_{\gamma,tot}$	all	6.51 ± 0.76 MeV		6.30 MeV	
$E_{\gamma,tot}$	1	1.99 MeV		3.00 MeV	
$E_{\gamma,tot}$	2	4.52 MeV		1.49 MeV	
$E_{\gamma,tot}$	unknown			1.81 MeV	

Even multipolarity-dependent spectra were generated [20], which however seem to show a dominating contribution of L=1 radiation. Apparently, this is in contrast with the experimental results presented in this work, although a considerable part of radiation consists of so-called "experimental transitions", to which no multipolarity was assigned. They were taken from the RIPL-3 database [21]. However, new FIFRELIN calculations were performed recently [22], whose results are shown on the right-hand side of Fig. 3. However, here it must be noted that, apart from γ rays of known multipolarity 1 and 2, there is a considerable amount of photons, whose multipolarity is not known, since they correspond to transitions between levels taken from the RIPL-3 database (see above). In fact, they constitute about 44% of the calculated total average PFGS multiplicity, as shown in Table 1. Since a dominating contribution of quadrupole radiation has been observed in all PFGS measurements, as referred to above, we tend to conclude that most of these transitions of unknown multipolarity should lead to quadrupole radiation. Therefore, the sum of these "unknown" transitions and the calculated ones with L=2 should be compared to the experimentally obtained contribution for L=2. With this condition comparison of the values given Table 1 shows indeed a rather good agreement.

This agreement would have been even better, if a probably more realistic assumption for the coefficient for pure dipole radiation, $\{A_2/A_0\} \approx -0.2$, had been made, considering the value $\{A_2/A_0\} = -0.212(26)$ from the example given in Ref. [15]. With this value, $\overline{M}_{\gamma,L=1} \approx 2.8$ and $\overline{M}_{\gamma,L=2} \approx 5.5$ would have been obtained, corresponding to relative contributions of 34% and 66%, respectively. Of course, these results are still considered preliminary and must be compared with other related information on fission fragments, such as average neutron separation energies and average moments of inertia, in order to verify the deduced energies carried away by dipole and quadrupole photons during the de-excitation process of fission fragments. Hence, these activities are to be continued.

One of the authors (A. O.) acknowledges the support from the Extreme Light Infrastructure Nuclear Physics (ELI-NP) Phase II, a project co-financed by the Romanian Government and the European Union through the European Regional Development Fund - the Competitiveness Operational Programme (1/07.07.2016, COP, ID 1334), with which this work had been finalized.

References

[1] C. Wagemans (editor), The Nuclear Fission Process, CRC Press Boca Raton (1991)

[2] A. Hotzel, P. Thirolf, Ch. Ender, D. Schwalm, M. Mutterer, P. Singer, M. Klemens, J.P. Theobald, M. Hesse, F. Gönnenwein, H.v.d. Ploeg, Z. Phys. A **356**, 299 (1996)

[3] P. Talou, T. Kawano, I. Stetcu, J.P. Lestone, E. McKigney, and M.B. Chadwick, Phys. Rev. C **94**, 064613 (2016)

[4] R. Vandenbosch and J.R. Huizenga, Nuclear Fission, Academic Press New York and London (1973)

[5] R. Billnert, F.-J. Hambsch, A. Oberstedt, and S. Oberstedt, Phys. Rev. C **87**, 024601 (2013)

[6] A. Oberstedt, R. Billnert, F.-J. Hambsch, and S. Oberstedt, Phys. Rev. C **92**, 014618 (2015)

[7] S. Oberstedt, A. Oberstedt, A. Gatera, A. Göök, F.-J. Hambsch, A. Moens, G. Sibbens, D. Vanleeuw, and M. Vidali, Phys. Rev. C **93**, 054603 (2016)

[8] A. Oberstedt, T. Belgya, R. Billnert, R. Borcea, T. Bryś, W. Geerts, A. Göök, F.-J. Hambsch, Z. Kis, T. Martinez, S. Oberstedt, L. Szentmiklosi, K. Takács, and M. Vidali, Phys. Rev. C **87**, 051602(R) (2013)

[9] S. Oberstedt, R. Billnert, T. Belgya, T. Bryś, W. Geerts, C. Guerrero, F.-J. Hambsch, Z. Kis, A. Moens, A. Oberstedt, G. Sibbens, L. Szentmiklosi, D. Vanleeuw, and M. Vidali, Phys. Rev. C **90**, 024618 (2014)

[10] A. Gatera, T. Belgya, A. Göök, F.-J. Hambsch, M. Lebois, A. Oberstedt, S. Oberstedt, L. Qi, L. Szentmiklósi, G. Sibbens, and M. Vidali, Phys. Rev. C **95**, 064609 (2017)

[11] O. Litaize and O. Serot, Phys. Rev. C **82**, 054616 (2010)

[12] C. Budtz-Jørgensen, H.-H. Knitter, Ch. Straede, F.-J. Hambsch and R. Vogt, Nucl. Instr. and Meth. A **258**, 209 (1987)

[13] `http://www.oecd-nea.org/tools/abstract/detail/nea-1525`

[14] S. Agostinelli and the Geant4 collaboration, Nucl. Inst. Meth. A **506** (2003) 250

[15] E.S. Paul, UK Nuclear Physics Summer School, Bristol, 27. August – 6. September 2013, `http://ns.ph.liv.ac.uk/ ajb/summerschool/files/ESP-Lecture4.pdf`

[16] M.M. Hoffman, Phys. Rev. C **133**, B714 (1964)

[17] Yu.N. Kopach, P. Singer, M. Mutterer, M. Klemens, A. Hotzel, D. Schwalm, P. Thirolf, M. Hesse, and F. Gönnenwein, Phys. Rev. Lett. **82**, 303 (1999)

[18] A. Bogachev, L. Krupa, O. Dorvaux, E. Kozulin, M. Itkis, M.-G. Porquet, A. Astier, D. Curien, I. Deloncle, G. Duchene, B.J.P. Gall, F. Hanappe, F. Khalfallah, M. Rousseau, L. Stuttge, N. Redon, and O. Stezowski, Eur. Phys. J. A **34**, 23 (2007)

[19] K.-H. Schmidt, B. Jurado, C. Amouroux and C. Schmitt, Nuclear Data Sheets **131**, 207 (2016)

[20] D. Regnier, "Contribution à l'étude des gammas prompts de fission", PhD thesis, Université de Grenoble, France (2013)

[21] R. Capote, M. Herman, P. Oblozinsky, P. Young, S. Goriely, T. Belgya, A. Ignatyuk, A. Koning, S. Hilaire, V. Plujko, M. Avrigeanu, O. Bersillon, M. Chadwick, T. Fukahori, Z. Ge, Y. Han, S. Kailas, J. Kopecky, V. Maslov, G. Reffo, M. Sin, E. Soukhovitskii, P. Talou, RIPL - Reference Input Parameter Library for Calculation of Nuclear Reactions and Nuclear Data Evaluations, Nucl. Data Sheets **110**(12) (2009) 3107

[22] A. Chebboubi, private communication (2017)

Role of the excitation energy of the compound nucleus in binary decay processes

H. Paşca[1,2,⋆], A.V. Andreev[1], G.G. Adamian[1], and N.V. Antonenko[1]

[1] *Joint Institute for Nuclear Research, 141980 Dubna, Russian Federation*
[2] *"Babeş-Bolyai" University, Faculty of Physics, 400084 Cluj-Napoca, Romania*

Abstract. Employing the improved scission-point model, the isotopic and excitation energy trends of the charge distribution of fission fragments are studied in fission of even-even Th isotopes at low and high excitation energies.

1 Introduction

As known, with increasing neutron number of fissioning even-even Th isotopes the transition from one-peaked to two-peaked charge distribution of fission fragments occurs through transient three-peaked shape in which the symmetric and asymmetric components of the distribution have almost equal weights [1]. The general opinion is that the competition between symmetric and asymmetric fission modes is related to the deformed shells. However, in the case of high-energy neutron-induced fission of ^{238}U [2], the experimental data show that the mass distribution maintains two-peaked shape even at neutron energies of 50–60 MeV. At these excitation energies, the shell effects are expected to be washed out, and the nucleus is supposedly left with a dominant symmetric mode, that is contrary to the experiments. The experimental data on fission of ^{180}Hg [3] also suggest that the asymmetric mass distribution can not be anticipated only from the microscopic effects in the fragments. Our aim is to explain the transformation of the charge distribution with neutron number in the electro-magnetic-induced fission ($E_\gamma = 11$ MeV) of Th isotopes and to predict the charge distributions at large excitation energies. The fission is described within the improved scission-point model [4]. The most important ingredient of the model is the potential energy of system as a function of charge (mass) asymmetry, deformations of the fission fragments, and internuclear distance. The knowledge of the deformations of nascent fragments at scission point is also crucial.

2 Statistical scission-point model

The statistical scission-point model relies on the assumption that the statistical equilibrium is established at scission where the observable characteristics of fission process are formed. At the scission point the decaying system is modeled by two uniformly charged, nearly touching coaxial ellipsoids - the fission fragments forming the dinuclear system (DNS). By taking into consideration the volume conservation, the system is completely defined by the mass and charge numbers (A_i, Z_i)

⋆e-mail: pasca@theor.jinr.ru

of the fragments, deformation parameters of fragments, β_i, and interfragment distance R. Here, the index i designates the light or heavy fragment of the DNS ($i = L, H$). The potential energy $U = B_L + B_H + V$ of the DNS consists of the binding energies B_i of the fragments and the mutual fragment-fragment interaction potential V [4]. The binding energy $B_i = U_i^{LD} + \delta U_i^{shell}$ of each fragment consists of the excitation-energy-dependent liquid-drop energy U_i^{LD} and deformation-dependent shell-correction term δU_i^{shell} calculated with the two-center shell model. The damping of the shell corrections with excitation energy is taken into account. The interaction potential $V = V^N + V^C$ consists of the Coulomb interaction potential V^C of the two uniformly charged ellipsoids and nuclear interaction potential V^N in the double-folding form [4]. The interaction potential presents an inner potential pocket and external barrier which prevents the system from decay. The barrier, $B_{qf}(A_i, Z_i, \beta_i) = V(A_i, Z_i, \beta_i, R_b) - V(A_i, Z_i, \beta_i, R_m)$, calculated as the difference of the interaction potentials at the top of the external barrier $[R = R_b(A_i, Z_i, \beta_i)]$ and at the bottom of the potential pocket $[R = R_m(A_i, Z_i, \beta_i)]$, prevents the decay of the DNS in R. The height B_{qf} of barrier has also an impact on the yields. With increasing elongation and decreasing charge (mass) asymmetry the value of B_{qf} decreases, the system becomes more unstable and decays. Because the dynamical treatment is not explicitly performed here, we simulate the dynamical effects by restricting the minimum value of the height B_{qf} of the decay barrier. In the calculations, we take into consideration only those configurations for which B_{qf} is larger than ~1 MeV. This condition restricts the highly deformed unstable configurations in the (β_L, β_H) plane.

Because the thermal equilibrium is assumed at the scission point, the excitation energy $E^*(A_i, Z_i, \beta_i, R_m) = E^*_{CN} + [U_{CN}(A, Z, \beta) - U(A_i, Z_i, \beta_i, R_m)]$ at scission is calculated as the initial excitation energy of the fissioning nucleus E^*_{CN} plus the difference between the potential energies of fissioning compound nucleus $U_{CN}(A, Z, \beta)$ and of the system at the scission point $U(A_i, Z_i, \beta_i, R_m)$ [4]. Note that the value of R_m is taken directly from the interaction potential and it is deformation-dependent, which allows us to have a precise definition of the scission configuration and excitation energy. The relative formation probability of the DNS with particular masses, charges and deformations of the fragments is calculated as [4]:

$$w(A_i, Z_i, \beta_i, E^*) = N_0 \exp\left[-\frac{U(A_i, Z_i, \beta_i, R_m) + B_{qf}(A_i, Z_i, \beta_i)}{T}\right], \tag{1}$$

where N_0 is the normalization factor. In Eq. (1), the temperature is calculated as $T = \sqrt{E^*/a}$, where $a = (A_L + A_H)/12$ MeV^{-1} is the level density parameter, which seems to be a good approximation for the systems studied here [4]. Note that in the literature various values for a are used (e.g. $a = (A_L + A_H)/12$ to $a = (A_L + A_H)/8$). In our calculations, a single value of the temperatureis used at the global potential minimum of U before the shell damping. Variations of the level density parameter in this work lead to minute changes in the results.

In order to obtain the mass-charge distribution $Y(A_i, Z_i, E^*)$ of fission fragments, one should integrate (1) over β_L and β_H. For the calculations of mass and charge distributions, the following expressions are used:

$$Y(A_i, E^*) = \frac{\sum_{Z_i} Y(A_i, Z_i, E^*)}{\sum_{Z_i, A_i} Y(A_i, Z_i, E^*)}, Y(Z_i, E^*) = \frac{\sum_{A_i} Y(A_i, Z_i, E^*)}{\sum_{Z_i, A_i} Y(A_i, Z_i, E^*)}. \tag{2}$$

3 Results and discussions

In Fig. 1 the calculated mass distributions resulting from the spontaneous fission of even-even nuclei Cf and Fm with $A = 250 - 256$ are depicted. The results seem to be in good agreement with the experimental data compiled and shown schematically in Fig. 8 of Ref. [5]. The calculated

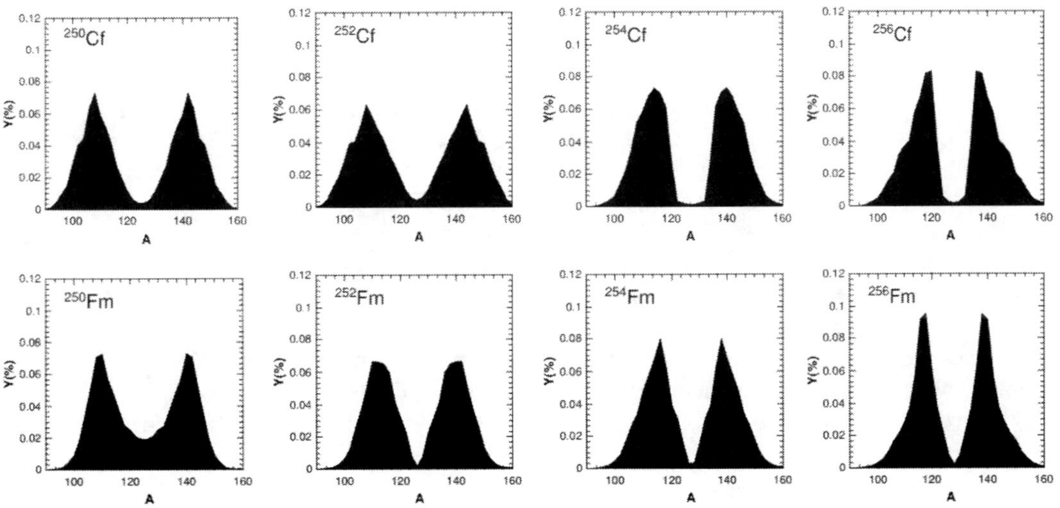

Figure 1. The calculated mass distributions in the spontaneous fission of the indicated even-even actinides.

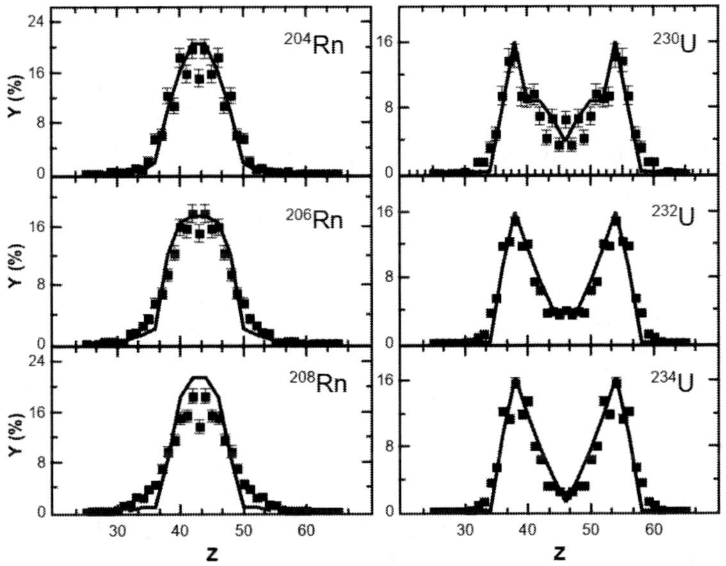

Figure 2. The calculated charge distributions (lines) for electro-magnetic-induced fission of the indicated radon and uranium isotopes at 11 MeV excitation energy are compared with the experimental data [1] (symbols). The lines connect the calculated points for even-even fission fragments.

charge distributions resulting from the electro-magnetic-induced (E_γ = 11 MeV) fission of even-even ^{204,206,208}Rn, ^{230,232,234}U, and ^{222,224,226,228}Th nuclei are compared with the experimental data of Ref. [1] in Figs. 2 and 3. For the isotopes of Rn (Fig. 2) and ^{222}Th (Fig. 3), the fragment charge distributions are clearly symmetric, with a single prominent peak around $Z/2$, while the charge distributions of fissioning nuclei ^{230,232,234}U (Fig. 2) and ^{228}Th (Fig. 3) have two distinct asymmetric peaks. The fission of the ^{226}Th nucleus shows a three-peaked distribution (Fig. 3). As seen, our model is suitable for describing both symmetric and asymmetric charge distributions. The transition from one-peaked to two-peaked distribution, going through transient three-peaked shape, can be explained by studying the driving potential:

$$\langle U(Z_i)\rangle = \frac{\sum_{A_i}\int d\beta_L d\beta_H U(A_i,Z_i,\beta_i,R_m)w(A_i,Z_i,\beta_i,E^*)}{\sum_{A_i}\int d\beta_L d\beta_H w(A_i,Z_i,\beta_i,E^*)} \quad (3)$$

averaged over the mass number A_i and deformations $\beta_{L,H}$. In the case of ^{222}Th isotope the driving potential $\langle U(Z_i)\rangle$ shows a deep minimum at $Z/2$=45. For the ^{224}Th nucleus, $\langle U(Z_i)\rangle$ starts displaying the formation of the second minimum around Kr–Sr, while for the ^{226}Th isotope, the second minimum becomes almost as deep as the central (symmetric) one. Even though that the values of $\langle U(Z_L = 36{-}38)\rangle$ is slightly higher than $\langle U(Z_L = 44)\rangle$, the corresponding yields are the same. This is easily explained by the fact that in (β_L, β_H) plane, the minima for Kr–Sr are wider than those for Ru–Pd [4]. For the ^{228}Th nucleus, the asymmetric minima become deeper, facilitating the asymmetric fragmentations, but the central minimum still remains.

To understand the conditions under which the mass-asymmetric minimum appears in the driving potential, we analyze all components of the driving potential: the liquid-drop energy $\langle U^{LD}(Z_i)\rangle = \langle U_L^{LD}\rangle + \langle U_H^{LD}\rangle$, the interaction potential $\langle V(Z_i)\rangle = \langle V^C\rangle + \langle V^N\rangle$, and the shell-correction energy $\langle\delta U^{shell}(Z_i)\rangle = \langle\delta U_L^{shell}\rangle + \langle\delta U_H^{shell}\rangle$ (Fig. 4).

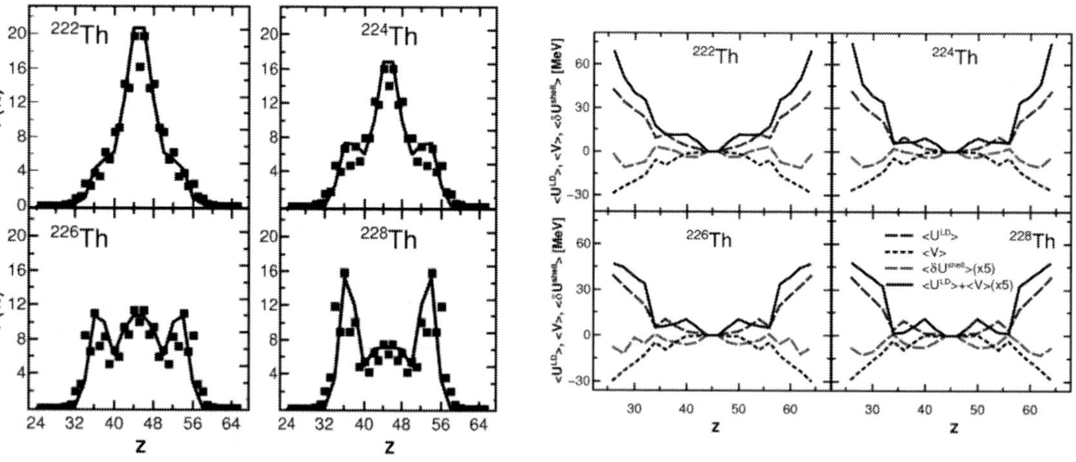

Figure 3. *Left side*: the same as in Fig. 2, but for Th isotopes. *Right side*: the calculated components of the driving potentials $\langle U(Z_i)\rangle$ at 11 MeV (solid black lines) excitation energy of the initial compound nucleus for indicated fissioning thorium isotopes. The liquid-drop energy $\langle U^{LD}\rangle$ (dashed blue lines), the nucleus-nucleus interaction potential $\langle V\rangle$ (dotted lines), and the shell-correction energy $\langle\delta U^{shell}\rangle$ (dashed red lines) are normalized to the zero energy at Z_L = 44. The values of $\langle\delta U^{shell}\rangle$ and $\langle U^{LD}\rangle + \langle V\rangle$ are scaled by a factor of 5, for visibility.

The liquid-drop energy increases globally when the charge (mass) number deviates from $Z_{CN}/2$ ($A_{CN}/2$). The interaction potential has the opposite global behavior. These two components depend on deformations of the DNS nuclei that create local minima in the $\langle U^{LD}(Z_i)\rangle$ and $\langle V(Z_i)\rangle$. The larger deformations result in the smaller interaction energy and the larger liquid-drop energy. The position of minimum in the (β_L, β_H) plane depends on the shell corrections. For the nuclei with $Z_L = 40-44$, the shell effects are relatively weak because Z_L (N_L) is midway between proton (neutron) shell closures. This leads to an almost pure liquid drop behavior of the scission configuration and larger deformations of the fragments. Thus, the asymmetric minimum in the driving potential appears due to the interplay between the liquid-drop energy and the nucleus-nucleus interaction potential. The shell effects also

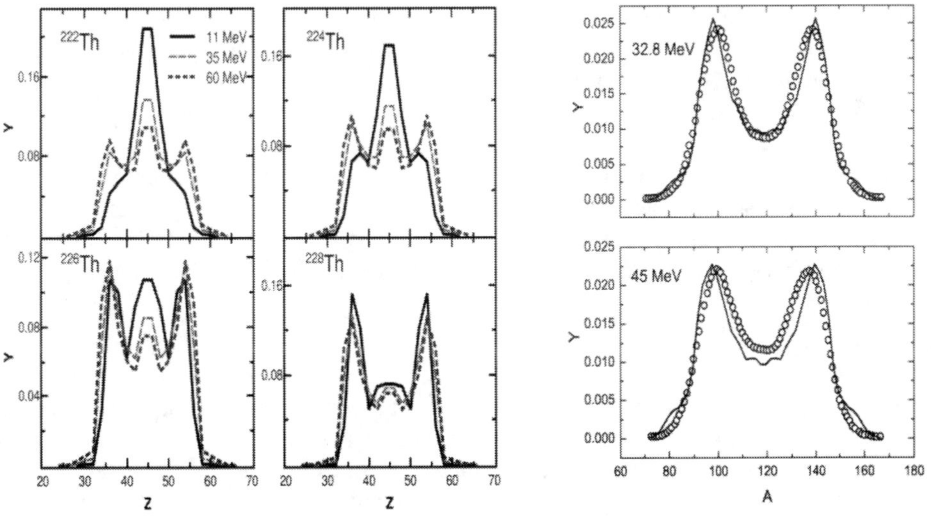

Figure 4. *Left side*: the calculated charge distributions for ^{222,224,226,228}Th isotopes at 11, 35, and 60 MeV excitation energies of the initial compound nucleus. *Right side*: the calculated mass distribution resulting from the $^{238}U(n, f)$ reaction at the indicated energies. The experimental data are from [2].

indirectly affect (through the deformations of nuclei) the appearance of the asymmetric minimum. The direct role of the shell effects is expressed by changing the position and depth of this minimum in the driving potential. For example, for ^{226}Th, at Z_L=34, 36, and 38 the DNS nuclei are near magic nuclei with $N_L = 50$ and $N_H = 82$. This implies high values of their stiffness parameter and so the minimum in the (β_L, β_H) plane is situated at small deformations. As a result, at $Z_L = 34$–38 the $\langle V\rangle$ decreases faster than the $\langle U^{LD}\rangle$ increases. At Z_L<34 the increase of the $\langle U^{LD}\rangle$ overcompensates the decrease of the $\langle V\rangle$, and, correspondingly, the asymmetric minimum appears at $Z_L = 34$. The shell corrections cause the shift of the minimum from $Z_L = 34$ to $Z_L = 36$ and make this minimum deeper in the driving potential (Fig. 3). It is interesting to note that only in the cases of ^{222,224}Th the shell-correction energy $\langle\delta U^{shell}(Z_i)\rangle$ diminishes the asymmetric minimum of the liquid-drop energy plus the interaction potential in Fig. 3. With increasing $N - Z$ of the fissioning nucleus the liquid-drop energy/interaction potential increases slower/faster with charge asymmetry. This is the reason of the appearance of the pronounced asymmetric minimum in the cases of ^{226,228}Th.

The predicted charge distributions of fissioning nuclei ^{222,224,226,228}Th at large excitation energies are given in the left part of Fig. 4. The calculated mass distributions of fissioning ^{239}U (lines) at two excitation energies (mentioned in the figure legends), describing the experimental data [2] (open circles) are plotted in the right part of Fig. 4. The distributions widens with increasing excitation energy for all nuclei. For ^{222,224}Th isotopes, the central peak recedes, and two asymmetric peaks appear suppressing the symmetric mode. The charge distributions of ^{226,228}Th at higher excitation energies (Fig. 4) emphasize the asymmetric aspect of the distribution, making the symmetric peak smaller and narrower, and the asymmetric maxima wider and taller. This is easily explained by the fact that with higher excitation energy the previously unaccessible asymmetric configurations are involved. The average driving potentials match the yields at 35 and 60 MeV excitation energy. At 60 MeV excitation energy $\langle U(Z_i)\rangle$ has asymmetric minima resulting in larger yields of asymmetric mode. The excitation energy reduces the shell effects that causes a widening of the minima in the deformation space (β_L, β_H) and a migration of the minima in the potential-energy surface. In the case of fissioning Th isotopes for Se–Kr charge fragmentations the minima widen with excitation energy

 https://doi.org/10.1051/epjconf/201816900015

faster than for the symmetric fragmentations. However, the influence of the excitation energy is not so dramatic. The deformations $\beta_L(Z_L)$ and $\beta_H(Z_L)$ of fragments, corresponding to the asymmetric minimum of the driving potential, do not change strongly, especially for the configurations with magic or close to magic nuclei Se–Kr [N_L=50 and N_H=82]. So, the deformations of the light nuclei with Z_L = 34−36 remain to be smaller than the deformations of the DNS nuclei with Z_L = 38−44. As a result, the $\langle V(Z_i)\rangle$ decreases faster than the $\langle U^{LD}(Z_i)\rangle$ increases from Z_L=38 to Z_L=34. At Z_L<34 their behaviors are opposite, and the minimum at Z = 34 is produced. So, the asymmetric minimum in the driving potential is produced due to the deformation effect. With increasing excitation energy of the fissioning nucleus, the liquid-drop energy $\langle U^{LD}(Z_i)\rangle$ increases slower with charge asymmetry. This leads to the pronounced asymmetric minimum in the cases of ^{222,224}Th.

4 Conclusions

In this paper it is shown that the improved scission-point model can be a suited tool for the description of the symmetric and asymmetric charge (mass) fragmentations resulting from the low-energy fission of different nuclei. The transition from a symmetric to an asymmetric fragmentation of Th nuclei is attributed to the interplay between the competing liquid-drop surface energy and the interaction energy in the driving potential of these systems at the scission point. This interplay depends on the excitation energy and the isospin of the fissioning nucleus. The shell effects affect indirectly through the deformations of nuclei on the appearance of the asymmetric minimum. The three-equal-peaked charge distribution is not the specific property of the ^{226}Th nucleus. The shape of the charge distribution depends on the excitation energy. For the fissioning nucleus ^{226}Th at high excitation energies, the charge distributions are well asymmetric but not the three-equal-peaked ones. The experimental verification of this unexpected effect is desirable.

References

[1] K.-H. Schmidt et al., Nucl. Phys. A **665**, 221 (2000)
[2] I.V. Ryzhov et al., Phys. Rev. C **83**, 054603 (2011)
[3] A.N. Andreyev, M. Huyse, P. Van Duppen, Rev. Mod. Phys. **85**, 1541 (2013)
[4] H. Paşca, A.V. Andreev, G.G. Adamian, N.V. Antonenko, Phys. Rev. C **94**, 064614 (2016); Phys. Lett. B **760**, 800 (2016); Eur. Phys. J. A **52**, 369 (2016)
[5] S. Hofmann and G. Münzenberg, Rev. Mod. Phys. **72** , 733 (2000)

Fission fragment mass and total kinetic energy distributions of spontaneously fissioning plutonium isotopes

K. Pomorski*, B. Nerlo-Pomorska, J. Bartel**, and C. Schmitt**,***

*UMCS, Lublin, Poland, ** IPHC, Strasbourg, France, *** GANIL, Caen, France*

Abstract. The fission-fragment mass and total kinetic energy (TKE) distributions are evaluated in a quantum mechanical framework using elongation, mass asymmetry, neck degree of freedom as the relevant collective parameters in the Fourier shape parametrization recently developed by us. The potential energy surfaces (PES) are calculated within the macroscopic-microscopic model based on the Lublin-Strasbourg Drop (LSD), the Yukawa-folded (YF) single-particle potential and a monopole pairing force. The PES are presented and analysed in detail for even-even Plutonium isotopes with $A = 236$–246. They reveal deep asymmetric valleys. The fission-fragment mass and TKE distributions are obtained from the ground state of a collective Hamiltonian computed within the Born-Oppenheimer approximation, in the WKB approach by introducing a neck-dependent fission probability. The calculated mass and total kinetic energy distributions are found in good agreement with the data.

1 Introduction

Our present understanding of nuclear fission is still based on the idea of Lisa Meitner and Otto Frisch [1] of a deformed charged liquid drop. In a seminal paper [2] Niels Bohr and John A. Wheeler fully developed the concept of the energy surface of a nucleus as function of a set of deformation parameters. The height of the minimal energy barrier that the nucleus has to overcome in the multidimensional energy hyper-surface as function of the deformation parameters on its decay path determines the stability of the nucleus against fission. Unfortunately, following the Bohr-Wheeler paper, one sometimes still uses up to now the Lord Rayleigh expansion of the nuclear surface into spherical harmonics. It was shown (e.g. in Ref. [3]) that such an expansion is not rapidly converging for large deformations and one needs to include terms up to multipolarity 16 in order to describe shapes close to the scission point. Fortunately, some better parametrizations of nuclear shapes exist. Among the most popular are the Quadratic Surfaces of Revolution (QSR) originally proposed by Nix [4], the Funny-Hills parametrization and its extension [5, 6], and the Cassini ovals with modifications due to Pashkevich [7]. The first two of the above quoted parametrizations are not analytical and closed (e.g. do no allow for the inclusion of higher order terms) while the last one is not easy to handle and its parameters do not have a clear physical interpretation.

The recently developed Fourier parametrization (see Appendix) of deformed nuclear shapes [8, 9] is free of these ambiguities and is rapidly converging. Using this parametrisation we have made an

*e-mail: `Krzysztof.Pomorski@umcs.pl`

attempt to obtain the fission-fragment mass and total kinetic energy distribution of even-even Plutonium isotopes with mass numbers $236 \leq A \leq 246$. The deformation-energy landscapes that we are going to discuss in Sect. 4 are described by three Fourier deformation parameters q_2, q_3, q_4 which are respectively elongation, mass-asymmetry and neck degree of freedom of the fissioning nucleus [8, 9]. Nonaxial shapes were not included in the present research since at large nuclear deformations they turn out to play a minor role. Thanks the fast convergence of the Fourier series, we believe that the above defined three deformation parameters are sufficient to describe the large variety of nuclear shapes up to very large deformations [9].The potential energy surfaces of different fissioning nuclei were calculated within the macroscopic-microscopic method [5, 10]. The fragment mass distribution obtained in low-energy fission of light actinides was evaluated in a quantum mechanics framework within the Born-Oppenheimer approximation (BOA) [11].

Within the aforementioned deformation space the fission yield is obtained from the probability distribution of the collective wave function on the (q_3, q_4) plane in the vicinity of the scission configuration ($q_2 \approx 2.3$). A neck-size dependent fission probability [12] was used to evaluate the fission-fragment mass and kinetic energy yields from the distribution probability at different elongations of the fissioning nucleus as will be explained in Sect. 3.

2 Macroscopic-microscopic model of the potential energy

The nuclear deformation energies of our analysis were determined in the macroscopic-microscopic approach, where the Lublin Strasbourg Drop (LSD) model [13] has been used for the macroscopic part of the potential-energy surface. Microscopic effects have been evaluated through a Yukawa-folded (YF) single-particle potential [14] with the parameters listed in Ref. [15], where also our way of solving of the eigenproblem of the YF Hamiltonian is described. Eighteen deformed harmonic oscillator shells were taken into account when diagonalising the YF Hamiltonian. The Strutinsky shell-correction method [5, 16] with a 8th order correctional polynomial and a smearing width $\gamma_S = 1.2\,\hbar\omega_0$ is used, where $\hbar\omega_0 = 41/A^{1/3}$ MeV is the spherical harmonic-oscillator frequency. The BCS [10] theory, including an approximate GCM+GOA particle-number projection as described in Refs. [17] was used for the pairing correlations. The pairing strength equal to $G \cdot \mathcal{N}^{2/3} = 0.28\,\hbar\omega_0$, (with $\mathcal{N} = Z, N$ for protons or neutrons) was adjusted to the experimental mass differences of nuclei in this region using a pairing window composed of $2\sqrt{15\mathcal{N}}$ single-particle levels closest to the Fermi surface [18].

3 Collective model of fission

In any coordinates q_n the collective Hamiltonian has the following form:

$$\widehat{H}_{\text{coll}} = -\frac{\hbar^2}{2}\sum_{i,j} |M|^{-1/2}\frac{\partial}{\partial q_i}|M|^{-1/2}M^{-1}_{ij}\frac{\partial}{\partial q_j} + V\,, \tag{1}$$

where $M_{ij}(\{q_i\})$ and $V(\{q_i\})$ denote the inertia tensor and the potential energy, respectively and $|M| = \det(M_{ij})$.

The eigenproblem of this Hamiltonian can be solved in the BOA in which one assumes that the motion towards fission, i.e. in the q_2 coordinate, is much slower than the motion in the q_3 and q_4 collective variables. This implies that the eigenfunction of $\widehat{H}_{\text{coll}}$ can be approximated in our case in the following form:

$$\Psi_{nE}(q_2, q_3, q_4) = u_{nE}(q_2)\,\varphi_n(q_3, q_4; q_2)\,. \tag{2}$$

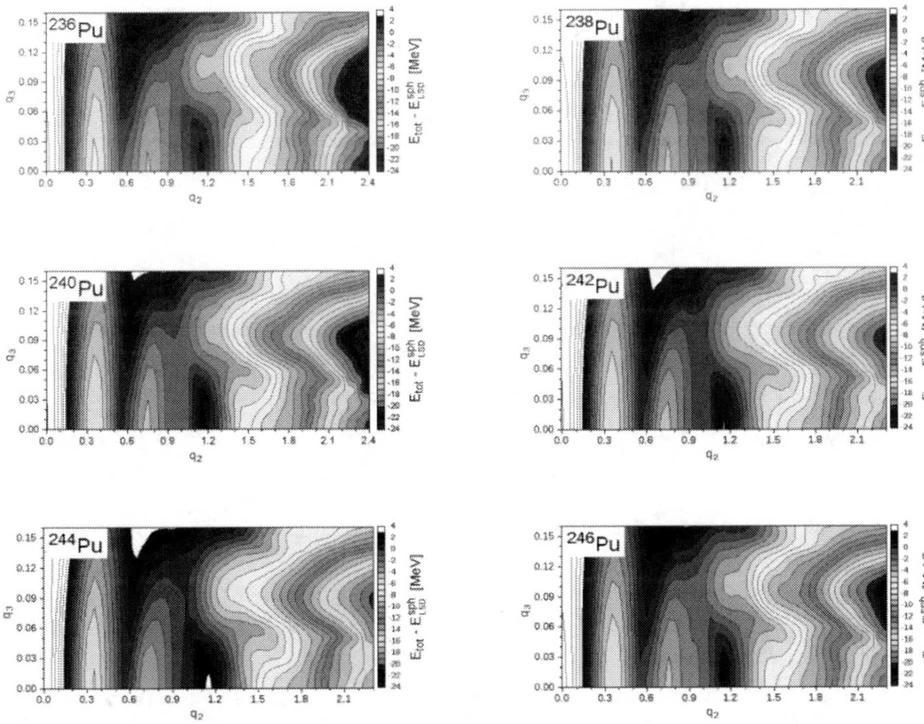

Figure 1. Potential energy surfaces on the (q_2, q_3) plane of even-even Pu isotopes minimized with respect to q_4.

Here $u_{nE}(q_2)$ is the wave function for the fission (elongation) mode and φ_n are the eigenfunctions of the Hamiltonian which describe the collective motion in the perpendicular directions. In the following we shall use the WKB approximation for the $u_{nE}(q_2)$ wave function and consider only the lowest energy eigenstate $\varphi_{n=0}$ since we are interested in fission at very low excitation energy. The effect of taking into account higher states was discussed in Ref. [11].

The probability of finding the system, for a given elongation q_2, in a deformation defined by (q_3, q_4) is equal to

$$W(q3, q4; q2) = |\Psi(q_3, q_4; q_2)|^2 = |\varphi_0(q_3, q_4; q_2)|^2 \quad (3)$$

Our model is still simplified further. Instead of the square of the collective wave function (3), we take the following Wigner function:

$$W(q3, q4; q2) \sim \exp\left\{-\frac{V(q_3, q_4; q_2) - V_{\rm eq}(q_2)}{E_0 + T}\right\} \quad (4)$$

where $V_{\rm eq}(q_2)$ is the potential minimum for a given q_2, E_0 is the zero-point energy, treated here as a free parameter and T is the temperature of the nucleus.

The probability distribution integrated over q_4

$$w(q_3; q_2) = \int W(q_3, q_4; q_2)\, dq_4 \,, \quad (5)$$

is directly related to the fragment mass yield at given elongation q_2.

It is obvious that the fission probability should also depend on the neck radius. Following Ref. [12] we assume a neck-rupture probability P in the form:

$$P(q_2, q_3, q_4) = \frac{k_0}{k} P_{\text{neck}}(R_{\text{neck}}) \,, \tag{6}$$

where k is the momentum in the direction towards fission (or simply the velocity along q_2), while $R_{\text{neck}}(q_2, q_3, q_4)$ is the deformation-dependent neck radius. k_0 plays the role of a scaling parameter. The neck rupture probability [12] can be taken e.g. in one of the following forms:

$$P_{\text{neck}}(R_{\text{neck}}) = \begin{cases} \exp[-\ln 2(R_{\text{neck}}/d)^2] & \textit{Gauss} \\ 1/\cosh[(2+\sqrt{3})R_{\text{neck}}/d] & \textit{1/cosh} \\ 1/[1+(R_{\text{neck}}/d)^2] & \textit{Lorentz} \end{cases} \tag{7}$$

The parameter d is the *half-width* of the probability and is treated here as a free adjustable parameter.

The momentum k in Eq. (6) has to ensure that the probability depends on the time in which one crosses the subsequent interval in q_2: $\Delta t = \Delta q_2 / v(q_2)$, where

$$v(q_2) = \hbar k / \overline{M}(q_2) \tag{8}$$

is the velocity towards fission. The inertia $\overline{M}(q_2)$ is evaluated using the approximation proposed in Ref. [21]:

$$\overline{M}(q_2) = \mu \left[1 + 11.5 \cdot (B_{\text{irr}}/\mu - 1) \right] \left(\frac{\partial R_{12}}{\partial q_2} \right)^2 \,, \tag{9}$$

where B_{irr} is the irrotational inertia corresponding to the distance R_{12} between the fragments and μ is the reduced mass. The value of k in Eq. (8) depends on the difference $E - V(q_2)$ and on the part of the collective energy which is converted into heat Q:

$$\frac{\hbar^2 k^2}{2\overline{M}(q_2)} = E_{kin} = E - Q - V(q_2) \,. \tag{10}$$

In order to introduce the neck-rupture probability P, in our formalism one has to rewrite the probability distribution $w(q_3; q_2)$ in the form

$$w(q_3; q_2) = \int W(q3, q4; q2) P(q_3, q_4, q_2) \, dq_4 \,. \tag{11}$$

Such an approximation means that the fission process is spread over some region of q_2 and that for a given q_2, at fixed mass asymmetry, one has to take into account the probability to fission at a previous q_2 value, i.e. one has to replace $w(q_3; q_2)$ by

$$w'(q_3; q_2) = w(q_3; q_2) \, \frac{1 - \int\limits_{q_2' \le q_2} w(q_3; q_2') \, dq_2'}{\int w(q_3; q_2') \, dq_2'} \,. \tag{12}$$

The integral mass yield can then be written as the sum of all partial yields at different q_2:

$$Y(q_3) = \int w'(q_3; q_2) \, dq_2 \; / \int w'(q_3; q_2) \, dq_2 \, dq_3 \,. \tag{13}$$

As one notices from (13), the scaling factor k_0 in the expression for P, Eq. (6), has vanished and does no longer appear in the definition of the mass yield. Our model thus has only two adjustable parameters, E_0 in Eq. (4) and the width parameter d, that appears in the neck-rupture probability (7).

4 Results

The potential energy surfaces (PES), relative to the corresponding spherical LSD energy, were calculated for even-even Plutonium isotopes $^{236-246}$Pu in the 3D (q_2, q_3, q_4) deformation space, where q_2 describes the elongation of the nucleus, q_3 its reflection asymmetry, and q_4 controls the neck size [8,9]. The $(q_2, q_3; q_4^{\min})$ cross-sections of the PES minimized with respect to q_4 are presented in Fig. 1. Pronounced mass asymmetric ($q_3 \neq 0$) fission valleys are visible in all isotopes beyond $q_2 \approx 1$. Symmetric fission valleys are generally present but lay always higher by a few MeV, and with a substantial ridge separating them from the asymmetric valleys.

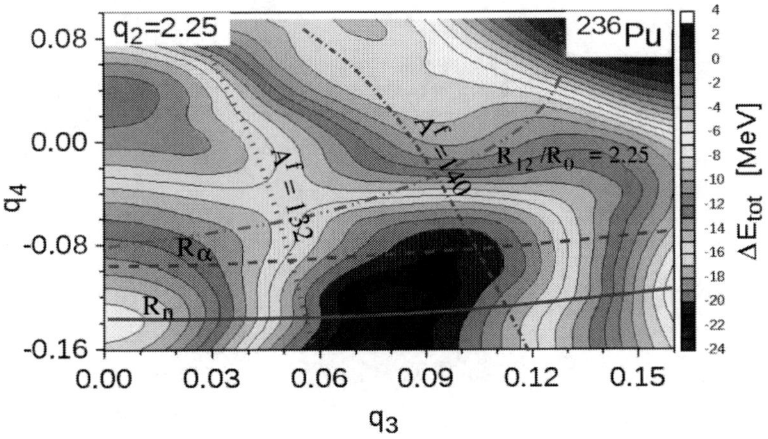

Figure 2. (q_3, q_4) PES cross-section of ^{236}Pu for $q_2 = 2.25$.

It is obvious that 2D maps do not give a complete information about the way in which a system goes to fission. The neck dimension, which for a given elongation q_2 and asymmetry q_3 depends on q_4, decides when fission occurs. The map of the PES of ^{236}Pu on the (q_3, q_4) plane for $q_2 = 2.25$ is presented as an example in Fig. 2. Lines corresponding to a neck radius equal to the one (R_n) of a nucleon, an alpha-particle (R_α), as well as of the heavier mass fragment $A_h^f = 132$ and $A_h^f = 140$ are marked in the plot. It is seen that the deformation parameter q_3 governs mostly the fragment mass, while q_4 is related to the neck size. The line corresponding to the distance $R_{12}/R_0 = 2.25$ between the fragment mass centres is also shown. Note that in all other points of the map R_{12} differs from this mean value by less than 5% what proves that at large elongations q_2 is roughly a measure of the distance between the fragments.

More informative are the projections of the PES's for different elongations q_2 onto the $(A_h^f, R_{\rm neck})$ maps presented in Fig. 3 for ^{236}Pu. It is seen that the structure of the bottom of the fission valley changes with increasing elongation of the nucleus. At smaller deformations $q_2 \leq 2.25$, the bottom of the valley in the vicinity of $R_{\rm neck} = R_n$, the size of a nucleon (thin solid), corresponds to the mass of the heavier fragment $A_h^f = 140$. At $q_2 = 2.3$ already two valleys at 138 and 144 are visible, while at $q_2 = 2.35$ three fission valleys corresponding to 136, 141 and 146 are formed. The sizeable dependence of the PES in the $(A_h^f, R_{\rm neck})$ plane on the elongation coordinate q_2 shows how important is to take into account the subsequent steps in the fission process as explained above.

 https://doi.org/10.1051/epjconf/201816900016

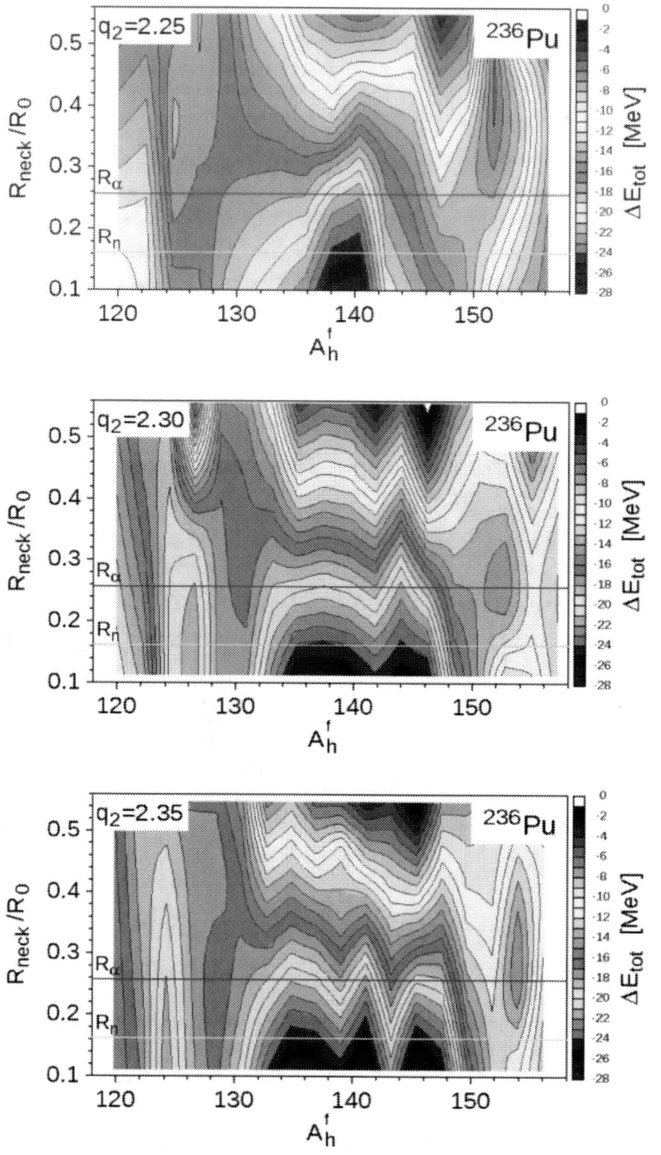

Figure 3. Potential energy surfaces for different elongations q_2 projected onto the $(A_h^f, R_{\rm neck})$ plane.

The Wigner function (5) integrated over q_4 is presented in Fig. 4 for different elongations q_2 and projected onto the A_h^f axis, instead of the asymmetry parameter q_3. Its maximum changes accordingly to the properties of the fission valleys shown in Fig. 3. This set of Wigner functions weighted with the neck-size dependent fission probability gives access to the mass yields (11) for different elongations q_2.

The final heavy-fragment mass yield (13), obtained for three different neck-breaking probabilities (7) is shown in Fig. 5 for ^{240}Pu, and compared to the yield measured in spontaneous fission [19]. The differences between the theoretical yields are small as can be seen in the figure. We shall therefore present in the following the results obtained with the Gauss distribution only keeping the half-width fixed at $d/R_0 = 0.15$ which roughly corresponds to the nucleon radius. The width-parameter E_0 of the

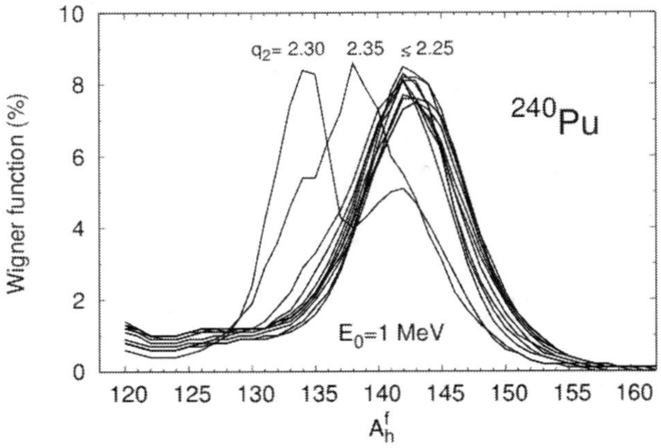

Figure 4. Wigner function for different elongations q_2 as a function of the heavy-fragment mass number A_h^f.

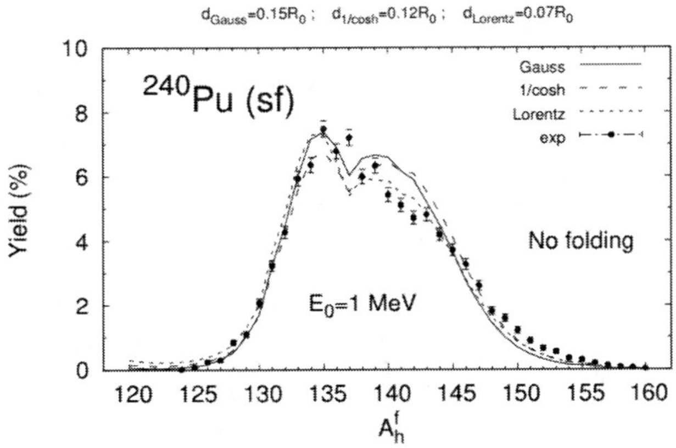

Figure 5. Heavy fragment mass yield for fission of ^{240}Pu as obtained with three different neck-breaking probability functions (7). The half-width of each function is written above the plot. The calculation is compared to the yield measured in spontaneous fission in Ref. [19].

Wigner function was taken as E_0=1 MeV what corresponds to a typical energy of collective modes in two dimensions (q_3 and q_4).

The spontaneous fission yields for six even-even Pu isotopes are compared in Fig. 6 with the experimental data whenever available [19]. Please note that these theoretical estimates were obtained with only two adjustable parameters $d/R_0 = 0.15$ and $E_0 = 1$ MeV. All other parameters of the model are unchanged as compared to those adjusted years ago when considering still different data.

The total kinetic energy (TKE) of the fragments is evaluated in the point-charge approximation of the Coulomb interaction energy between the fragments. No pre-fission kinetic energy is used in the present calculation, *i.e.* we assume that the whole energy gain on the slope to scission is transferred into heat. One has to bear in mind that for a given elongation q_2 the mass of the fragment $A_h^f(q_3, q_4; q_2)$ and the TKE$(q_3, q_4; q_2)$ are functions of q_3 and q_4. So, it is easy to map the yields from the (q_3, q_4) onto the (TKE, A_h^f) plane. The experimental data have a finite resolution in mass and in TKE, which in not negligible for the present cases [19]. Typical values are $\sigma_{\rm TKE}$=10 MeV and a variance of $\sigma_{A_h^f}$=1.5 a.m.u. Hence, the theoretical A_h^f and TKE yields have to be folded according to these values before they can be compared to the experiment. The resulting folded TKE yield is presented in Fig. 7 (top).

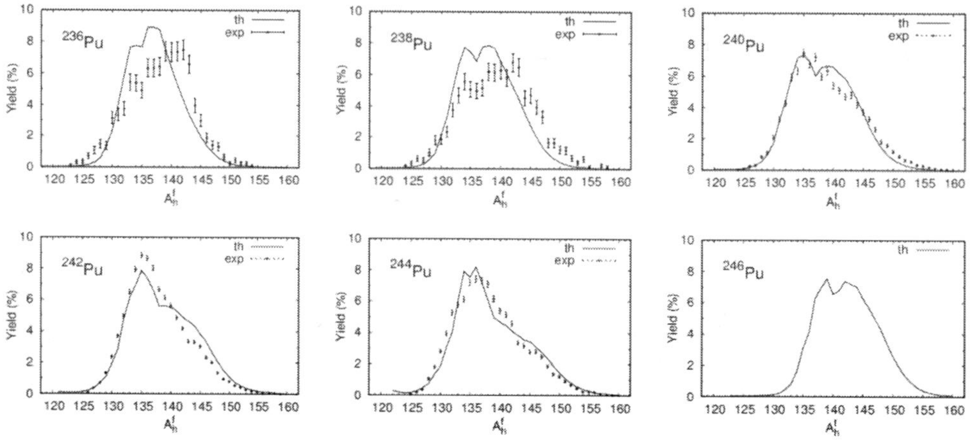

Figure 6. Same as in Fig. 5 for fission of $^{236-246}$Pu but for the optimal calculation (see the text). The theoretical results, obtained here without any folding procedure, are compared to the yield measured in spontaneous fission in Ref. [19]) for $^{236-244}$Pu.

One can see in the figure that the maximum of the TKE yield for ^{240}Pu corresponds to TKE $\approx$ 179 MeV what is in line with the measured data [20]. The calculated correlation between the heavy-fragment mass and the TKE is presented in Fig. 7 (bottom). Although a quantitative analysis is still to be done, it is already noted that the theoretical correlation shows a trend similar to the experiment [22], with a mean TKE larger for the lighest ($A_h^f \approx 132$) heavy-fragment group, than for the heaviest ($A_h^f \approx 140$) group. The relatively good agreement of the TKE distribution and mean value with the experiment can be considered as quite encouraging. Keep in mind, however, that our theoretical approach is still quite crude, with the use of a Wigner functions for the probability distribution instead of the true quantum mechanical density distributions, the absence in our calculations of extremely elongated shapes, since it was impossible to construct for them a reliable Wigner function, the inclusion of only the lowest energy phonon in the direction perpendicular to the scission mode, i.e. neglecting the dynamical coupling of the fission and the perpendicular modes.

5 Summary

Summing up the following conclusion can be drawn from our investigation:

- an effective three-dimensional set of Fourier deformation parameters is used to describe the fission process,
- a collective 3D model describing elongation, mass asymmetry and neck modes was introduced,
- a Wigner distribution is assumed for the neck and mass asymmetry degrees of freedom,
- a neck-breaking probability depending on the neck-size has been adopted,
- our 3D model which couples fission, neck and mass asymmetry modes is able to describe the main features of the fragment mass and total kinetic energy yields.

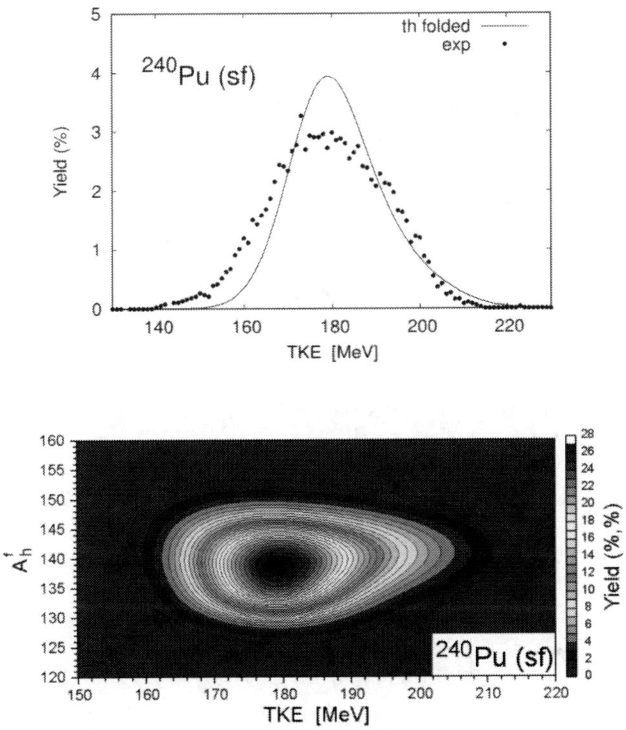

Figure 7. *Top*: calculated and experimental (as measured in spontaneous-fission [20]) TKE distribution for ^{240}Pu. *Bottom*: calculated correlation between the heavy-fragment mass and the TKE. Calculated mass and TKE values were folded with the experimental resolution (see text).

This contribution presents the present status of our research. Similar calculations are going to be performed to describe low-energy fission of other isotopes. In parallel, we aim to develop a Langevin code to study fission at higher energies with the new Fourier deformations space.

This work has been partly supported by the Polish-French COPIN-IN2P3 collaboration agreement project No. 08-131 and by the Polish National Science Centre, grant No. 2013/11/B/ST2/04087.

References

[1] L. Meitner, O.R. Frisch, Nature **143**, 239 (1939)
[2] N. Bohr, J.A. Wheeler, Phys. Rev. **56**, 426 (1939)
[3] A. Dobrowolski, K. Pomorski, J. Bartel, Phys. Rev. C **75**, 024613 (2007)
[4] J.R. Nix, Nucl. Phys. A **130**, 241 (1969)
[5] M. Brack, J. Damgaard, A.S. Jensen, H.C. Pauli, V.M. Strutinsky, and C.Y. Wong, Rev. Mod. Phys. **44**, 320 (1972)
[6] J. Bartel, B. Nerlo-Pomorska, K. Pomorski, and C. Schmitt, Phys. Scr. **89**, 054003 (2014)
[7] V.V. Pashkevich, Nucl. Phys. A **169**, 275 (1971)
[8] K. Pomorski, B. Nerlo Pomorska, J. Bartel, C. Schmitt, Acta. Phys. Polon. **B** Supl. **8**, 667 (2015)

[9] C. Schmitt, B. Nerlo Pomorska, K. Pomorski, J. Bartel, Phys. Rev. C **17**, 034612 (2017)
[10] S.G. Nilsson et al., Nucl. Phys. A **131**, 1 (1969)
[11] B. Nerlo-Pomorska, K. Pomorski and F.A. Ivanyuk, Acta Phys. Pol. B Supl. **8**, 669 (2015)
[12] K. Pomorski, B. Nerlo-Pomorska and F.A. Ivanyuk, Eur. Phys. J. A **53**, 59 (2017)
[13] K. Pomorski and J. Dudek, Phys. Rev. C **67**, 044316 (2003)
[14] K.T.R. Davies and J.R. Nix, Phys. Rev. C **14**, 1977 (1976)
[15] A. Dobrowolski, K. Pomorski, and J. Bartel, Comp. Phys. Comm. **199**, 118 (2016)
[16] V.M. Strutinsky, Yad. Fiz. **3**, 614 (1966); Nucl. Phys. A **95**, 420 (1967)
[17] A. Góźdź and K. Pomorski, Nucl. Phys. A **451**, 1 (1986)
[18] S. Pilat, K. Pomorski, A. Staszczak, Zeit. Physik A **332**, 259 (1989)
[19] L. Dématté et al, Nucl. Phys. A **617**, 331 (1997)
[20] C. Wagemans, E. Allaert, A. Deruytter, R. Barthelemy, and P. Schillebeeckx, Phys. Rev. C **30**, 218 (1984)
[21] J. Randrup et al., Phys. Rev. C **13**, 229 (1976)
[22] P. Schillebeeckx, C. Wagemans, A.J. Deruytter, and R. Barthelemy, Nucl. Phys. A **545**, 623 (1992), and S. Oberstedt, private communication.

Appendix

A new nuclear shape parametrization [8, 9] is used that gives an expansion of the nuclear surface in the form of a Fourier analysis in cylindrical coordinates

$$\frac{\rho_s^2(z)}{R_0^2} = \sum_{n=1}^{\infty} \left[a_{2n} \cos\left(\frac{(2n-1)\pi}{2} \frac{z - z_{\rm sh}}{z_0} \right) + a_{2n+1} \sin\left(\frac{2n\pi}{2} \frac{z - z_{\rm sh}}{z_0} \right) \right] ,$$

where, similarly to the famous "Funny-Hills" (FH) shape parametrization [5], $\rho_s(z)$ defines the distance of the equivalent sharp surface from the symmetry z axis, and z_0 is half the elongation of the nuclear shape along that axis with end points located at $z_{\rm min} = z_{\rm sh} - z_0$ and $z_{\rm max} = z_{\rm sh} + z_0$. The coordinate $z_{\rm sh}$ is chosen so as to locate the centre of mass of the shape at the origin of the coordinate system. R_0 represents the radius of the corresponding spherical shape having the same volume and $z_0 = c\,R_0$.

The parameters a_2, a_3, a_4 are related to elongation, left-right asymmetry, and neck degree of freedom, respectively. More and more elongated prolate shapes correspond to decreasing values of a_2, while oblate ones are described by a_2 larger than one, which is somehow contrary to the traditional definition of a quadrupole deformation parameter. So, it was proposed in Ref. [9] to use a more *natural* definition of deformation parameters which in addition ensures that along the LD fission path these parameters show only a small variation around zero:

$$q_2 = \frac{a_2^{(0)}}{a_2} - \frac{a_2}{a_2^{(0)}} , \qquad q_3 = a_3 , \qquad q_4 = a_4 + \sqrt{\left(\frac{q_2}{9}\right)^2 + \left(a_4^{(0)}\right)^2} ,$$

$$q_5 = a_5 - (q_2 - 2)\,\frac{a_3}{10} , \quad q_6 = a_6 - \sqrt{\left(\frac{q_2}{100}\right)^2 + \left(a_6^{(0)}\right)^2} .$$

Here $a_n^{(0)}$ stands for the value of the a_n coefficient in the spherically symmetric case:
$a_2^{(0)} = 1.03205$, $a_4^{(0)} = -0.03822$, $a_6^{(0)} = 0.00826$.

Measurement of fission yields and isomeric yield ratios at IGISOL

Stephan Pomp[1,*], *Andrea* Mattera[1], *Vasileios* Rakopoulos[1], *Ali* Al-Adili[1], *Mattias* Lantz[1], *Andreas* Solders[1], *Kaj* Jansson[1], *Alexander V.* Prokofiev[1], *Tommi* Eronen[2], *Dimitri* Gorelov[2], *Ari* Jokinen[2], *Anu* Kankainen[2], *Iain D.* Moore[2], *Heikki* Penttilä[2], and *Sami* Rinta-Antila[2]

[1] *Uppsala University, Department of Physics and Astronomy, Box 516, 75120 Uppsala, Sweden*
[2] *University of Jyväskylä, Department of Physics, P.O. Box 35 (YFL), 40014 Jyväskylä, Finland*

Abstract. Data on fission yields and isomeric yield ratios (IYR) are tools to study the fission process, in particular the generation of angular momentum. We use the IGISOL facility with the Penning trap JYFLTRAP in Jyväskylä, Finland, for such measurements on ^{232}Th and ^{nat}U targets. Previously published fission yield data from IGISOL concern the ^{232}Th(p,f) and ^{238}U(p,f) reactions at 25 and 50 MeV. Recently, a neutron source, using the Be(p,n) reaction, has been developed, installed and tested. We summarize the results for (p,f) focusing on the first measurement of IYR by direct ion counting. We also present first results for IYR and relative yields for Sn and Sb isotopes in the 128-133 mass range from ^{nat}U(n,f) based on γ-spectrometry. We find a staggering behaviour in the cumulative yields for Sn and a shift in the independent fission yields for Sb as compared to current evaluations. Plans for the future experimental program on fission yields and IYR measurements are discussed.

1 Introduction

Fission yields (FY) are a key observable for studies of the fission process. They are measured for example using the 2E or the 2E-2v-technique [1], techniques involving conventional mass spectrometers like LOHENGRIN [2,3] or using irradiation techniques in combination with γ-spectrometry [4]. The challenges in the mentioned techniques are connected to the achievable mass resolution (2E and 2E-2v technique), limits in the achievable excitation energies in the fissioning nucleus (in the case of LOHENGRIN), or the necessity of sufficiently long life times of the studied fission products in combination with the requirement of well-known decay data (when using γ-spectrometry).

An alternative method to measure fission yields is provided by the usage of the ISOL-technique (Isotope Separation On-Line) in combination with a Penning trap for mass separation. With this approach independent isotopic FY can be measured for a range of fissioning systems with different excitation energies and, since the method is based on direct ion counting, without requirements on γ-decay schemes.

Furthermore, the high mass resolving power provided by Penning traps make it possible to measure isomeric yield ratios (IYR). This is then, again, not based on the measurement of gamma decays

*e-mail: stephan.pomp@physics.uu.se

but on direct ion counting, resolving the ground state mass from the mass of the isomeric state. Routinely achievable mass resolving powers correspond to a resolution of about 0.5 MeV in excitation energy. New developments like the Phase-Imaging Ion-Cyclotron-Resonance technique [5] can achieve a resolution of 50 keV. This opens the door to both high-quality FY data and measurements of a wide range of IYR for various fission systems.

The IYR may be used to deduce the angular momentum distributions of the fission fragments. This allows thus to study the amount of energy available in the fission process that is converted into rotational energy. The initial spin also influences the production of fission neutrons and gamma rays in the de-excitation process of the fragment [6]. Therefore, IYR, measured for a range of fissioning systems and across all masses, provide important information and constraints for modelling the fission process and its dynamics.

It has also been pointed out that the differences between calculated and measured antineutrino production at Daya Bay may be due to a lack of knowledge on IYR [7,8].

Last but not least, it should be mentioned that, due to high N/Z ratio and low average prompt-neutron emission multiplicity, neutron-induced fission of ^{232}Th and ^{238}U is an efficient way to produce exotic nuclei. Using a neutron-production target within an ISOL facility and the ^{232}Th(n,f) reaction is the most promising way to produce very neutron-rich nuclei especially in the light mass region, close to ^{78}Ni, for further studies with a Penning trap.

2 Experimental

The measurements discussed here have been performed at the IGISOL facility, located at the Accelerator Laboratory of the University of Jyväskylä, Finland. Here we give a brief outline of the facility. Details can be found in Refs. [9,10].

The K-130 cyclotron delivers a proton beam to the fission chamber. In the presented cases this beam is used for both (p,f) and (n,f) reaction studies. In the first case a 25 MeV beam directly interacts with ^{232}Th or ^{nat}U targets to induce (p,f) reactions. In the (n,f) case, a 30 MeV beam is used to create neutrons in a thick Beryllium target via the Be(p,n) reaction. The Be target is located inside the fission chamber and the neutrons then induce (n,f) reactions in the ^{nat}U target located a few cm downstream the neutron production target. Further details on the neutron converter are given below and in Ref. [11].

Helium is used as buffer gas to slow down the fission products and flush them towards a radiofrequency (RF) ion guide where they are stepwise accelerated to 30 kV towards a magnetic mass separator. The time spent by the ions in the buffer gas until extraction towards the magnet is on the order of a few tens of milliseconds.

The ions produced in this way are dominantly in the 1+ charge state. The relative fraction of ions in the 1+ state may vary from element to element in a non-trivial way and can be measured in dedicated studies. An example is given below. Details of the behaviour of the target cell that is used in the (p,f) case, i.e., the stopping efficiency for the different ions and potential bias for different fission product masses or their total kinetic energies (TKE), have been studied in Ref. [12]. It was found that such dependences are weak and the associated uncertainties are small compared to other systematic uncertainties. A similar study for the (n,f) case is currently in progress.

After extraction from the fission chamber by the ion guide and a second vacuum chamber with extraction electrodes, a dipole magnet is used to select ions for an isobaric mass chain. The mass resolving power, $M/\Delta M$ of the magnet is about 500.

After the magnet, the ions pass through a switch yard where they can either be deflected to a γ-ray measurement station, or towards the radiofrequency cooler and buncher (RFQ). After deceleration and cooling, short ion bunches of about 10 μs duration are finally released towards a setup of two

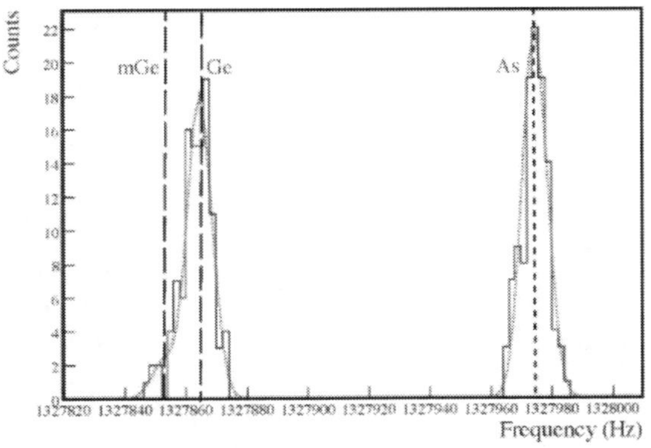

Figure 1. Experimental frequency distribution spectrum for the region of mass A=81. The identification of the peaks is based on a frequency to mass calibration. The figure shows the positions of ^{81}As, ^{81}Ge, and ^{81m}Ge (dashed black), and the Gaussian fits used to extract the IYR (dotted red).

consecutive Penning traps, JYFLTRAP. In the traps the ions are stored inside a magnetic field and their eigenmotion is manipulated by applying a RF field that selects different q/m ratios in the static 7.0 T magnetic field. The selected ions then pass through a 1.5 mm wide aperture and are extracted towards a multichannel plate (MCP). The last step allows for additional time-of-flight measurements which can supress impurities with the same q/m ratio. The total time from production of the fission product until registration on the MCP varies depending on the setting of the experiment, e.g., collection and cooling times in the RFQ. These settings can be changed depending on the measurement needs, e.g., desired mass resolving power. Typically this total time is on the order of a few hundred ms. Therefore, decay corrections have to be applied to the raw data.

The case of ^{81}Ge is the most difficult measurement discussed in this paper. The mass difference between isomer and ground state is only 679 keV and the IYR is small. The isomeric state thus shows up as a small bump almost completely immersed in the mass peak related to the ground state of ^{81}Ge. Peak positions are calculated relative to a reference peak, ^{81}As in this case, and all peaks are fitted with Gaussian distributions (see Fig. 1).

It turns out that applying Gaussian fits does not always reproduce the experimental data sufficiently well and the peaks might have systematic distortions depending on settings of the Penning trap. While this has, in general, a negligible effect on the IYR, it is desirable to apply the correct shapes in the fitting process, especially in cases with overlapping peaks. To this end, a method using Gaussian processes to fit imperfect models with non-linear least squares and using a prior distribution has been developed in Uppsala [13]. The method has, however, not yet been applied in the results presented in this paper.

3 Results and discussion

3.1 Proton-induced fission at 25 MeV

Over the past years a series of measurements of FY from the (p,f) reaction on ^{232}Th and ^{nat}U have been performed by the IGISOL group [14].

From some of these data IYR could be extracted for six nuclei, each from the two fissioning systems. These measurements represent the first IYR obtained by direct ion counting. Furthermore, nine

Table 1. IYR for the six studied nuclei from (p,f) at 25 MeV for the two studied fissioning systems. Uncertainties are statistical only. The IYR are given as Ym/(Ym+Yg.s.).

	Spin g.s.	Spin isom.	ΔM (keV)	IYR for ^{nat}U from [17]	IYR for ^{nat}U	IYR for ^{232}Th
^{81}**Ge**	(9/2$^+$)	(1/2$^+$)	679	-	**0.03±0.01**	**0.08±0.02**
^{96}**Y**	0$^-$	8$^+$	1140	-	**0.55±0.01**	**0.61±0.02**
^{97}**Y**	(1/2$^-$)	(9/2)$^+$	667	0.71	**0.73±0.02**	**0.78±0.02**
^{128}**Sn**	0$^+$	(7$^-$)	2091	0.52	**0.58±0.02**	**0.46±0.01**
^{129}**Sb**	7/2$^+$	(19/2$^-$)	1851	-	**0.47±0.04**	**0.46±0.01**
^{130}**Sn**	0$^+$	(7$^-$)	1946	0.34	**0.54±0.02**	**0.38±0.01**

out of the twelve measurements are the first IYR measurements for the studied system. Preliminary results were presented recently [15]. Here we briefly summarize revised results. A more detailed publication is in progress [16].

In Table 1 we give the spin assignments for the six nuclei in their ground state and their studied isomer, their mass difference, and our measured IYR with their statistical uncertainties. Data from the only other measurement of IYR in the (p,f) reaction obtained by Tanikawa et al. [17] by means of γ-spectrometry are also included in the table. The IYR are given as the yield for the metastable state over the total yield.

The ^{81}Ge case is the only one out of the six cases where the isomeric state has lower spin compared to the ground state. This is consistent with the comparatively low observed IYR. When comparing the two Yttrium cases, we note that ^{96}Y shows a larger difference in spin between ground state and isomer than ^{97}Y and has as small IYR. In the case of ^{128}Sn and ^{130}Sn we find that, at least for the ^{nat}U reaction, similar IYR may be expected since ground state and isomeric state have the same spin assignments in both cases. These observations indicate that our results are consistent with each other. Our data are in reasonable agreement with the data from Tanikawa et al. [17] except in the ^{130}Sn case where Ref. [17] gives a value that is much lower and closer to our result for the Thorium case but also much lower than their IYR for ^{128}Sn.

Our most striking observation is related to differences in the IYR between the two studied fissioning systems. We see a tendency that the ^{232}Th(p,f) reaction gives higher IYR, compared to the ^{nat}U(p,f) reaction, for nuclei in the light-mass region whereas the opposite trend is seen in the heavy-mass region. Most notably the IYR for ^{128}Sn and ^{130}Sn are significantly lower for ^{232}Th(p,f) compared to ^{nat}U(p,f). A possible explanation for this observation might be linked to different scission configurations due to differences in fission mode ratios; for ^{232}Th the super-long mode is found to be stronger compared to ^{238}U at comparable excitation energies in the compound system [18].

3.2 Relative yields and isomeric ratios for (n,f) reactions

The neutron-induced fission data have been obtained using a thick (6 mm) Beryllium target. This implies only minor differences in the energy spectra compared to the characterized 5-mm thick water-cooled Be target [11,19]. The idea behind the development of this target is motivated by the newly installed MCC30/15 cyclotron that is designed to deliver a 30 MeV proton beam with currents up to 100 μA. This would lead to a power of 3 kW on the target that needs to be cooled, but could also be used to create a neutron flux that induces a fission rate that is high enough to ensure reasonable measurement times for IFY measurements over a wide mass range. Characterization campaigns at The Svedberg Laboratory in Uppsala and in Jyväskylä showed that the design goal of reaching a total

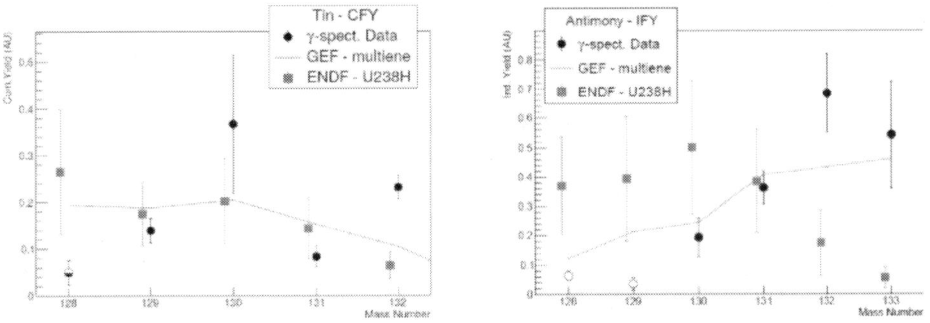

Figure 2. Preliminary results for relative cumulative FY for Sn (left) and independent FY for Sb (right) in the ^{nat}U(n,f) reaction. The semi-open circle indicates that only the ground state is included in the data point, and the open circles indicate that only the isomer could be measured. The solid red line represents calculations using the GEF code weighted with the neutron energy spectrum. The blue squares are evaluated data from ENDF/B-VII at 14 MeV.

flux of more than $2{\cdot}10^{12}$ n/sr/s can be achieved with a 100 μA proton current. The neutron flux with energies above 1 MeV would actually be $2\text{–}3{\cdot}10^{12}$ n/sr/s [11].

The data presented here were, however, obtained with the K-130 cyclotron and a proton current of only 1 μA since the MCC30/15 was not available during the measurement campaign in December 2016. This meant that we could not achieve sufficient intensities to get good measurement statistics using JYFLTRAP. Instead the fission products were directed to a γ-spectrometry station where the presented results were obtained. Nevertheless, the measured data here are the first (n,f) data measured at IGISOL.

According to the characterization data the effective neutron energy spectrum is rather wide. The effective neutron energy, determined by weighted averaging of the neutron energy spectrum with the fission cross section for the ^{nat}U target, is 12.4±8.8 MeV. About (97±1)% of all fission events are from ^{238}U(n,f). According to calculations with the GEF code [20], first chance fission contributes with 55%, second chance with 31%, third and fourth chance fission with 8 and 6%, respectively. The pre-fission emission of other particles is negligible. We could also measure the ratio of the 1+/2+ charge states. For Sn we obtain 1.19±0.17 and for Sb 2.78±0.42. It should be noted, however, that the different probabilities to reach the 1+ state do not affect the results presented below.

Here we summarize the preliminary results from the December 2016 campaign. A more detailed publication has been submitted [21]. Figure 2 shows the measured relative fission yields for Sn and Sb in the mass range between 128 and 133.

The measured FY for Sn are cumulative. The measured FY for Sn data can be subtracted from the Sb data and, therefore, the Sb yields are independent FY. The figure shows a comparison of the measurement results to evaluated data and calculations using the GEF code. Note that we normalize to the area under the respective distributions since our measured data are not on an absolute scale. The discussion is, therefore, based on the observed shapes only. In the Sn case, we observe a structure with a relatively low yield for ^{131}Sn and a high yield in ^{132}Sn. This is not predicted by GEF or present in the evaluated data. The yield for the Sb isotopes is peaked at mass 132. The trend in the yields is roughly reproduced by GEF but is in sharp contrast to the data in the ENDF evaluation which shows a peak in the yield at mass 130.

Table 2. Isomeric yield ratios in the ^{nat}U(n,f) reaction as obtained in the present measurement compared to calculations with the GEF code and the ENDF/B-VII and JENDL-4.0 evaluations.

	Spin g.s.	Spin isom.	**This work**	GEF	ENDF/B-VII.1	JENDL 4.0
^{129}Sn	$3/2^+$	$11/2^-$	**0.37±0.13**	0.63	0.15±0.12	0.68±0.43
^{130}Sn	0^+	(7^-)	**0.64±0.48**	0.49	0.48±0.49	0.70±0.46
^{131}Sn	$(3/2^+)$	$(11/2^-)$	**0.43±0.19**	0.69	0.48±0.45	0.81±0.51
^{130}Sb	(8^-)	$(4,5)^+$	**0.81±0.43**	0.68	0.43±0.41	0.43±0.28
^{132}Sb	(4^+)	(8^-)	**0.25±0.10**	0.49	0.61±0.41	0.61±0.20

Table 2 shows the derived IYR ratio for five cases compared to GEF calculations and evaluated data. Compared to the IYR data in the (p,f) case, the statistical uncertainties are rather large. We expect this to improve considerably when using JYFLTRAP in forthcoming campaigns.

We note that our data are consistent when comparing, e.g., ^{129}Sn and ^{131}Sn where the ground state and the isomer carry the same spin assignments. We also note that the IYR is inversed when comparing the two Sb isotopes. This is consistent with the spin assignments in these two cases which are also inversed between their respective ground state and isomer. Since the evaluations also give large uncertainties it is not possible to observe any clear discrepancy. Also the results from the GEF code are in rough agreement with the measured data. The only exceptions might be ^{129}Sn and ^{132}Sb where the measured IYR seem to be significantly lower than the GEF predictions.

4 Outlook

Future work will focus on continued studies on the stopping efficiencies in the fission ion guide and measurement of the transport efficiency of the system using, e.g., ^{252}Cf(SF). This will allow the FY data to be discussed on an absolute scale. We also plan to implement the Phase-Imaging Ion-Cyclotron-Resonance technique to achieve 50 keV mass resolution. This will open up for measurement of a wide range of IYR for many different fissioning systems and a detailed study of the origin angular momentum generation in nuclear fission.

This work has been supported by the European Commission within the CHANDA project (contract 605203), the Swedish Radiation Safety Authority (SSM), and by the Swedish Nuclear Fuel and Waste Management Co. (SKB).

References

[1] A. Al-Adili et al., these proceedings
[2] S. Julien-Laferrière et al., these proceedings
[3] A. Bail, et al., Phys. Rev. C **84**, 034605 (2011)
[4] H. Naik, et al., Nucl. Phys. A **941**, 16 (2015)
[5] S. Eliseev, et al., Phys. Rev. Lett. **110**, 082501 (2013)
[6] I. Stetcu, et al., Phys. Rev. C **88**, 044603 (2013)
[7] A.A. Sonzogni, et al., Phys. Rev. Lett. **116**, 132502 (2016)
[8] F.P. An, et al., Phys. Rev. Lett. **118**, 251801 (2017)
[9] H. Penttilä, et al., Eur. Phys. J. A **44**, 147 (2010)
[10] D. Gorelov, et al., Nucl. Inst. Meth. Phys. Res. B **376**, 46 (2016)

[11] A. Mattera, et al., Eur. Phys. J. A **53**, 173 (2017)
[12] A. Al-Adili, et al., Eur. Phys. J. A **51**, 59 (2015)
[13] P. Helgesson and H. Sjöstrand, in manuscript
[14] H. Penttilä, et al., Eur. Phys. J. A **52**, 104 (2016)
[15] V. Rakopoulos, et al., to appear in Eur. Phys. J. (Proceedings of ND2016)
[16] V. Rakopoulos, et al., in manuscript
[17] M. Tanikawa, et al., Z. Phys. A **347**, 53 (1993)
[18] I.V. Ryzhov, et al., Phys. Rev. C **83**, 054603 (2011)
[19] A. Solders, et al., Nucl. Data Sheets **119**, 338 (2014)
[20] K.-H. Schmidt, et al., Nucl. Data Sheets **131**, 107 (2016)
[21] A. Mattera, et al., submitted to Eur. Phys. J.

Prompt fission gamma-ray emission spectral data for ^{239}Pu(n,f) using fast directional neutrons from the LICORNE neutron source

L. Qi[1], J.N. Wilson[1,⋆], M. Lebois[1], A. Al-Adili[2], A. Chatillon[3], D. Choudhury[4], A. Gatera[5], G. Georgiev[6], A. Göök[5], B. Laurent[3], A. Maj[7], I. Matea[1], A. Oberstedt[4], S. Oberstedt[5], S.J. Rose[8], C. Schmitt[9], B. Wasilewska[7], and F. Zeiser[8]

[1]*Institut de Physique Nucléaire, CNRS-IN2P3, Univ. Paris-Sud, Université Paris-Saclay, 91406 Orsay Cedex, France*
[2]*Department of Physics and Astronomy, Uppsala University, Box 516, 751 20 Uppsala, Sweden*
[3]*CEA, DAM, DIF, 91297 Arpajon, France*
[4]*Extreme Light Infrastructure-Nuclear Physics (ELI-NP)/Horia Hulubei National Institute for Physics and Nuclear Engineering (IFIN-HH), 077125 Bucharest-Magurele, Romania*
[5]*European Commission, Joint Research Centre, Directorate G for Nuclear Safety and Security, Unit G.2, 2440 Geel, Belgium*
[6]*CSNSM Orsay, CNRS-IN2P3, Univ. Paris-Sud, Université Paris-Saclay, 91405 Orsay Cedex, France*
[7]*The Henryk Niewodniczański Institute of Nuclear Physics, 31-342 Krakow, Poland*
[8]*University of Oslo, Department of Physics, P.O. Box 1048, Blindern 0316 Oslo, Norway*
[9]*Grand Accélérateur National d'Ions Lourd, Bd. Henri Becquerel, BP. 55027, 14076 Caen Cedex 05, France*

Abstract. Prompt fission gamma-ray spectra (PFGS) have been measured for the ^{239}Pu(n,f) reaction using fast neutrons at $\bar{E}_n$=1.81 MeV produced by the LICORNE directional neutron source. The setup makes use of $LaBr_3$ scintillation detectors and PARIS phoswich detectors to measure the emitted prompt fission gamma rays (PFG). The mean multiplicity, average total energy release per fission and average energy of photons are extracted from the unfolded PFGS. These new measurements provide complementary information to other recent work on thermal neutron induced fission of ^{239}Pu and spontaneous fission of ^{252}Cf.

1 Introduction

One of the key isotopes for present and future energy applications is the fissile nucleus ^{239}Pu. As such it occurs frequently in the NEA Nuclear Data High Priority Request List (HPRL). In conventional Light Water Reactors (LWR), up to one third of the energy released is from the fission of ^{239}Pu. ^{239}Pu is produced through neutron capture reaction on ^{238}U, followed by two beta decays. For many of the Generation IV reactors or current LWR using MOX fuel, fission of plutonium dominates the energy production.

Gamma-ray emission in nuclear fission is one of the least known fission observables and for this reason measurement requests of PFGS for ^{239}Pu in the thermal and fast neutron region are in the NEA

⋆e-mail: wilson@ipno.in2p3.fr

Nuclear Data HPRL. Recent results on PFG emission in thermal neutron induced fission of ^{239}Pu have been published in Ref [1]. Recent reactor experiments indicate heating effects from gamma-rays are underestimated up to 28% [2] when compared with the simulations based on the currently available experimental data.

Here we report on the PFG measurement for fission of ^{239}Pu induced by fast neutrons at an energy very relevant for future Gen IV fast reactors ($\bar{E}_n$=1.8 MeV). An experiment was performed at the ALTO facility at the IPN Orsay, using the LICORNE neutron source [3]. Results from the experiment will provide important nuclear data for reactor physics, e.g. as an input for gamma heating calculations and for a better understanding of gamma emission in the fission process.

2 LICORNE

The LICORNE neutron source generates forward focused neutron cones by exploiting neutron-producing reactions in inverse kinematics. The major advantage is that detectors can be placed around the sample to be irradiated without the need for shielding from neutrons emitted by the source. A further consequence is a very low thermal neutron background in the experimental room. This feature of LICORNE makes the study of fissile isotopes in the fast neutron region possible. In the experiment reported here, PFGS characteristics of ^{239}Pu at an incident neutron energy of 1.81±0.52 MeV were studied.

In order to estimate the neutron flux in the space after the LICORNE gas cell, a GEANT4 [4] code was developed to simulate the physical processes in LICORNE (see Fig. 1). The program incorporates user-defined physical processes, which are p(^{7}Li,n)^{7}Be and p(^{11}B,n)^{11}C reactions, into the standard physics list provided by GEANT4. They are based on relativistic kinematic calculations and consider the experimental differential cross section data from literature [5, 6]. Since the cross sections of these reactions are low (less than 1 barn), event biasing techniques are used to have more sampling in the region of interest to decrease the calculation time. There are mainly two categories of event biasing techniques, importance biasing [7] and multi-level splitting. Importance biasing is to increase the cross section of the process of interest. Multi-level splitting includes force collision (widely used in MCNP) and increasing the final neutron yields (encouraged by the GEANT4 example: bremsstrahlung splitting). All these un-physical techniques will need a weight to be corrected. They were implemented in the G4WrapperProcess without changing the tool-kit itself. Analogue simulation is also compared to the biased simulation. The final results agree with each other while the Figure-Of-Merit (FOM) has been improved more than 20 times with the event biasing techniques. The final strategy is to make full use of these two techniques, which is to increase the cross section of the reaction of interest and at the same time increase neutron yields once the reaction is invoked. This code has been validated by several experiments, e.g. the neutron time-of-flight (TOF) spectra measurement with liquid scintillation detectors from the EDEN array [8].

3 Experimental methods

3.1 Fission events tagging

A fission chamber, developed at JRC-Geel, was used to discriminate α emissions and fission fragments (FF) of the fissile samples. The chamber contained high purity (99.97%) ^{239}Pu samples of 3.519 mg in a compact geometry. The counting gas was pure Tetrafluoromethane (CF_4). Due to high activity of the sample (7 MBq), Pulse Shape Discrimination (PSD) was applied to have a better discrimination (see Fig. 2) by eliminating α pile-up events.

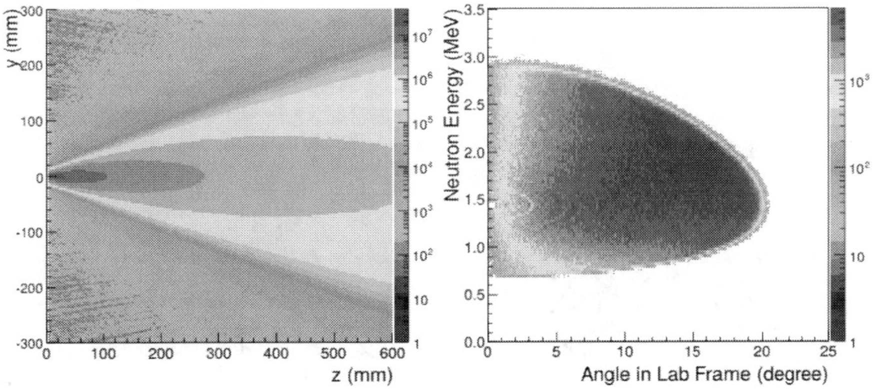

Figure 1. (a) Neutron flux distribution, normalized to the beam intensity 100 nA. The total neutron flux is of the order of 10^{6-7}n/(cm^2 s) a few centimeters after the LICORNE gas cell; (b) Kinematic curve correlating the outgoing neutron energy and the angle in the laboratory frame for the p(^{7}Li,n)^{7}Be reaction.

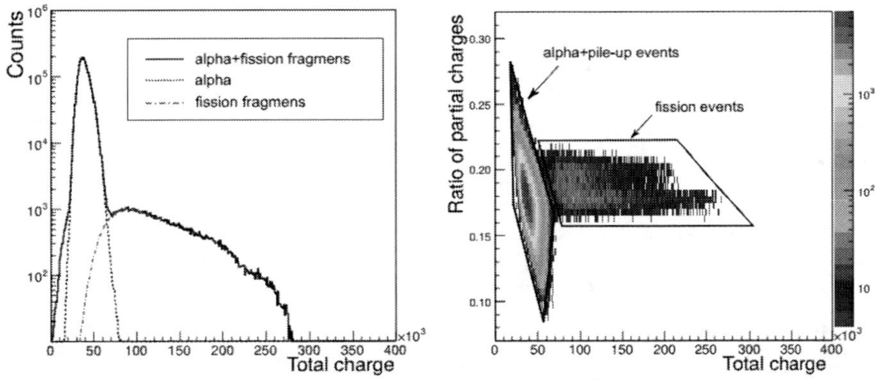

Figure 2. (a) Total integrated charge spectrum, seperating α particles and FF; (b) Correlation between the ratio of partial charges and the total integrated charge. PSD improves the discrimination between α particles and FF.

3.2 PFG measurements

The prompt gamma rays were measured, in coincidence with the fission fragments tagged in the ionization chamber, by scintillation detectors: 6 individual $LaBr_3$(Ce) detectors with two sizes (50.8 mm × 50.8 mm and 76.2 mm × 76.2 mm in diameter and length) and a cluster of 9 phoswiches detectors from the PARIS collaboration [9]. $LaBr_3$ detectors were selected due to their state-of-the-art energy and time resolution. Detection efficiency for high energetic gamma-rays is favourable in terms of the PARIS phoswiches. The calibration was performed by using some conventional gamma-ray sources, including ^{137}Cs, ^{60}Co,^{152}Eu and thorium series radionuclides. In addition, a practical and convenient source of 9 MeV gamma rays (see Fig. 3) [10] was used. It is able to check the linearity of the detectors and the dependance of resolution versus energy, i.e. $(\frac{dE}{E}) = \sqrt{\alpha^2 + \frac{\beta^2}{E} + (\frac{\gamma}{E})^2}$ [11], up to 9 MeV. It also provides an opportunity to validate the simulation of the experimental setup in the high-energy region.

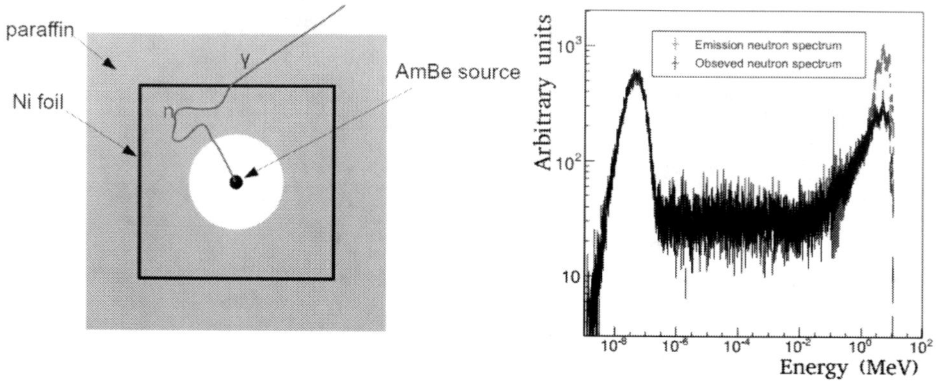

Figure 3. (a) Schematic of the 9 MeV gamma-ray source, consisting of an AmBe source, paraffin and nickel foils; (b) Emission neutron spectrum from the AmBe source and observed neutron spectrum in the nickel foils. The fast neutrons from the AmBe source are thermalized by the surrounding paraffin, and then captured in the nickel foils.

3.3 Deconvolution

Figure 4 represents the matrix of TOF between the fission chamber and the gamma detector versus the observed gamma-ray energies. The first structure in the zero time reference corresponds to the PFG and the second structure corresponds to the prompt neutron response. The prompt neutrons are well separated from the prompt gamma rays. The time resolution, expressed in terms of the full width half maximum (FWHM), of the prompt gamma-ray peak is 2 ns. The measured PFGS is extracted with the time window set to be [−3 ns, +3 ns]. This time window was chosen due to the convenience in comparison with another experiment [1], which measured the PFGS characteristics with thermal neutrons.

Since gamma rays have a complicated response in a single detector, the measured PFGS is folded by the response function. The aim is to unfold the measured PFGS with the simulated response matrix in GEANT4. The unfolding algorithm is by linear iteration [12]. The comparison of different unfolding techniques has been presented in another work [8]. The unfolding process highly depends on the simulated response matrix. In order to validate the response matrix, the efficiency of each gamma detector has to been verified with the calibration results in the range between 100 keV and 9 MeV. In the low energy region, the gamma spectrometer is very sensitive to the threshold effect, including electronic threshold and physical threshold. Preliminary results from one $LaBr_3$ have been extracted, with values for the mean multiplicity of $\bar{M}_\gamma = 7.71 \pm 0.09$, average total energy release per fission $E_{\gamma,tot} = 6.67 \pm 0.05$ MeV and average energy of photons $\epsilon_\gamma = 0.87 \pm 0.01$ MeV. A full analysis of the data is ongoing and final results will be published as a journal article in the near future, so these numbers are liable to change.

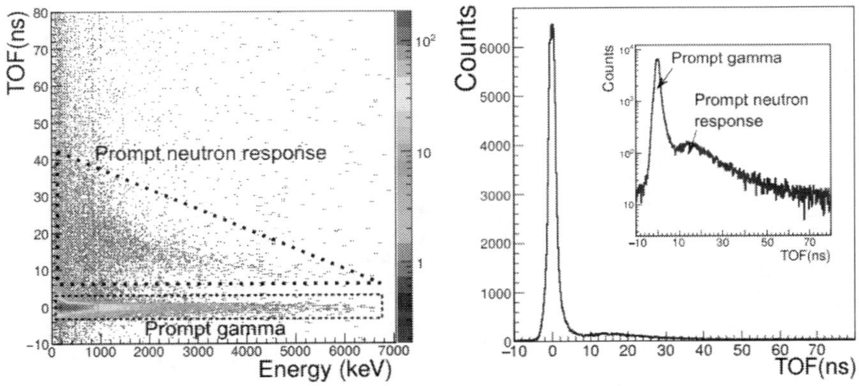

Figure 4. (a) Correlation between the gamma-ray energy detected in one gamma detector and the TOF, where the ionization chamber gives the start signal and the gamma detector gives the stop signal; (b) TOF spectra in linear scale; the inset shows the same distribution in logarithmic scale.

4 Conclusions

PFGS have been measured for the ^{239}Pu(n,f) reaction using fast neutrons produced by the LICORNE directional neutron source. Preliminary results from one $LaBr_3$ have been extracted, including the mean multiplicity, average total energy release per fission and average energy of photons. These results provide complementary information to recent thermal neutron induced fission and spontaneous fission measurements. Further analysis concerning all the detectors is still ongoing.

We would like to thank the staff at the ALTO facility for the production of high intensity beams and the PARIS Collaboration. We also acknowledge the support from the ENSAR2 project and the CHANDA program of the European Commission.

References

[1] A. Gatera et al., Phys. Rev. C **95**, 064609 (2017)
[2] Nuclear Data High Priority Request List of the NEA (Req. ID: H.3, H.4), `http://www.nea.fr/html/dbdata/hprl/hprlview.pl?ID=421` and `http://www.nea.fr/html/dbdata/hprl/hprlview.pl?ID=422`
[3] M. Lebois et al., Nucl. Instrum. Methods Phys. Res. A **735**, 145 (2014)
[4] S. Agostinelli et al., Nucl. Instrum. Methods Phys. Res. A **506**, 250 (2003)
[5] S.A. Elbakr et al., Nucl. Instrum. Methods **105**, 519 (1972)
[6] L. Van der Zwan and K.W. Geiger, Nucl. Phys. A **306**, 45 (1978)
[7] M.H. Mendenhall and R.A. Weller, Nucl. Instrum. Methods Phys. Res. A **667**, 38 (2012)
[8] L. Qi et al., submitted to Phys. Rev. C.
[9] A. Maj et al., Acta Phys. Pol. B **40**, 565, (2009)
[10] J.G. Rogers et al., Nucl. Instrum. Methods Phys. Res. A **413**, 249 (1998)
[11] N.V. Kornilov et al., Nucl. Instrum. Methods Phys. Res. A **599**, 226 (2009)
[12] A. László, J. Phys. Conf. Ser. **368**, 012043, (2012)

Fission dynamics with microscopic level densities

Jørgen Randrup[1,*], *Daniel* Ward[2], *Gillis* Carlsson[2], *Thomas* Døssing[3], *Peter* Möller[4], and *Sven* Åberg[2]

[1] *Lawrence Berkeley National Laboratory, Berkeley, California 94720, USA*
[2] *Mathematical Physics, Lund University, 221 00 Lund, Sweden*
[3] *Niels Bohr Institute, 2100 Copenhagen Ø, Denmark*
[4] *Los Alamos National Laboratory, Los Alamos, New Mexico 87545, USA*

Abstract. Working within the Langevin framework of nuclear shape dynamics, we study the dependence of the evolution on the degree of excitation. As the excitation energy of the fissioning system is increased, the pairing correlations and the shell effects diminish and the effective potential-energy surface becomes ever more liquid-drop like. This feature can be included in the treatment in a formally well-founded manner by using the local level densities as a basis for the shape evolution. This is particularly easy to understand and implement in the Metropolis treatment where the evolution is simulated by means of a random walk on the five-dimensional lattice of shapes for which the potential energy has been tabulated. Because the individual steps between two neighboring lattice sites are decided on the basis of the ratio of the statistical weights, what is needed is the ratio of the local level densities for those shapes, evaluated at the associated local excitation energies. For this purpose, we adapt a recently developed combinatorial method for calculating level densities which employs the same single-particle levels as those that were used for the calculation of the pairing and shell contributions to the macroscopic-microscopic deformation-energy surface. For each nucleus under consideration, the level density (for a fixed total angular momentum) is calculated microscopically for each of the over five million shapes given in the three-quadratic-surface parametrization. This novel treatment, which introduces no new parameters, is illustrated for the fission fragment mass distributions for selected uranium and plutonium cases.

1 Introduction

Soon after the discovery of nuclear fission in 1938 [1], it was recognized that the process can be viewed qualitatively as an evolution of the nuclear shape from that of a single compound nucleus to two receding fragments [2, 3] and that Langevin transport theory provides an appropriate model framework [3, 4]. A number of Langevin treatments of fission have been successfully developed and applied for excitations high enough to render the dynamics macroscopic, see for example Refs. [5–7].

If the collective shape dynamics is idealized as being highly dissipative, then the Langevin equation reduces to the Smoluchowski equation in which the evolution depends on the balance between the driving force provided by the potential energy of deformation and the dissipative force resulting from the coupling of the collective degrees of freedom to the remaining system. In this limit, it

*e-mail: `JRandrup@LBL.gov`

has proven possible to describe the Brownian shape motion as a Metropolis walk on the associated multi-dimensional deformation-energy surface [8]. This calculationally simple method has yielded remarkably good results for the fission-fragment mass distributions [9] and it has made it possible to predict fission-fragment mass distributions for poorly explored regions of the nuclear chart [10, 11].

2 Microscopic shape-dependent level density

In Ref. [12] a method was developed for microscopic calculations of level densities for deformed nuclei and it has been adapted to the fission process [13]. For a specified shape χ, the single-particle levels for protons and neutrons needed for the combinatorial calculation of the level density are obtained by solving the Schrödinger equation in the associated folded-Yukawa potential. These are the same levels as those previously used in Ref. [11] to calculate the microscopic shell and pairing energies in the construction of the deformation energy surface, thus guaranteeing consistency of the treatment. The corresponding local many-body vacuum state $|0;\chi\rangle$ has N neutrons and Z protons filling the lowest single-particle states and the uncorrelated excited states consist of all multiple particle-hole excitations,

$$|i;\chi\rangle \;=\; \prod_{n\geq 1} a^{\dagger}_{\nu_n^{(i)}} a_{\mu_n^{(i)}} |0;\chi\rangle \,, \tag{1}$$

For each many-body state $|i;\chi\rangle$, blocked BCS calculations for neutrons and protons separately provide the state-dependent pairing gaps, $\Delta_i^{\rm n}(\chi)$ and $\Delta_i^{\rm p}(\chi)$, respectively, and the energy of the correlated intrinsic many-body state, $E_i(\chi) = E_i^{\rm n}(\chi) + E_i^{\rm p}(\chi)$.

For each such intrinsic state, the angular momentum along the nuclear symmetry axis is denoted by K and it is assume that it forms a rotational band head. The resulting total angular momenta I may then take on the values $I = K, (K+1), (K+2), \ldots$ and the corresponding rotational energies of the band members are

$$E_i^{\rm rot}(I;\chi) \;=\; \frac{I(I+1) - K_i(\chi)^2}{2\mathcal{J}_\perp(\chi, \Delta_i^{\rm n}(\chi), \Delta_i^{\rm p}(\chi))} \,. \tag{2}$$

The moment of inertia $\mathcal{J}_\perp$ is approximated by the moment of inertia of a rigid body with the shape χ, modified by the calculated pairing gaps for the state [14].

In Ref. [12] it was found that the enhancements arisng from collective vibrational modes are unimportant and they are therefore ignored in the present fission applications. The total energy of a state is then given by

$$E_i(I,\chi) \;=\; E_i^{\rm n}(\chi) + E_i^{\rm p}(\chi) + E_i^{\rm rot}(I;\chi) \,. \tag{3}$$

For each shape χ, the resulting states are binned according to their energy E_i and their total angular momentum I; the bin width was taken as $\Delta E = 200\,$keV. The sensitivity of our results to the bin size has been tested and it was found that a doubling or tripling of ΔE produces negligible changes in the calculated mass distributions.

In Fig. 1, we show calculated level densities versus the backshifted excitation energy for three special shapes for which the shell and apiring energies have different magnitudes. At the second minimum, the gaps in the single-particle spectra for neutrons and protons provide the special stability of this shape region, quantitatively expressed by a considerable negative shell correction energy. On the other hand, the same gaps render it costly in energy to make particle-hole excitations, resulting in a slow increase of the level density when the nucleus is excited. The opposite situation is encountered at the symmetric barrier, where a high density of single-particle states around the Fermi energy results in a large positive shell correction energy, while at the same time delivering particle-hole states at a low cost in energy, providing a rapid increase of the level density with local excitation energy.

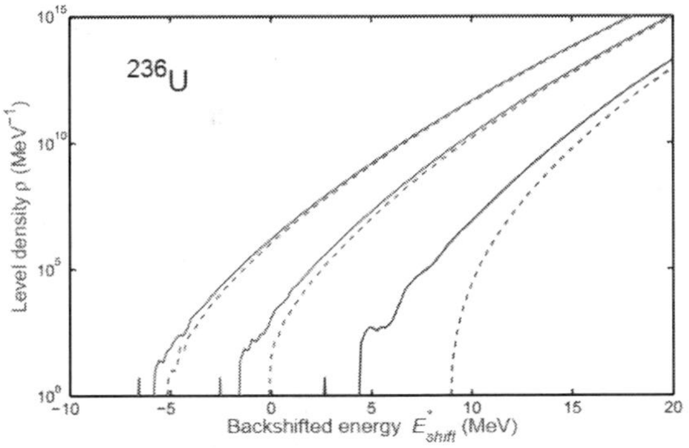

Figure 1. The microscopic level density, $\rho(E^*)$, calculated for ^{236}U at angular momentum $I = 4\hbar$, plotted versus the backshifted energy defined as $E^*_{\rm shift} = E^* + E_{\rm sh} + E_{\rm pc}$ where $E_{\rm sh}$ is the shell energy and $E_{\rm pc}$ is the pairing correlation energy; three selected shapes are illustrated, either with (solid) or without (dashed) pairing: (i) the second minimum (red) ($E_{\rm sh} = -5.4$ MeV, $E_{\rm pc} = -1.5$ MeV), (ii) the outer symmetric fission barrier (blue) ($E_{\rm sh} = +8.7$ MeV, $E_{\rm pc} = -6.1$ MeV), and (iii) an intermediate shape (green) ($E_{\rm sh} = -0.3$ MeV, $E_{\rm pc} = -2.5$ MeV).

These opposite effects of the shell structure on the potential energy and on the level density cause the no-pairing results (dashed curves on Fig. 1) to converge towards the level density with zero shell energy, illustrating the suppression of shell energy with increasing exitation energy, as discussed in Refs. [15–17].

The effects on the level density from shell structure and pairing correlations subside steadily as the excitation energy is increased. Two different energy scales are involved, with the pairing disappearing at a lower excitation energy than the shell structure. The present study of fission shape evolution considers a large number ($> 5 \cdot 10^6$) of nuclear shapes and, consequently, a large variety of shell and pairing energies appear. It is therefore important to employ a level density that accounts for the changes in shell and pairing effects when the shape is varied. This can be ensured by employing a microscopic calculation of the level density up to sufficiently high excitation energies for all shapes considered. To reduce the required computation times, we employ an extrapolation procedure that takes account of the changes in pairing and shell effects with excitation energy for each shape.

Because the computational effort required by the combinatorial method grows exponentially with excitation, it is practically important to develop a simple way of extending the results to high energy. We do that by switching to an analytical expression at $E^*(\chi) \approx 6\,\text{MeV}$, below which most of the specific structure effects, such as spectral gaps or non-monotonic behavior, have disappeared. In this way, important structures in the level density are maintained, while the numerical calculations are kept to a manageable level (about 3 CPU seconds per shape). For each particular shape χ, the extrapolation makes use of its specific shell and pairing energies, $E_{\rm sh}(\chi)$ and $E_{\rm pc}(\chi)$. Their influence diminishes with increasing energy and the level density approaches the analytical expression for a Fermi gas.

We employ the following simple analytical Fermi-gas expression [21], which is suitable for deformed nuclei with a fixed small angular momentum,

$$\rho(N, Z, E^*(\chi), I) = C(\chi)\, \tilde{E}_{\rm intr}^{-3/2} \exp\left(2\sqrt{a_0 \tilde{E}_{\rm intr}}\right), \tag{4}$$

where $E_{\rm intr} = E^*(\chi) - I(I+1)\hbar^2/2\mathcal{J}_\perp(\chi)$ is the approximate intrinsic energy. Accounting for the different energy scales of shell effects and pairing effects, we introduce a backshifted intrinsic excitation energy $\tilde{E}_{\rm intr}$, which is similar to the effective excitation $E^*_{\rm eff}$ of Ref. [9] and can be considered as a generalization of the prescription originally employed by Ignatyuk *et al.* [17],

$$\tilde{E}_{\rm intr}(\chi) = E_{\rm intr}(\chi) + (1 - {\rm e}^{-E_{\rm intr}(\chi)/E_{\rm d,sh}})E_{\rm sh} + (1 - {\rm e}^{-E_{\rm intr}(\chi)/E_{\rm d,pc}})E_{\rm pc}\,. \tag{5}$$

Here $E_{sh}(\chi)$ is the shell-correction energy and $E_{pc}(\chi)$ is the pairing condensation energy, both calculated for the lowest state at the specified shape χ. The full backshift, $\tilde{E}_{intr} = E_{intr} + E_{sh} + E_{pc}$ emerges when E_{intr} is much larger than both damping scales.

The parameter $E_{d,sh}$ sets the energy scale for the melting of shell structure and the parameter $E_{d,pc}$ is the corresponding energy scale for the melting of pairing correlations. For simplicity, both quautities are assumed to be common to all shapes. By fitting the analytic expression for a number of typical shapes, optimal values of $E_{d,sh}$ and $E_{d,pc}$ were determined then applied for all shapes. The normalization constant $C(\chi)$ in Eq. (4) is determined by continuity with the corresponding microscopic value at the matching energy. The validity of the analytical extrapolation is illustrated in Fig. 2.

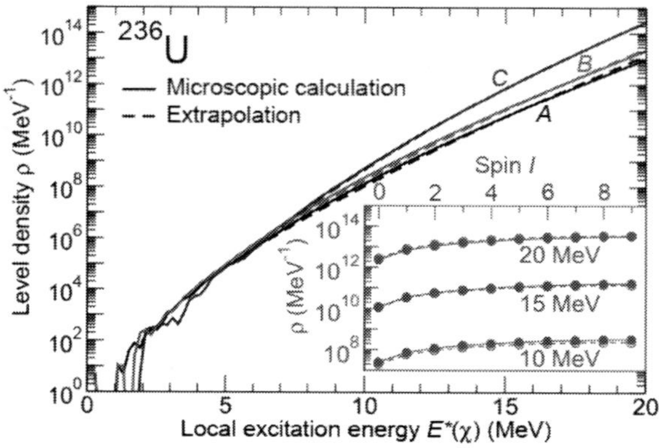

Figure 2. Microscopic level densities (solid lines) compared to extrapolated values (dashed lines). Three different deformations for ^{236}U are considered: the second minimum (A), the asymmetric second saddle (B), and an elongated symmetric shape close to the outer barrier (C). Extrapolated values are shown for local excitation energies $E^*(\chi) \geq 6$ MeV. The inset shows the angular momentum dependence of the level density at the asymmetric saddle (B) for different excitation energies.

3 Brownian shape evolution

The description of nuclear fission as a generalized Brownian motion builds on the assumption that the evolving nucleus can be characterized by its shape degrees of freedom χ. The associated shape parameters are treated as classical variables that are coupled dissipatively to the remaining nuclear degrees of freedom which have a primarily microscopic character and are assumed to have a short relaxation time. The resulting large-amplitude collective motion then exhibits a strongly damped diffusive evolution that can be described as a random walk on the associated multi-dimensional potential-energy surface $U(\chi)$. When the values of the potential are available on a discrete lattice of shapes (as is the case for the 5D tabulations in Ref. [18]), the Smoluchowski equation can be approximately simulated by means of a random walk on this lattice [8].

Because the random walk must satisfy detailed balance, the following relation must hold between the rates of transition between one lattice site χ and another χ', $\nu(\chi \to \chi')$, and the corresponding statistical weights which are proportional to the local level densities, $\rho(\chi)$,

$$\nu(\chi \to \chi')/\nu(\chi' \to \chi) = \rho(\chi')/\rho(\chi) . \tag{6}$$

This condition can be satisfied in many ways and the Metropolis procedure [19] is merely one particularly simple possibility: A proposed shape change from χ to χ' is accepted unconditionally if

$\rho(\chi') > \rho(\chi)$ (corresponding to a "downhill" step), whereas it it accepted only with the probabilty $\rho(\chi')/\rho(\chi)$ otherwise (i.e. if the proposed step is "uphill"); it is readily verified that this procedure satisfies detailed balance.

In Ref. [8] and all subsequent applications until recently [13], simplified Fermi-gas level-density expressions were used, $\rho_E(\chi) \approx \exp[2\sqrt{E - U(\chi)}]$. Because there is no explicit dependence on the shape, the ratio between neighboring level densities can be approximated by means of a logarithmic expansion,

$$\frac{\rho(\chi')}{\rho(\chi)} \approx \exp\left[-\frac{\partial \ln \rho(E^*)}{\partial E^*}\frac{\partial U(\chi)}{\partial \chi}\Delta\chi\right] \approx e^{-\Delta U/T}, \tag{7}$$

where $\Delta U = U(\chi') - U(\chi)$ is the change in the potential energy associated with the proposed shape change $\chi \to \chi' = \chi + \Delta\chi$ and $T(\chi) = 1/[\partial \ln \rho(E^*)/\partial E^*]$ is the local nuclear temperature. In the new treatment [13], we wish to account for pairing correlations and shell effects and we therefore employ the microscopic level densities (see Sect. 2) to evaluate the required ratios $\rho(\chi')/\rho(\chi)$.

To obtain a general understanding of the Brownian shape evolution towards fission, we have studied trajectories for the case ^{236}U at E^*=6.55 MeV. The aggregate number of visits to different sites on the shape lattice is shown in Fig. 3, projected onto the plane of mass asymmetry α_g and elongation q_2. The solid black curve shows the mean "scission" elongation, i.e. the average elongation, q_2, for random walks terminating with a particular mass asymmetry α_g.

The potential-energy landscape plays a key role in the shape evolution. The Metropolis walks all start in the second potential-energy minimum at $\alpha_g = 0$ and $q_2 = 2.36$. From this starting point, they exhibit a diffusive track that probes the potential landscape at ever larger deformations, including the

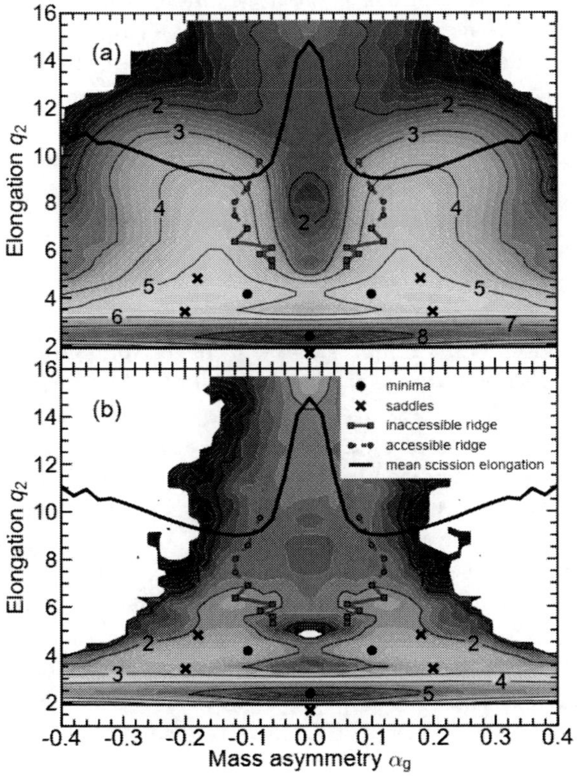

Figure 3. Contour diagrams (on a logarithmic scale) of the cumulative number of visits to shape lattice sites with asymmetry α_g and elongation q_2 for ^{236}U at $E^* = S_n$ =6.55 MeV. Panel (a) includes all tracks, whereas panel (b) contains only tracks that end in symmetric fission (α_g=0). Energy minima are shown as solid circles, while saddle points are shown as crosses. The solid black curve shows the mean "scission" elongation (where the random walk is halted and the mass split binned). The ridge separating the symmetric valley from the asymmetric valleys is shown as a solid red/dashed green curve, where green/red indicates that the ridge is below/above the total energy E^* and thus is either accessible or inaccessible, respectively. Symmetric fission events generally occur at larger elongations. This bimodal character of the mass distribution reflects the asymmetry dependence of the shell-correction energy in the scission region: Wheeras for asymmetric splits the combined shell correction of the two prefragments encourages early neck formation, the more symmetric scission shapes tend to have a smaller shell energy.

third minimum of potential energy located at an asymmetric shape having $\alpha_g \approx \pm 0.1$ and $q_2 \approx 4$. The distribution of site visits exhibits maxima in the regions around the second and third minima.

Beyond the outmost saddle point(s), around $\alpha_g \sim \pm 0.2$ and $q_2 \sim 5$, the potential starts to slope steadily down towards scission. Most often, a decreasing potential is accompanied by an increase in the level density and the Metropolis walk then changes its character from being diffusive in character to being drift dominated. As a result, "backwards" steps become unlikely and the average number of site visits tend to become relatively constant with elongation.

Concurrent with this steady drift in elongation, sideways diffusive steps are taken, thus providing some possibility for probing the potential with regard to changes in the fragment deformations, the neck radius, and, most important, the mass asymmetry α_g. A prominent region of shapes where this kind of evolution occurs is the interval $6 < q_2 < 9$, $0.1 < \alpha_g < 0.24$ (displaying a rather uniform light green color on Fig. 3a).

Turning now to symmetric shapes, $\alpha_g \approx 0$, we note that for the considered energy, $E^* = 6.55$ MeV, it is not possible to pass directly through the barrier region while keeping α_g close to zero. Therefore, in order to reach elongated symmetric shapes, the nucleus must first acquire some degree of asymmetry. A comparison between the upper and lower parts of Fig. 3 shows that the diffusion process indeed proceeds in identical ways up to elongations of $q_2 \approx 5$. But for elongations lager than $q_2 \approx 5$, the potential landscape displays sloping asymmetric and symmetric valleys separated by a (sloping) ridge. The ridge location in the (q_2, α_g) plane is indicated in Fig. 3, with the color of the ridge line indicating whether the ridge lies below (green) or above (red) the total energy. This implies that the distribution of tracks leading to symmetry is markedly different from the overall distribution.

4 Energy dependence of the fragment yields

The Metropolis walks were performed on a discrete lattice of more than five million shapes given in the three-quadratic-surface parametrization [20]. The five independent parameters characterizing a shape are the overall elongation of the nucleus (in terms of the reduced quadrupole moment q_2), the constriction of the central part (in terms of the neck radius c), the spheroidal deformations ε_{f1} and ε_{f2} of the two nascent fragments, and the reflection (mass) asymmetry α_g.

The use of microscopic level densities is particularly well-suited for studying the dependence of the shape evolution on the total excitation energy of the fissioning system because the microscopic pairing and shell effects automatically disappear the energy is raised.

The preliminary treatment in Ref. [9] used an energy-dependent effective potential,

$$U_E(\chi) = U_{\rm macro}(\chi) + \mathcal{S}(E^*(\chi))\, E_{\rm micro}(\chi)\,, \tag{8}$$

which is obtained by reducing the microscopic term with a suppression factor of the form $\mathcal{S}(E^*) = (1+c)/[\exp(E^*/E_{\rm damp}) + c]$, which decreases from one to zero as the local excitation $E^* = E - U(\chi)$ increases. The damping energy in the suppression fucntion was fixed to be $E_{\rm damp} = 15\,$MeV and the constant $c = 20\,$MeV was adjusted to ensure reproduction of the experimental yields for fission of ^{234}U at $E^* = 11\,$MeV [23]. The local statistical weights are then proportional to a simplified Fermi-gas level density, $\rho_E(\chi) \sim \exp[2\sqrt{E - U_E(\chi)}]$, and the Metropolis walk can thus be carried out as before without any other modifications of the code which is obviously very convenient. Although this approach yields quite reasonable results [9], it is unsatisfactory from a theoretical point of view because it fails to take account of the different energy scales of the pairing and shell effect as well as of their dependence on the nuclear shape.

Figure 4 shows the calculated charge yields for ^{234}U, as obtained by either using the simplified Fermi-gas level density based on the effective potential Eq. (8) introduced in Ref. [9] or the microscopic level densities [13]. The experimental data for ^{233}U(n_{th},f) [22] and ^{234}U(γ,f) [23] are also

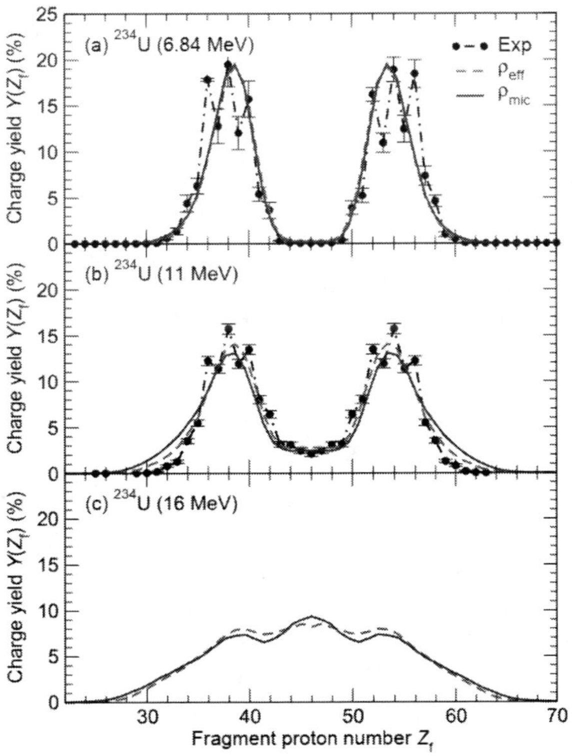

Figure 4. Fission-fragment charge distributions for ^{234}U at three different excitation energies E^* (MeV): 6.84 (a), 11 (b), and 16 (c). The blue solid curves have been obtained with the microscopic level densities, while the dashed red curves were calculated with the effective level density $\rho_{\rm eff}$ introduced in Ref. [9]. The results for $E^* = 6.84$ MeV are compared to (n_{th}, f) data [22], while those for $E^* = 11$ MeV are compared to (γ, f) data [23]. The calculated results for $E^* = 6.84$ MeV are practically identical and both are in good agreement with the experimental data. For $E^* = 11$ MeV the present approach reproduces the symmetric yield of around 2%, while the very asymmetric yields are too large. The calculated yields for $E^* = 16$ MeV are also quite similar.

shown. The (n,f) results are similar (and in good agreement with the data). For $E^* = 11$ MeV, the current approach correctly reproduces the symmetric yield of around 2%, whereas the yields for highly asymmetric splits are too large. The yields for $E^* = 16$ MeV come out quite similar in both approaches.

The transition from asymmetric to symmetric fission with increasing energy in the actinides is usually associated with the "melting" of shell structure, with the yield gradually assuming the symmetric form characteristic of the macroscopic limit. This feature appears automatically when the microscopic level densities are used [13], as illustrated in Fig. 5 for fission of ^{236}U at various excitations. As the energy is increased from 6.55 to 11.55 MeV there is a marked increase in the symmetric component because the shape diffuses ever more readily into the symmetric valley as the microscopic effects attenuate. At still higher energies the pairing and shell effects have largely subsided and the level density is determined by the excitation above the macroscopic potential, favoring symmetry.

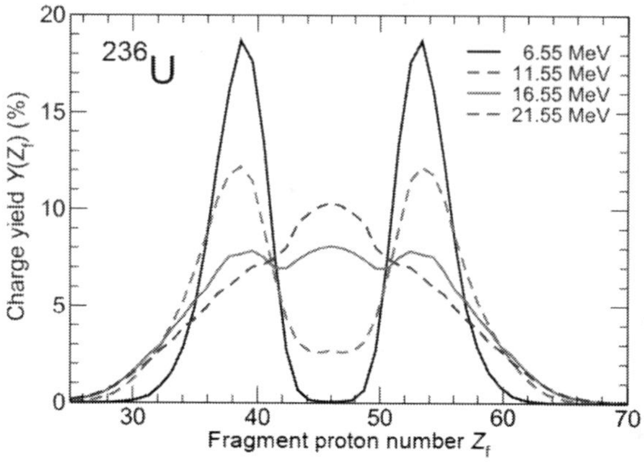

Figure 5. Calculated charge yields for fission of the nucleus ^{236}U at four values of the excitation energy extending from the neutron separation energy up to $E^* \approx 22$ MeV. The calculated yields correspond to first-chance fission. At excitation energies above $\approx$11 MeV the system is above the second-chance fission threshold, and it has an increasing probability of multi-chance fission as the energy is further increased. Consequently, the displayed results cannot be compared directly to experimental yields obtained with higher-energy neutrons impinging on ^{235}U.

The importance of taking account of nuclear structure effects in the level density is well illustrated by the energy dependence of the fission-fragment yield for symmetric splits. Figure 6 displays results for the nucleus ^{236}U obtained with different strengths of the pairing interaction entering into the microscopic level densities. The standard calculation leads to a local maximum at $E^* = 7$ MeV, followed by a local minimum at 8 MeV; this bump is followed by a second bump ending with an inflection at around 10 MeV. The experimental data exhibit a qualitatively similar behavior, in particular a local maximum at $E^* = 7$ MeV. Without pairing, the absence of a gap produces a larger level density, especially for shapes with positive shell corrections, where both the gap and the single-particle level density are larger. The non-monotonic behavior of the symmetric yield has its origin in the structure of the microscopic level density related to pairing. It is thus evident that the symmetric yield is highly sensitive to the pairing properties of the level density.

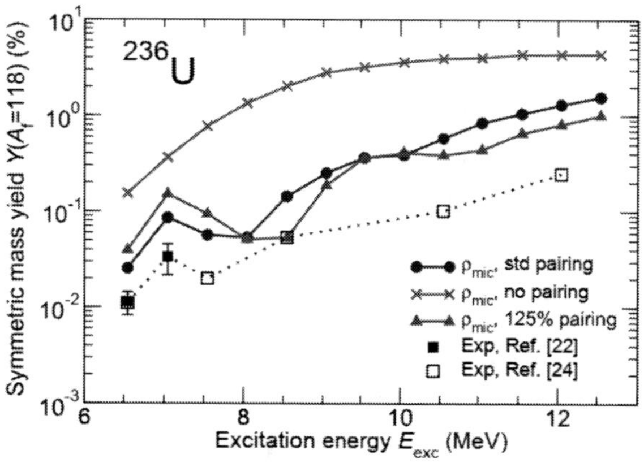

Figure 6. The dependence of the symmetric mass yield for ^{236}U on the excitation energy. In addition to the standard calculation (filled blue circles) are shown two additional calculations: one where the level densities are obtained without pairing (green crosses) and one where the pairing strength is 25% above the standard value (purple triangles). (It is the same potential-energy surface that has been used in all three calculations, namely the surface obtained with the standard pairing strength [18].) Also shown are the experimental data from Refs. [22] (solid squares) and [24] (open squares).

5 Conclusions and outlook

We have refined the Metropolis-walk approximation to the Brownian-motion treatment of fission dynamics [8] by employing microscopic local level densities obtained by the recently developed combinatorial method [12]. Because the single-particle levels used are the same as those employed for the macroscopic-microscopic calculation of the potential-energy surfaces, the approach is consistent and no new parameters need be introduced. Moreover, the combinatorial procedure provides access to the level density for a fixed value of angular momentum.

In order to reduce the considerable computing effort required for calculating the level density up to sufficiently high excitations, we employ an analytical extension above ≈6 MeV where the combinatorial level density has become smooth. This procedure uses a damped backshift of the excitation energy based on the local pairing and shell energies. In this manner, we have obtained the local level density $\rho(N, Z, E, I; \chi)$ for over 5 million shapes of several nuclei in the uranium region and for angular momenta up to $I = 9$.

Using these parameter-free microscopic level densities with the Metropolis-walk method, we calculate fission fragment charge distributions for ^{226}Th, ^{234,236}U, and ^{240}Pu. The agreement with experimental data was on par with or better than what was obtained previously with a phenomenological level-density parametrization [9]. The angular-momentum dependence of the fission yields was found to be relatively small and decreasing with increasing excitation energy.

Because the microscopic level densities automatically contain the diminishing effects of pairing correlations and shell structure, the present refined model makes it possible to make more detailed predictions for the energy dependence of the fission yields. The gradual transition from asymmetric to symmetric fission and the detailed energy dependence of the symmetric yield were studied.

Particularly interesting is the finding that the symmetric fragment yield is not monotonically increasing with excitation energy. This perhaps counter intuitive effect appears to be a result of the large pairing correlations for shapes with positive shell-correction energies separating the asymmetric-fission path from symmetric shapes. It is in this connection intriguing that a recent experiment has reported a non-monotonic energy dependence of the asymmetric peak shape in the fragment mass yields for fission of ^{240}Pu [25]. We plan to investigate these phenomena in more detail.

The present refined model provides a consistent and computationally manageable theoretical framework for studying large-scale collective motion of warm nuclear systems far from equilibrium and, in particular, it provides a unique tool for calculating energy-dependent fission fragment mass distributions. The present work extends the use of microscopic level densities from nuclei in shape equilibrium to arbitrary shapes and our present studies have revealed the intriguing possibility that the pairing interaction in shapes far from equilibrium. may manifest itself in a measurable manner through the energy dependence of the fisssion yields.

We are grateful to N. Schunck, A. Tonchev, and R. Vogt for discussions, helpful comments, and valuable suggestions. This work was supported by the Swedish Natural Science Research Council (BGC and SÅ), by the National Nuclear Security Administration of the U.S. Department of Energy at Los Alamos National Laboratory under Contract No. DE-AC52-06NA25396 (PM), and by the Office of Nuclear Physics in the U.S. Department of Energy(tm)s Office of Science under Contract No. DE-AC02-05CH11231 (JR).

References

[1] O. Hahn and F. Straßmann, Naturwiss. **27**, 11 (1939)
[2] L. Meitner and O.R. Frisch, Nature **143**, 239 (1939)
[3] N. Bohr and J.A. Wheeler, Phys. Rev. **56**, 426 (1939)
[4] H.A. Kramers, Physica **VII**, 284 (1940)
[5] A.V. Karpov et al., Phys. Rev. C **63**, 054610 (2001)
[6] P.N. Nadtochy et al., Phys. Rev. C **65**, 064615 (2002)
[7] E.G. Ryabov et al., Phys. Rev. C **78**, 044614 (2008)
[8] J. Randrup and P. Möller, Phys. Rev. Lett. **106**, 132503 (2011)

[9] J. Randrup and P. Möller, Phys. Rev. C **88**, 064606 (2013)
[10] P. Möller, J. Randrup, and A.J. Sierk, Phys. Rev. C **85**, 024306 (2012)
[11] P. Möller and J. Randrup, Phys. Rev. C **91**, 044316 (2015)
[12] H. Uhrenholt, S. Åberg, A. Dobrowolski, Th. Døssing, T. Ichikawa, and P. Möller, Nucl. Phys. A **913**, 127 (2013)
[13] D. Ward, G. Carlsson, Th. Døssing, P. Möller, J. Randrup, and S. Åberg, Phys. Rev. C **95**, 024618 (2017)
[14] R. Bengtsson and S. Åberg, Phys. Lett. B **172**, 277 (1986)
[15] J.R. Huizenga and L.G. Moretto, Ann. Rev. Nucl. Sci. **22**, 427 (1972)
[16] A. Bohr and B.R. Mottelson, Nuclear Structure II, Benjamin, Reading, MA (1974)
[17] A.V. Ignatyuk, K.K. Istekov, and G.N. Smirenkin, Sov. J. Nucl. Phys. **29**, 450 (1979)
[18] P. Möller, A. J. Sierk, T. Ichikawa, A. Iwamoto, R. Bengtsson, H. Uhrenholt, and S. Åberg, Phys. Rev. C **79**, 064304 (2009)
[19] N. Metropolis, A.W. Rosenbluth, M.N. Rosenbluth, A.H. Teller, and E. Teller, J. Chem. Phys. **21**, 1087 (1953)
[20] J.R. Nix, Nucl. Phys. A **130**, 241 (1969)
[21] T. Ericsson, Adv. Phys. **9**, 425 (1960)
[22] M.B. Chadwick et al. Nucl. Data Sheets, **112**, 2887 (2011)
[23] K.-H. Schmidt et al., Nucl. Phys. A **665**, 221 (2000)
[24] L.E. Glendenin, J.E. Gindler, D.J. Henderson and J.W. Meadows, Phys. Rev. C **24**, 2600 (1981)
[25] M.E. Gooden et al., Nucl. Data Sheets **131**, 319 (2016)

Fourier transform of single particle wave functions in extremely deformed nuclei: Towards the momentum distribution of scission neutrons

M. Rizea[1,⋆] and N. Carjan[1,2,⋆⋆]

[1] *"Horia Hulubei" National Institute of Physics and Nuclear Engineering, Bucharest, Romania*
[2] *Joint Institute for Nuclear Research, FLNR, Dubna, Moscow Region, Russia*

Abstract. The Fourier transform of single particle wave functions in cylindrical coordinates is applied to the study of neutrons released during scission. We propagate the neutron wave packets in time through the bi-dimensional time dependent Schrödinger equation with time dependent potential. We separate the parts of these wave packets that are in the continuum and calculate their Fourier transforms at different times: immediately after scission ($T = 1 \times 10^{-22}$ s) and at several intervals afterwards (until $T = 50 \times 10^{-22}$ s). The momentum distributions corresponding to these Fourier transforms are then estimated. The evolution of these distributions in time provides an insight into the separation of the neutron from the fissioning system and asymptotically gives the kinetic energy spectrum of that particular neutron.

1 Introduction

The Fourier transform (FT) appears in many nuclear physics applications. For example, FT is required in theoretical approaches of direct nuclear reactions of stripping, pick-up and knock-out types. In particular, the FTs of the single-particle wave-function (WF) represent the momentum transfers that occur during the reaction and give the probability that such an event takes place. Another application of the FT of a WF is in the field of nuclear fission. The momentum distribution of the scission neutrons (those that are emitted during the neck rupture) is given by the FT of the tail of their wave packets that is in the continuum. From the momentum distribution one can deduce an important observable: the kinetic energy distribution. This may allow us to distinguish the scission neutrons from the neutrons evaporated from fully accelerated fragments. Although the nuclei involved in the direct reactions are often deformed, the calculations have been performed in spherical coordinates neglecting the, undoubtedly important, role of the deformation. This is mainly because a deformed WF is commonly expressed in cylindrical coordinates and the procedure to calculate its FT is quite complicated. Recently, we have developed adequate numerical procedures to calculate the FT of a WF in cylindrical (ρ, z) coordinates [1]. After a short description of the numerical methods used, we apply them to scission neutrons and present some physical results in the frame of the Dynamical Scission Model [2, 3].

⋆e-mail: `rizea@theory.nipne.ro`
⋆⋆e-mail: `carjan@theory.nipne.ro`

2 Formalism

2.1 Continuous case

The standard form of the Fourier transform of a function $\Psi(\vec{r})$ in three dimensions is:

$$\Phi(\vec{k}) = \frac{1}{(2\pi)^{3/2}} \int_{-\infty}^{\infty} \Psi(\vec{r}) e^{-i\vec{k}\vec{r}} d^3\vec{r}, \quad \vec{k} = \vec{p}/\hbar. \tag{1}$$

It gives the probability in the momentum space i.e., the probability that the nucleon has its momentum $\vec{p}$ in the volume element $d^3\vec{p}$.

In our applications we consider nuclear forms with axial symmetry so that we use the cylindrical coordinates (ρ, z) and omit the dependence on the angle θ. It can be shown that the Fourier transform in this case has the form:

$$F(R, Z) = 2\pi \int_{-\infty}^{\infty} \left[\int_0^{\infty} f(\rho, z) J_0(2\pi\rho R) \rho d\rho \right] e^{-2\pi i z Z} dz \tag{2}$$

where J_0 is the zero-order Bessel function of the first kind. The transform with respect to the variables ρ, R is called the zero-order Hankel transform. Thus, the Fourier transform in cylindrical coordinates implies a combination of Hankel and one-dimensional Fourier transforms. The form used in quantum mechanics is based to the relations $R = k_\rho/(2\pi)$ and $Z = k_z/(2\pi)$. The variables ρ, z belong to the position space, while k_ρ, k_z belong to the momentum space. The factor $1/2\pi$ is introduced as a phase convention.

2.2 Discrete grid

Suppose now that the function $f(\rho, z)$ is known only on the nodes of a discrete grid. Then, the Fourier transform will be also a discrete function, approximation of the continuous transform. Let $\rho_j = \rho_0 + j\Delta\rho$, $j = 0, 1, \ldots, M-1$ and $z_k = z_0 + k\Delta z$, $k = 0, 1, \ldots, N-1$ be the points (uniformly spaced) defining the grid for which $f(\rho_j, z_k)$ are given. The function $f(\rho, z)$ is supposed to be zero or negligible for $\rho > \rho_{M-1}$, $z < z_0$ and $z > z_{N-1}$. The Fourier transform will be also calculated on a grid with the mesh points R_m, Z_n. For any R_m from the discrete grid, let us consider the integral:

$$H(R_m, z) = 2\pi \int_0^{\infty} f(\rho, z) J_0(2\pi\rho R_m) \rho d\rho \tag{3}$$

For a position $z = z_k$, since f is negligible beyond ρ_{M-1}, one has

$$H(R_m, z) \approx 2\pi \int_{\rho_0}^{\rho_{M-1}} f(\rho, z) J_0(2\pi\rho R_m) \rho d\rho.$$

The integral on the interval $[\rho_0, \rho_{M-1}]$ is represented as a sum of integrals on subintervals of the size $2\Delta\rho$. To approximate these partial integrals we have deduced a quadrature formula (see [1]) whose coefficients are expressed in terms of Bessel and Struve functions [4] of order 0 and 1.

Let us now consider the integral:

$$F(R_m, Z) = \int_{-\infty}^{\infty} H(R_m, z) e^{-2\pi i z Z} dz. \tag{4}$$

The variable Z takes a set of N values defined as

$$Z_n = \frac{n - N/2}{N\Delta z}, \quad n = 0, 1, \ldots, N-1.$$

The value of $F(R_m, Z_n)$ is approximated by using the Fast Fourier Transform - FFT [5]. In conclusion, the values $F(R_m, Z_n)$, representing the Discrete Fourier Transform in cylindrical coordinates, are obtained by the discrete Hankel (ρ-axis) and Fourier transforms (z-axis).

3 Application to the scission neutrons

We apply the Fourier transform of single particle wave functions in cylindrical coordinates to the study of neutrons released during scission. The neutron wave packets are propagated through the bi-dimensional time dependent Schrödinger equation (TDSE) with time dependent potential (TDP):

$$i\hbar\frac{\partial\Psi(\rho,z,t)}{\partial t} = \mathcal{H}(\rho,z,t)\Psi(\rho,z,t), \tag{5}$$

$$\mathcal{H}\Psi = \begin{bmatrix} O_1 - CS_c & -CS_a \\ -CS_b & O_2 - CS_d \end{bmatrix}\begin{bmatrix} f^{(1)} \\ f^{(2)} \end{bmatrix}, \quad O_{1,2} = -\frac{\hbar^2}{2\mu}(\Delta - \frac{\Lambda_{1,2}^2}{\rho^2}) + V(\rho,z,t). \tag{6}$$

Δ is the Laplacean, V is the potential, C is a constant and the operators $S_a,\ldots,S_d$ represent the spin-orbit coupling. We use a realistic mean field potential of Woods-Saxon type adapted to nuclear shapes described by generalized Cassini ovals [6]. The wave functions have two components corresponding to spin "up" and spin "down". For a given projection Ω of the total angular momentum on the symmetry axis they assume the form:

$$|\Psi\rangle = f^{(1)}(\rho,z)e^{i\Lambda_1\theta}|\uparrow\rangle + f^{(2)}(\rho,z)e^{i\Lambda_2\theta}|\downarrow\rangle, \quad \Lambda_1 = \Omega - \frac{1}{2}, \quad \Lambda_2 = \Omega + \frac{1}{2}. \tag{7}$$

Due to the axial symmetry, only the coordinates ρ and z remain.

3.1 Dynamical scission model

According to the Dynamical Scission Model [2, 3], we separate the calculation in two stages:

1) The scission process itself, i.e., the neck rupture and its absorption by the fragments. The corresponding nuclear configurations are defined by a set of deformations $\{\alpha^i\}$ (just before scission) and $\{\alpha^f\}$ (immediately after scission). The duration of this stage is relatively short ($\Delta T \approx 10^{-22}$ s) and therefore the potential in which the neutrons move changes rapidly.

2) The detachment from the fragments of the fraction of the neutrons that are left unbound at the end of the previous stage. Since, at this stage, their motion is much faster than that of the just separated fragments we maintain the fragments at the configuration $\{\alpha^f\}$ and therefore keep the potential constant during the subsequent propagation in time.

In the 1st stage we follow the time evolution of the wave functions corresponding to each neutron state that is occupied at $\{\alpha^i\}$. At each time step the nuclear shape and the potential is changing as the deformation parameter set varies uniformly from $\{\alpha^i\}$ to $\{\alpha^f\}$. The initial wavefunctions (at $t = 0$) are taken as eigensolutions of the eigenvalue problem: $\mathcal{H}\Psi = E\Psi$. To solve it, the infinite physical domain is replaced by a finite one, which is discretized on a grid with the mesh points ρ_j and z_k. At each point the partial derivatives in $\mathcal{H}$ are approximated by finite difference formulas [7]. The eigenvalue problem is transformed into an algebraic eigenvalue problem with a large sparse matrix, which is solved by an appropiate method (Arnoldi) [8]. The propagation in time is performed by a numerical scheme of Crank-Nicolson type [9]. At the end of this stage we calculate quantities like:

- The probability that a neutron occupying the state $|\Psi^i\rangle$ before scission populates a state $|\Psi^f\rangle$:

$$a_{if} = \langle\Psi^i(\Delta T)|\Psi^f\rangle = 2\pi\int\int\rho\Psi^i(\Delta T)\Psi^f d\rho dz.$$

- The probability that the neutron is emitted, given by:

$$P^i_{em} = v_i^2(\sum_{unbound}|a_{if}|^2) = v_i^2(1 - \sum_{bound}|a_{if}|^2)$$

where v_i^2 are the occupation probabilities of the initial eigenstates.

• The emitted part of the wave packet:

$$|\Psi^i_{em}\rangle = |\Psi^i(\Delta T)\rangle - \sum_{bound} a_{if}|\Psi^f\rangle.$$

In the 2nd stage we follow these wave packets (that describe the unbound neutrons) in time (by TDSE with the potential fixed for the set $\{\alpha^f\}$) and calculate their FT at certain time intervals T.

3.2 The evolution of the emitted wave functions and of the momentum distributions

We have investigated the reaction $^{235}U(n_{th}, f)$ with the light fragment mass $A_L = 96$ (which is the most probable experimental mass asymmetry). Two parameters of the Cassini representation have been considered: α (elongation) and α_1 (mass asymmetry). We have chosen: $\alpha = 0.985$ (before scission) and $\alpha = 1.001$ (after scission). The parameter α_1 results from the condition of fixed ratio of the fragment volumes that corresponds to A_L=96. We have performed the Fourier transform for several wavefunctions corresponding to $\Omega = 1/2$. At each point in the (k_ρ, k_z) plane are associated an absolute value $K = \sqrt{k_\rho^2 + k_z^2}$ and a probability $P = |F(k_\rho, k_z)|^2 k_\rho \Delta k_\rho \Delta k_z$. To represent the K-distributions as histograms, we divide the domain of K-values in equal intervals and group the points according to the interval to which they belong. Summing up the probabilities of the points in each group one obtains the probability $P(K)$ that a neutron has its K-value in the respective interval. In the Figs.1 and 2 are shown the emitted wavefunctions with the indices 30 and 33 (as sum of square moduli of the two components) aside with the momentum histograms at different times T. The time-dependent wavefunctions are represented relative to the function at $T = 20$, while the values on the ordinates of the histograms are the $P(K)$ probabilities multiplied by 10^2.

As one can see, at increasing time intervals the wavefunction values are diminishing, showing that the neutron is leaving the nucleus. At the same time, the K-distributions are shrinked to lower values, reflecting the fact that the neutrons are less and less present in the potential well. At large times they tend to a gaussian shape.

From the momentum distribution one can deduce the kinetic energy distribution (according to $E = \frac{\hbar^2}{2\mu}K^2$). To obtain the whole kinetic energy spectrum one has to repeat such calculations for all the states at each Ω value. Here we present results only for Ω=1/2 states and from WF_{11} to WF_{41}. The states lying at the bottom of the well (WF_1 to WF_{10}) have negligible P_{em} values. Note however that $\Omega = 1/2$ gives the dominant contribution (65%).

Figure 3 shows kinetic energy spectra for two individual states (30 and 33) around the Fermi level and the sum for all Ω=1/2 states with the index above 10 (having energies greater than −23 MeV). The total spectrum presents a maximum around 0.7 MeV and an exponentially decreasing tail till 5 MeV in qualitative agreement with the measured spectrum [10] of all prompt fission neutrons (PFN). For a quantitative comparison with experimental data, states with higher Ω values (3/2 to 9/2) has to be included in the calculations.

4 Summary

In this work we have presented numerical procedures and some applications of the Fourier transform on wave functions given in cylindrical coordinates. As initial functions for our procedure we have chosen single-particle states in extremely deformed nuclei associated with the wave packets of the neutrons released during scission. We propagate the neutron wave packets in time through the bi-dimensional TDSE with TDP. We select Ω=1/2 states and calculate their Fourier transforms

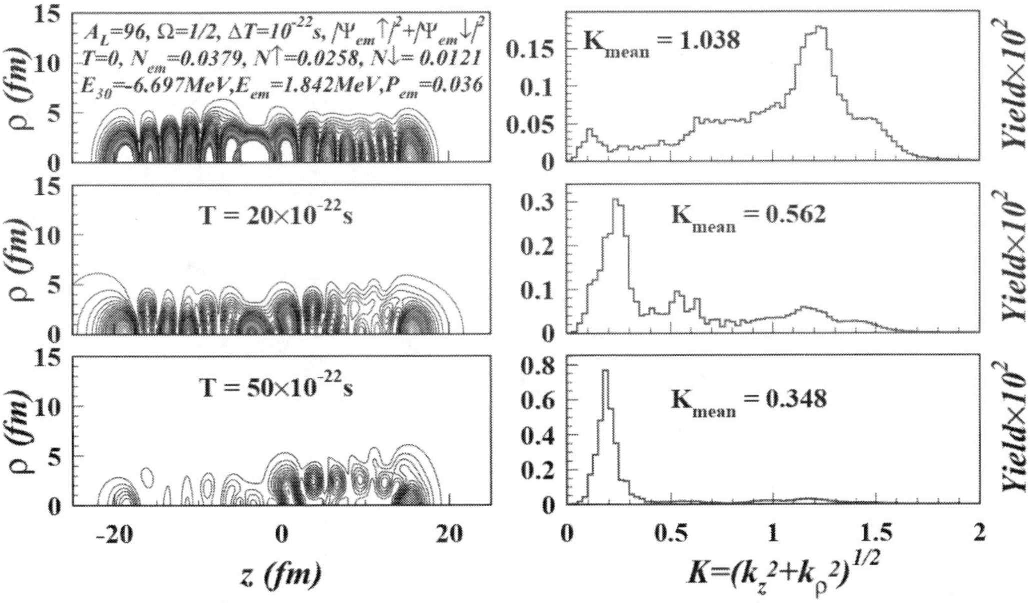

Figure 1. Square modulus of the emitted WF_{30} and momentum distribution at different times T. N_{em} is the norm of the emitted function, N↑, N↓ are the norms of the two components and $K_{mean} = \frac{\sum_{m,n} KP}{\sum_{m,n} P}$, $P = k_\rho |F|^2 dk_\rho dk_z$.

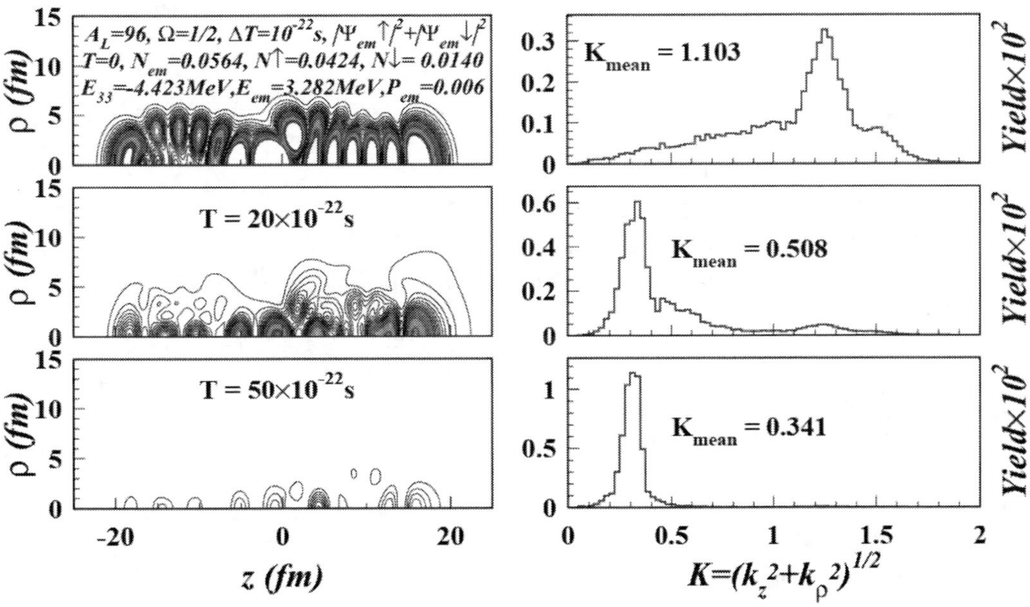

Figure 2. The same as in Fig. 1 but for the emitted WF_{33}.

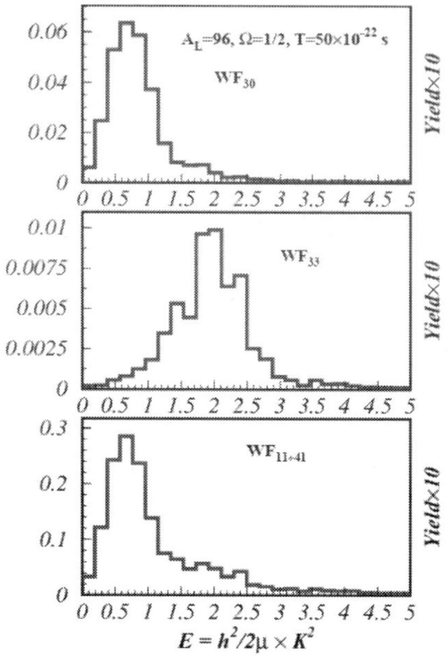

Figure 3. Energy distribution at $T = 50 \times 10^{-22}$ s (Yield = $P(K) \times v_i^2$).

at different times T and the momentum distributions corresponding to these transforms. From such distributions one can follow the separation of the neutron from the fissioning system and obtain the kinetic energy spectrum of each emitted neutron. The shape of the energy spectrum obtained summing over all emitted neutrons resembles the measured PFN spectrum. For a more accurate comparison the effect of reabsorption of the unbound neutrons by the imaginary potential has to be taken into account.

This work was done in the frame of the projects PN-III-P4-ID-PCE-2016-0649 (contract No. 194/2017) and PCE-2016-0014 (contract No. 7/2017).

References

[1] M. Rizea, N. Carjan, Eur. Phys. J. A **52**, 368 (2016)
[2] M. Rizea, N. Carjan, Nuc. Phys. A **909**, 50–68 (2013)
[3] N. Carjan, M. Rizea, Phys. Lett. B **747**, 178–181 (2015)
[4] M.Abramowitz, I.A.Stegun, *Handbook of Mathematical Functions* (8th ed., Dover, New York, 1972) 358, 496
[5] W.H. Press, B.P. Flannery, S.A. Teukolsky, W.T. Vetterling, *Numerical Recipes* (Cambridge Univ. Press, 1996) 451–453
[6] V.V. Pashkevich, Nucl. Phys. A **169**, 275–293 (1971)
[7] M. Rizea, V. Ledoux, M. Van Daele, G. Vanden Berghe, N. Carjan, Comp. Phys. Commun. **179**, 466–478 (2008)
[8] R.B. Lehoucq , D.C. Sorensen, C. Yang, *ARPACK Users Guide* (Philadelphia, SIAM, ISBN: 978-0-89871-407-4, 1998)
[9] M. Rizea, N. Carjan, Proc. Rom. Acad. A **16**, 176–183 (2015)
[10] N. Kornilov, F.-J. Hambsch et al., Nucl. Sci. Eng. **165**, 117–127 (2010)

Neutron induced fission of ^{237}Np – status, challenges and opportunities

Ivan Ruskov[1,2,⋆], *Andrei* Goverdovski[3], *Walter* Furman[1], *Yury* Kopatch[1], *Oleg* Shcherbakov[4], *Franz-Josef* Hambsch[5], *Stephan* Oberstedt[5], and *Andreas* Oberstedt[6]

[1] *Joint Institute for Nuclear Physics, 141980 Dubna, Moscow region, Russia*
[2] *Institute for Nuclear Research and Nuclear Energy of BAS, 1784 Sofia, Bulgaria*
[3] *JSC "SSC-RF-IPPE", 249033 Obninsk, Kaluga region, Russia*
[4] *NRC "Kurchatov Institute"-PNPI, 188300 Gatchina, Leningrad region, Russia*
[5] *European Commission, Joint Research Centre, Directorate G, 2440 Geel, Belgium*
[6] *ELI-NP/IFIN-HH, 077125 Bucharest-Magurele, Romania*

Abstract. Nowadays, there is an increased interest in a complete study of the neutron-induced fission of ^{237}Np. This is due to the need of accurate and reliable nuclear data for nuclear science and technology. ^{237}Np is generated (and accumulated) in the nuclear reactor core during reactor operation. As one of the most abundant long-lived isotopes in spent fuel ("waste"), the incineration of ^{237}Np becomes an important issue. One scenario for burning of ^{237}Np and other radio-toxic minor actinides suggests they are to be mixed into the fuel of future fast-neutron reactors, employing the so-called transmutation and partitioning technology. For testing present fission models, which are at the basis of new generation nuclear reactor developments, highly accurate and detailed neutron-induced nuclear reaction data is needed. However, the EXFOR nuclear database for ^{237}Np on neutron-induced capture cross-section, σ_γ, and fission cross-section, σ_f, as well as on the characteristics of capture and fission resonance parameters (Γ_γ, Γ_f, $\sigma_o\Gamma_f$, fragments mass-energy yield distributions, multiplicities of neutrons ν_n and γ-rays ν_γ), has not been updated for decades.

1 Introduction

For basic (fundamental) and applied nuclear science, ^{237}Np continue to be amongst the most important and interesting isotopes (minor actinides) for experimental investigation with neutrons of different energies. This is because, as it is a long-lived ($T_{1/2} = 2.144 \times 10^6$ years) isotope, which is present in spent fuel of light-water nuclear reactors, it contributes significantly to the long-term radio-toxicity of nuclear waste. For its successful incineration (transmutation, $\sigma_f \sim 0.01$–10 barn), neutrons of different energies can be used [1–7]. In this case, neutron-induced reaction cross sections (particularly for fission) should be known with high precision for a broad neutron energy range (from thermal to fast) for estimating the effectiveness of the transmutation process.

⋆e-mail: ivan.n.ruskov@gmail.com

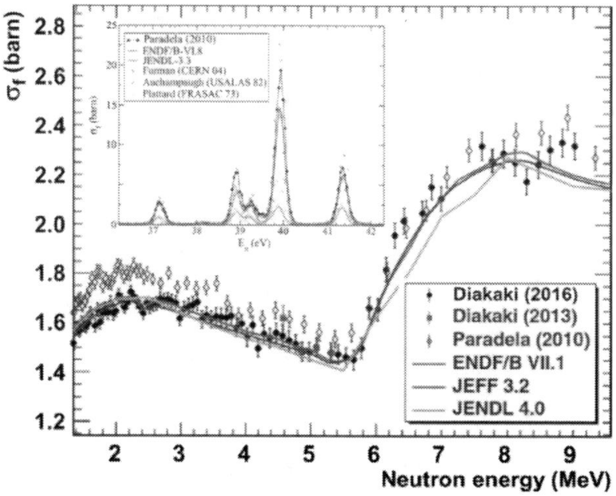

b

Figure 1. ^{237}Np(n,f) cross-section in the vicinity of the first resonance cluster around E_n = 40 eV (Fig. 6 [12]) and in the neutron energy range between 1.5 MeV and 9 MeV (Fig. 7 [14]); the depicted data are from different experiments, as well as from different evaluated nuclear data libraries (see the references in [12–14]).

According to one of the contemplated scenarios, a high-flux thermal reactor ($\sigma_f^{th} \sim 20$ mb) can be used as a transmutation device [7], because the majority of the operating commercial reactors in the world are based on a thermal spectrum.

2 Neutron-induced fission cross section of ^{237}Np

As ^{237}Np is a major component of spent nuclear fuel, the accurate knowledge of the capture and fission cross-sections is needed for studies of waste transmutation and advanced nuclear facilities (reactors, ADS, etc.). However, significant discrepancies exist in neutron-induced reaction experimental data repositories and in the present evaluated data libraries, based on the different sets of experimental data (see Fig. 1 and Ref. [8–14]). Similar discrepancy between different data libraries exists for fission-resonance parameters below the fission threshold [10]. Although some data sets for fast-neutron induced fission are consistent within the claimed accuracy, usually about 3–4%, some significant discrepancies show up between more recent data, when compared to several previous measurements. For example, several experiments based on the time-of-flight (TOF) technique were dedicated to the measurement of fission cross section data in a wide neutron energy range (0.7 eV–1 GeV), using different fission-fragment detectors [12–14].

While being in good agreement with previous measurements and evaluated data files at energies below and around the fission threshold, the data exhibit an excess of about 6–7% above E_n = 1 MeV [8–13]. The more recent measurements [14] were not able completelly to resolve the situation. That is why, to study the ^{237}Np(n,f) and ^{238}U(n,f) cross sections in a comprehensive manner, two experiments were performed at two different Van de Graaf (VDG) facilities: IRMM and NPL. Analysis of both experiments is in progress. Some preliminary results were reported at the ND2016 Conference on Nuclear Data for Science and Technology [15].

Very recently, two simultaneous experiments with PPACs (at EAR1) and Micromegas detectors (at EAR2) were conducted at the n_TOF facility to re-measure the ^{237}Np(n,f) cross-section [16], but the results have not published yet.

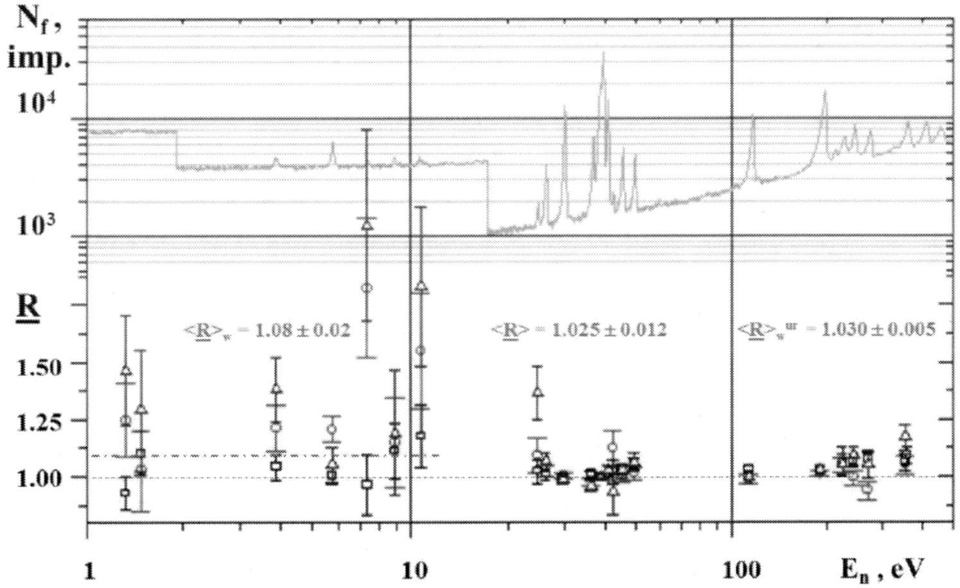

Figure 2. *Upper graph:* fission count rate in neutron energy interval from 1 eV up to 500 eV. The steps indicate the different value used for a single time-channel in the neutron time-of-flight (TOF) fission spectrum; *Lower graph:* variation of the relative (to the strongest resonance around 40 eV) fission γ-ray yield $\underline{R}$ from resonance to resonance at different thresholds of γ-quanta registration E_γ: ~0.2 MeV (circles), ~0.4 MeV (squares), ~0.6 MeV (triangles). $< \underline{R} >$, $< \underline{R} >_w$ and $< \underline{R} >_w^{ur}$ are explained in the text.

3 The (n,γf)-reaction

The investigation of the resonance neutron induced fission of ^{237}Np and the (n, γf) reaction provide important information about the compound nucleus (CN) internals:

- the height and structure of fission barriers;
- the transition states to fission and the degree of their damping;
- the strength of the coupling between the collective mode of motion with the internal excitations;
- the influence of the quantum characteristics of the excited states on the properties of the fission fragments;
- the scheme of γ-transitions between highly excited states and fission resonances;
- the connection strength between states with different deformations (classes I and II).

It aims at refining the quality of nuclear data, such as σ_γ and σ_f. The reaction (n, γf) leads to significant changes in Γ_γ, σ_γ, Γ_f and σ_f of the order of 10–15% (for e.g. ^{239}Pu at E_n = 1 keV) on σ_t and $\alpha = \sigma_\gamma/\sigma_f$, especially, at low excitation energy region of the CN. On the other hand, the database on fission resonance parameters for ^{237}Np is meagre (only few experimental datasets available in the EXFOR database).

The nature of the coupling between 1st and 2nd class states, as well as the existence and location of class II resonances remains unclear. The existence of very weak resonances in the sub-threshold fission of ^{237}Np, with $\Gamma_f \sim 1$ μeV and, consequently, relatively long lifetimes of excited states in the ^{238}Np* compound nucleus (t~1 ns) suggest that the probability of emission of a pre-fission g-quantum might be larger than the probability for this reaction to occur in the fission resonances of ^{235}U and ^{239}Pu. A detailed review on the investigation of the ^{235}U(n,γf) and ^{239}Pu(n,γf) until 1990 is given in reference [18].

Until now, the only experiment on the investigation of fluctuations of the γ-ray yield from resonance to resonance in ^{237}Np(n,f) (Fig. 2, [18–20]) shows that this fluctuation might be partly explained assuming the existence of a ^{237}Np(n, γf) reaction channel. In Fig. 2, $< \underline{R} >$ is the average relative γ-ray yield from the resonances in the first cluster around 40 eV, $< \underline{R} >_w$ is the weighted average relative value from the resonances below 10 eV and $< \underline{R} >_w^{ur}$ is the weighted average relative value from the unresolved resonance above 100 eV.

4 Conclusion and outlook

The EXFOR nuclear database for the ^{237}Np(n,f) reaction does not provide experimental data with uncertainties appropriate for nuclear reactor applications. This is true for all of the available incident neutron energy regions. The data from the sole experiment on the fluctuation of fission γ-ray yields from resonance to resonance does not exclude a (n,γf) reaction channel in the sub-threshold fission of ^{237}Np [20]. New experiments can be conducted, e.g. at the high-resolution neutron time-of-flight facility GELINA [24], with a focus on measuring $< \nu_n >$ and $< \nu_\gamma >$, prompt n/γ-ray competition, correlations r(<TKE>, Γ_f^{-1}) [21], n- and γ- yields as a function of resonance neutron energy. The assessment of those quantities in correlation with fission fragment properties (A, TKE), i.e. fission fragment mass-dependent γ- and n-emission is recommended but requires a highly efficient experimental set-up.

To minimize the influence of the statistical and systematic uncertainties, new experiments need to be conducted on the different existing resonance-neutron TOF facilities [22–24], using fission fragment and multi n-γ detector spectrometers.

References

[1] T. Iwasaki, Prog. Nucl. Energy **40**, 481 (2002)
[2] G. Fioni, O. Deruelle, J. Nucl. Sci. Technol. Suppl. **2**, 876 (2002)
[3] T. Wakabayashi, Prog. Nucl. Energy **40**, 457 (2002)
[4] T. Aoyama, S. Maeda, Y. Maeda, S. Suzuki, J. Nucl. Radiochem. Sci. **6**, 279 (2005)
[5] M. Kotschenreuther et al., Fusion Eng. Des. **84**, 83 (2009)
[6] B. Liu et al., J. Nucl. Materials **437**, 95 (2013)
[7] S. Chen, C. Yuan, J. Wu, Y. Zou, Proc. ICONE25-66250, 1 (2017)
[8] J. Meadows et al., Nucl. Sci. and Eng. **85**, 271-279 (1983)
[9] P. Lisowski et al., Proc. Int. Conf. on Nucl. Data for Sci. and Eng. **79** (1988)
[10] W. Furman, P. Cennini, V. Ketlerov, A. Goverdovski et al. (n_TOF Collab.), AIP Conference Proceedings **769**, 1039 (2005), `http://dx.doi.org/10.1063/1.1945184`
[11] F. Tovesson and T. Hill, Phys. Rev. C **75**, 034610 (2007)
[12] C. Paradela, R. Capote et al. (n_TOF Collab.), Phys. Rev. C **82**, 034601 (2010)
[13] L. Leong et al., EPJ Web of Conferences **42**, 05003 (2013)
[14] M. Diakaki, P. Dimitriou et al. (n_TOF Collab.), Phys. Rev. C **93**, 034614 (2016)
[15] P. Salvador-Castiñeira, F.-J. Hambsch, A. Göök, M. Vidali, N.P. Hawkes, D.J. Thomas, ND2016 Conf. Nucl. Data for Sci. Tech., Brugge, Belgium (2016)

[16] D. Macina, 55th INTC Meeting, CERN (2017)
[17] Yu. Ryabov, J. Trochon, and D. Shackleton, Nucl. Phys. A **216**, 395 (1973)
[18] O.A. Shcherbakov, Phys. Part. Nucl. **21**(2), 177 (1990)
[19] E. Dermendjiev, A.A. Goverdovsky, W.I. Furman et at., Proc. Int. Conf. on Nucl. Data for Science and Technology, Juelich, 147 (1992)
[20] S.B. Borzakov, E. Dermendjiev, A.A. Goverdovsky, A.I. Kalinin, V.Yu. Konovalov, I. Ruskov, S.M. Soloviev, Yu.S. Zamyatnin, Phys. Atomic Nuclei **59**, 1125 (1995); **62**(5), 872 (1999)
[21] F.-J. Hambsch, Luca Demattè, Hugo Bax, Ivan Ruskov, J. Nucl. Sci. and Tech. **2**, 307 (2002)
[22] I. Ruskov, Yu.N. Kopach, V. Skoy, F.-J. Hambsch, S. Oberstedt, Physics Procedia **59**, 107 (2014)
[23] O.A. Shcherbakov, A.S. Vorobyev et al., Proc. ISINN-23, Dubna, JINR, 214 (2016)
[24] S. Oberstedt, R. Billnert, F.-J. Hambsch, M. Lebois, A. Oberstedt, J.N. Wilson, European Physical Journal A, EDP Sciences, 51 (2015)

EPJ Web of Conferences **169**, 00022 (2018) https://doi.org/10.1051/epjconf/201816900022
Theory-4

The GEF model: Assessment of fission-fragment properties over an extended region

Christelle Schmitt[1,⋆], *Karl-Heinz* Schmidt[2,⋆⋆], and *Beatriz* Jurado[2,⋆⋆⋆]

[1] *IPHC, CNRS/IN2P3, 23 Rue du Loess, BP. 28, 67037 Strasbourg Cedex 2, France*
[2] *CENBG, CNRS/IN2P3, Chemin du Solarium, BP. 120, 33175 Gradignan, France*

Abstract. Calculations with the fission model GEF are performed over an extended region in fissioning-system mass A, charge Z, excitation energy E^* and angular momentum L. The present contribution focuses on fission-fragment mass and isotopic distributions, benefiting from recent innovative measurements. The influence of the system (A, Z, E^*, L) properties and of multi-chance fission is studied. The impact of accurate experimental knowledge about these properties and instrumental resolution is discussed. The need of as selective and accurate as possible experimental data to improve the model is demonstrated.

1 Introduction

In recent years, the need for codes characterized by high-predictive capabilities in anticipating consistently and simultaneously various fission observables has become more and more urgent for nuclear application purposes, but also for large-scale simulations in nuclear astrophysics as well as for guiding the development of purely microscopic theories. The General description of Fission observables (GEF) code [1] is a semi-empirical model responding to this demand. GEF is able to handle spontaneous and, more generally, induced fission of a compound nucleus formed in a nuclear reaction. The code gives access to a nearly-exhaustive set of fission observables, including fission probabilities, fragment properties, prompt and delayed neutrons and γ-rays. The present contribution focuses on fission-fragment mass and charge distributions. Recent innovative measurements are used to evaluate the accuracy of the GEF predictions for independently varied fissioning-system size, excitation energy and angular momentum, over a wide range beyond the region used in the model adjustement.

2 The GEF model

For fundamental research, assisting and interpreting measurements, and in various applications, fission models are essential. Models of different levels of sophistication have been developed over the years. Semi-classical models based on transport theory, as well as specific fully quantum mechanical frameworks, are currently encountering promising quantitative results (see e.g. Refs. [2, 3] and

⋆e-mail: christelle.schmitt@iphc.cnrs.fr
⋆⋆e-mail: schmidt-erzhausen@t-online.de
⋆⋆⋆e-mail: jurado@cenbg.in2p3.fr

Ref. [4], respectively). Though, there is no unified model available yet and which can be used for any fissioning system. In addition, the corresponding codes do not implement a full description of the fission process, due to the many exit channels. Definitively, computing time is an issue, and still prevents use in large-scale calculations. Empirical and phenomenological models - again more or less sophisticated, were developed in parallel.

GEF is a semi-empirical model that exploits several general laws of mathematics and physics, combined with empirical information. In particular, GEF is based on the characteristics of the potential-energy landscape, some principles of quantum mechanics, nuclear dynamics, and statistical mechanics. Although GEF does not describe quantities from the microscopic level, it is *not* a mathematical fit of experimental observations. Also, it preserves the link between different fission observables. This theoretical framework defines the model in a *qualitative* way. The parameters entering into the formalism (around 100) are connected to the underlying physics; they specify the model in a *quantitative* way. These parameters were adjusted once with benchmark experimental data, and are taken the same for all systems over a large range of excitation energy. That has to be compared with the popular systematic of Ref. [5] which involve a similar number of parameters but for one fissioning system, only.

In praxis, GEF is a Monte-Carlo code which, starting from a specified, either entrance-channel reaction, or initial compound nucleus, computes the sequential decay of the system, including: possibility of pre-equilibrium emission, pre-scission evaporation (i.e., multi-chance fission)[1], (A, Z) production of the primary (excited) fission fragments, post-scission evaporation and γ-ray emission, (A, Z) population of the final (secondary) products and their radioactive decay wherever suited by delayed neutrons and/or γ-rays. All along the calculation, the correlations between the different quantities are preserved. No parameter is to be adjusted. As mentioned above, these were fixed once, and are system-independent. The user is acting at two places: at the beginning, by specifying the reaction or initial compound nucleus, and at the end, by analyzing the output results. The latter can be provided upon request as an event-by-event list mode data file.

2.1 Theoretical ideas and hypotheses

A detailed presentation of the modeling of fission in GEF is reviewed in Ref. [1]. Those features which are crucial for the observables (primary and secondary fragment mass and charge) which this contribution focuses on are discussed briefly below.

2.1.1 Fission barriers

A first important aspect concerns the fission probability, which determines the initial properties of the nuclei that finally go to fission. In the calculation of this probability, the fission barrier plays a central role. In GEF, fission barriers are calculated within the idea of the macro-microscopic approach, and using of the topographic theorem [6]: The barrier is given by the sum of the macroscopic barrier (taken from the extended Thomas-Fermi prescription) and the additional binding energy by the empirical ground-state shell correction, plus enhanced pairing correlations at saddle. That implies that shell corrections at saddle are neglected, as suggested by the topographic theorem. The approach avoids the uncertainties of theoretical shell-correction energies and allows to discriminate different macroscopic models [7].

[1] In GEF, only pre-fission evaporation leads to multi-chance fission. Evaporation between saddle and scission occurs, when the nucleus is already "condemned" to fission.

2.1.2 Fission channels

The second aspect of main importance in this work is the (primary) fragment (A, Z) partition at scission. To model this, GEF exploits the basic ideas of the early manifestation of fragment shells [8], the concept of quantum oscillators for the fission modes, and dynamical freezing [11]. In this frame, the fission channels are related to the statistical population of quantum oscillators in the mass-asymmetry degree of freedom that form the fission valleys. The quantum oscillator of each channel is characterized by three parameters (position, depth, and curvature) that are traced back to the macroscopic potential and to shells in the proton and neutron subsystems of both fragments, which are assumed to be effective already little beyond the outer saddle [8]. These shells are assumed to be the same for all fissioning systems. It is the superposition of different shells and the interaction with the macroscopic potential that create the mass distributions which differ for different systems [9]. Note that these shells also determine the shape (mainly quadrupole deformation) of the nascent fragments at scission. According to Strutinsky-type calculations, the fragment shapes are found to be a linearly increasing function of proton, respectively neutron, number in regions between closed spherical shells [10]. Also the charge-polarization (deviation of the N/Z degree of freedom at scission - mean value and fluctuations - from the "UCD" value of the fissioning nucleus) is treated by the corresponding quantum oscillator [11].

Adjustement of the predictions by the above formalism to benchmark experimental mass and charge distributions over a wide region of the nuclear chart showed that four fission channels are necessary: the symmetric SL channel, and three asymmetric channels ("standard" S1 and S2, and very asymmetric SA). This empirical adjustement procedure has fixed the value of all parameters of the model.

2.1.3 Energy sorting

The third aspect which is important for this contribution concerns fission energetics, and in particular energy sharing between the fragments at scission. It determines the decay of the primary, and population of the secondary, fragments. By the influence of pairing correlations, the nuclear temperature below the critical pairing energy is assumed to be constant [12]. Therefore, the di-nuclear system between saddle and scission consists of two coupled microscopic thermostates [13]. This leads to a sorting process of the available intrinsic energy and of unpaired nucleons before scission [14–16].

3 Results

Within the aforementioned framework, the GEF code succeeded to describe a large spectrum of observables [1] over a wide range of nuclides, excitation energy and angular momentum. The simplicity and efficiency in handling and running GEF makes this code used by the community for various purposes including the treatment of experimental data, e.g. Ref. [17], guidance for theories, e.g. Ref. [18], interpretation of data, e.g. Refs. [19, 20]. Implementation in large-scale calculations at Radioactive Beam Facilities and in astrophysics (e.g. Refs. [21, 22]) has been done. Finally, the GEF code can be used for examining the consistency of experimental results and for evaluation [1]. For a more comprehensive validation, and possible improvement, of the GEF code, new and high-quality experimental data are mandatory [23]. A step in this direction is proposed in this contribution benefiting from recent innovative measurements.

 https://doi.org/10.1051/epjconf/201816900022

3.1 Dependence of fragment mass partition on the fissioning system properties

Using multi-nucleon transfer-induced-fission from $^{18}O+^{232}Th$ collisions, Leguillon et al. [24] measured in a single experiment the fragment pre-neutron mass distribution for about 15 fissioning isotopes between ^{231}Th and ^{238}U, over a wide excitation energy range, from threshold to about 60 MeV. The data set of Ref. [24] is particularly relevant for the present concern. First, it permits to probe the ability of GEF to describe the evolution with fissioning N and Z in a region not much explored yet, and, importantly, well beyond the region of the model adjustement. Second, the E^* domain covered in Ref. [24] is a critical one: *i)* multi-chance fission progressively sets in, and *ii)* shell effects progressively vanish. Finally, it shall be emphasized that all systems are produced, measured, and analyzed in the same way, what ensures consistency between the results obtained for the different fissioning nuclei.

The comparison of experiment and GEF calculation is displayed in Fig. 1. Note that, in experiment, due to limited resolution and statistics, the initial excitation energy is distributed over domains which are 10 MeV-wide about the value given in the figure. The calculation was done for a single E^* (quoted in the figure), which was taken as the central value of the experimental window. The survey of the chart of Fig. 1 shows that GEF describes rather well the evolution of the distribution, with initial system N and Z, and E^*. A detailed inspection, though, shows some deviation. The latter is attributed to several reasons. Transfer reactions can induce a non-negligible amount of angular momentum L in the fissioning compound nucleus, what affects the fragment mass distribution; in the calculation of Fig. 1, L was set to zero. Another aspect is the 10 MeV-wide E^* interval inherent to the experiment, and which influences the mass distribution also; the calculation assumed a well-defined E^* situated mid-way in this interval. Another effect (purely experimental) is the limited mass resolution (σ = 6 units quoted in [24]) which is not taken into account in the calculated curves of Fig. 1. Last, but not least, model deficiencies cannot be excluded.

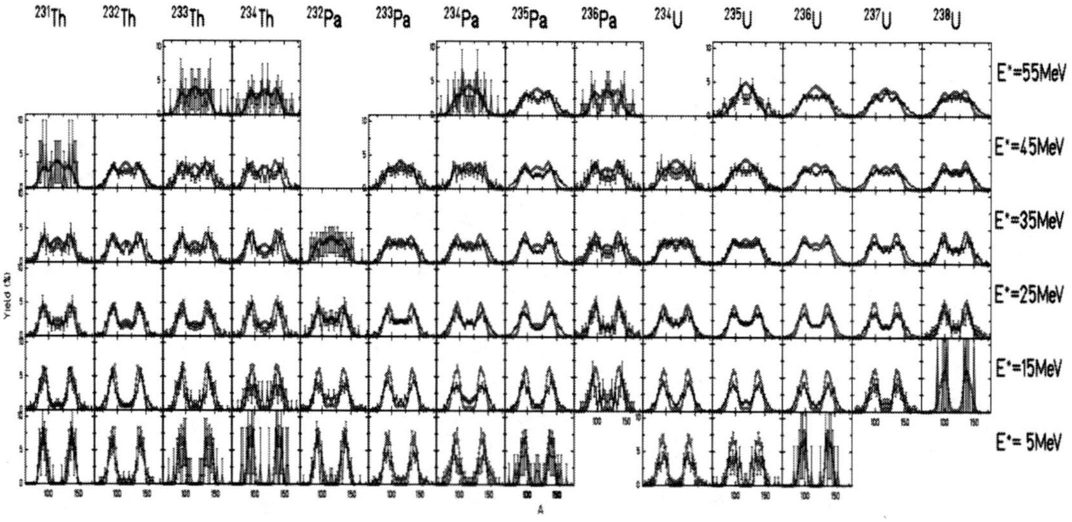

Figure 1. Fission-fragment pre-neutron mass distribution for various systems as measured in Ref. [24] (black) and compared with GEF calculations (red). Initial compound nucleus isotopes are ordered by columns, and initial mean E^* values are ordered by rows.

To investigate the magnitude of the aforementioned reasons of deviation between the measurement and GEF, the calculations were repeated at various L and E^* values. Also, they were folded with the experimental resolution. A sample of results is gathered in Fig. 2. The GEF model suggests that the influence of angular momentum is present, but not that strong for the present systems. The width of the E^* window can have a large impact. Experimental mass resolution has a non-negligible influence. Finally, the crucial need to account for multi-chance fission is also illustrated in Fig. 2, even at the "medium" E^* covered here. In conclusion, neglecting these four aspects can yield an un-proper comparison of experiment and theory, and may even lead to data mis-interpretation.

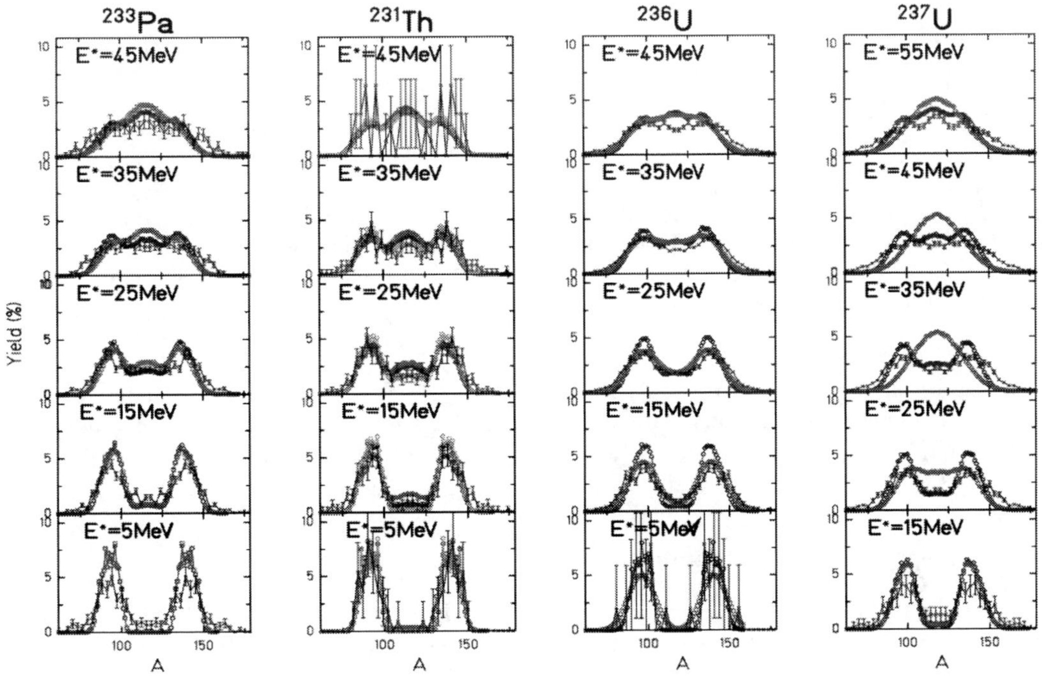

Figure 2. Fission-fragment pre-neutron mass distribution for selected systems from Ref. [24] (black) and compared with various GEF calculations. *First column:* calculations with L=0 (blue) and L=20 (red) as a typical maximum value in transfer; *second column:* calculations with E^* taken as the center ($<E^*>$) (blue) of the experimental interval, and as the lower ($<E^*>-5$ MeV) (red) and upper ($<E^*>+5$ MeV) (green) limit; *third column:* calculations assuming unique mass resolution (blue) and folded with the experimental resolution of σ = 6 units (red); *fourth column:* calculations without (red) and with (blue) multi-chance fission taken into account.

The potential of GEF for unfolding experimental observations from the aforementioned effects is demonstrated in Fig. 3. The mass distribution measured at Tokai [24] for one example is compared to a measurement performed in Jyvaskyla [25] for the same initial system. There, fission was induced by bombarding a heavy target with a proton beam. In such a configuration, the angular momentum imparted to the system is small, and the excitation energy is well defined. In addition, the set up of Ref. [25] permitted to achieve a mass resolution of σ around 1.5 unit. Also, the statistics collected was high. Unfortunately, according to the method, only one system could be measured. The deviation between the Tokai and Jyvaskyla measurements *and* the good agreement between GEF and the Jyvaskyla data set corroborate the non-negligible intricate interplay of L, E^* and resolution effects. Furthermore, it demonstrates that the GEF code is a relevant tool to trace back the different influences, and possibly unfold each of them in the measurement. In the present specific case, Fig. 2

would suggest that most of the discrepancy observed between the two experiments in Fig. 3 may be ascribed to resolution effects.

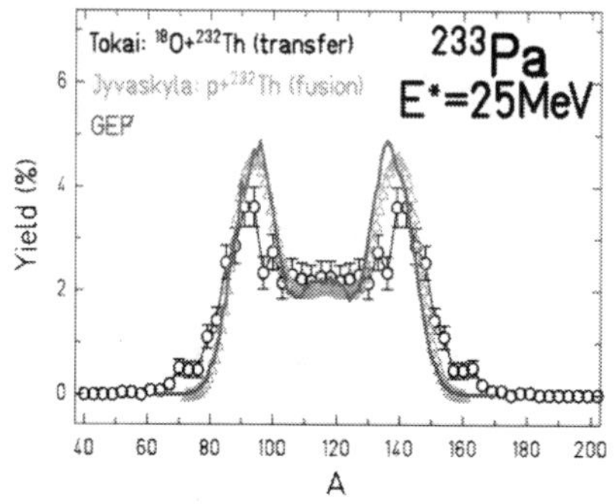

Figure 3. Fission-fragment pre-neutron mass distribution for ^{233}Pa at $E^* = 25$ MeV as measured at Tokai (black) and at Jyvaskyla (green) in transfer- and proton-induced fission, respectively. The GEF calculation (red) with $L =$ 0 and $E^* = 25$ MeV overlaps reasonably well with the distribution of Jyvaskyla. This experiment was performed under better-defined conditions, as inherent to the method. See also text.

The chart measured at Tokai, Fig. 1, allows to study the evolution with mass, charge, excitation energy, and possibly angular momentum, of the fissioning system. Yet, Figs. 1 to 3 also show that to probe the influence of E^* and L in detail the data set of Ref. [24] is still lacking experimental accuracy (in terms of fissioning system properties and fragment mass) and statistics.

3.2 Fragment isotopic distribution

In parallel to the above study scanning the (N, Z, E^*, L) fissioning hyperspace, the validation of GEF can still be performed on another front, thanks to the emergence of recent experiments at VAMOS/GANIL and SOFIA/GSI. These gave access for the first time to the complete nuclide production in low excitation-energy fission for several actinides. Mass and charge identification was achieved with unique precision for all the light and heavy fragments. Such *precise and complete* isotopic distributions certainly constitute a very valuable constraint for any model of fission. Yet, most models - among very sophisticated ones, restrict to either mass or charge partitioning. That is, a fixed ratio (usually the one of the compound system) is assumed between N and Z. In GEF, fragment N and Z are two independent degrees of freedom, and whose population is governed by the physics outlined in Sect. 2.

The approach at VAMOS/GANIL has some similarity with the one used at Tokai. Namely, fission is induced by transfer reactions around the Coulomb barrier with ^{238}U (6.1 MeV/nucleon) +^{12}C collisions, and the initial compound-nucleus mass, charge and excitation energy is deduced event-by-event from the measurement. As compared to Ref. [24], inverse kinematics is employed, and use of the VAMOS magnetic spectrometer permits to intercept and isotopically identify one of the fragments [26]. Unfortunately, statistical significance is available for a few systems only, between ^{238}U and ^{250}Cf. According to the detection method, post-neutron masses are measured at GANIL. In Ref. [27, 28] good agreement was observed between the experimental and GEF-predicted mass and charge integral distributions. A deeper survey is here proposed [29] by comparing the experimental and calculated isotopic distributions for the initial ^{240}Pu compound system excited at about 10 MeV (FWHM $\approx$ 7 MeV), see Fig. 4. The description by GEF is observed to be reasonable, over the entire

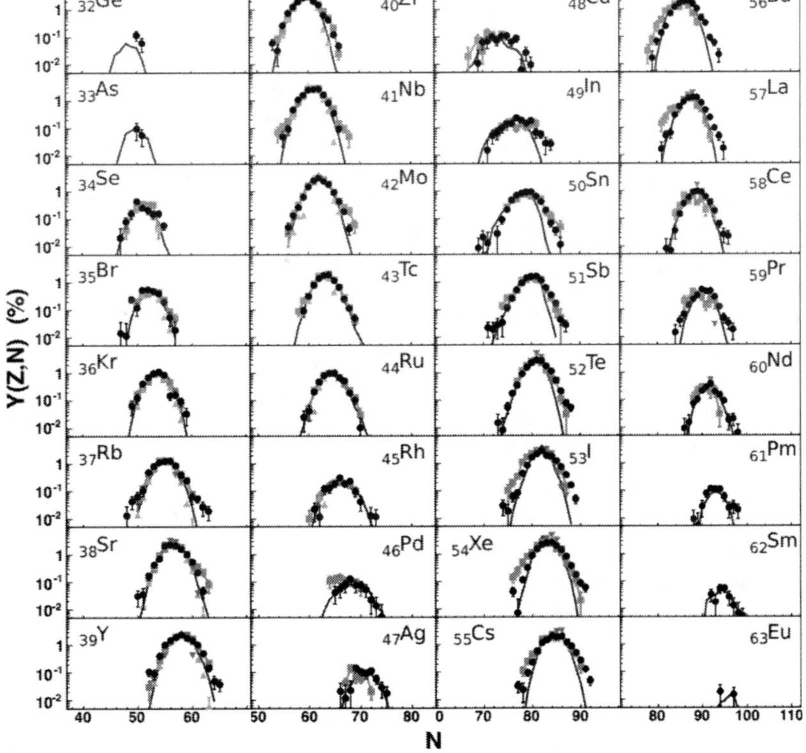

Figure 4. Fragment post-neutron N distribution for elements between Ge and Eu as measured at VAMOS (black and red) in low-energy fission of ^{240}Pu. Experiment is compared with GEF (blue). Figure is adapted from Ref. [29].

range from light to heavy fragments. Some deviation is visible at the edges of the distribution for a few elements. This is attributed so far to still not perfect Z identification in the experiment [30].

The approach at SOFIA/GSI is based on fission induced by electromagnetic excitation following heavy-ion collisions at relativistic energy [31]. To make available, in a single experiment, a wide variety of fissioning isotopes, a two-step scheme is employed. The reaction ^{238}U (1000 MeV/nucleon) +^{9}Be is used to produce by fragmentation a large set of stable and r adioactive products. These are used as secondary relativistic beams which bombard a heavy (^{238}U or ^{208}Pb) secondary target, what induces fission of the secondary beam by electromagnetic excitation with an average $E^{*} \approx 14$ MeV (extending from threshold up to 35 MeV). As compared to the GANIL run, inverse kinematics is employed at SOFIA also, but the set up gives access to the post-scission mass and charge, with unique precision, for *both* fragments in coincidence. However, the initial excitation energy is not measured event-by-event, yielding the issue of an intricate convolution of a complex E^{*} distribution [20]. Few systems were already published, among which the results for low-energy fission of ^{235,238}U. The experimental and calculated isotopic distributions are reported in Fig. 5 for the case of ^{238}U [20]. Overall the model describes the measurement well. It is noteworthy to emphasize that the isotopic distribution is seen to deviate from a Gaussian shape for many elements, and that the calculation is capable of explaining this observation [20]. Though, discrepancy between experiment and theory

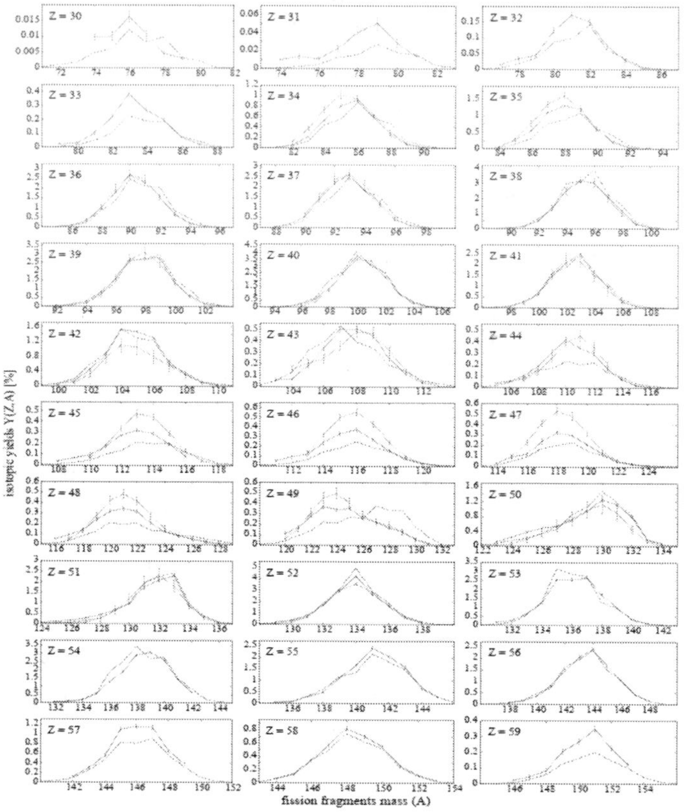

Figure 5. Fragment post-neutron mass distribution for elements between Ga and Pr as measured at SOFIA (red) and earlier at GSI (blue) in low-energy fission of ^{238}U. Experiment is compared with GEF (black). Figure is adapted from Ref. [20].

exists in the symmetric fission region, and whose origin is under investigation. Note that this is the region where the yields are smallest, and where accurate account of E^* may be critical. A slight underestimation by GEF of the yields around symmetry with increasing excitation energy was already seen for selected cases in Refs. [1, 32].

According to Figs. 4 and 5, and within the experimental control of E^*, GEF describes consistently the fragment (N, Z) population. Gaining further insight calls for more exclusive experimental information (in terms of e.g. E^* and L).

3.3 Neutron multiplicity

According to the direct identification of both fragments in mass, the SOFIA experiment gives also access to the total neutron multiplicity ν_{tot} for each (N, Z) fission partition. This multiplicity is reported in Fig. 6 for electromagnetic-induced fission of ^{235}U. The calculation (left) is in agreement with the measurement (right). That could already be inferred from Fig. 5 where the experimental and calculated peak localizations coincide well. Comparing the measured and predicted ν_{tot} quantity can be used to study fission energetics. Though, to get deeper insight, and namely investigate the debated issue of energy sharing at scission, requires to consider the neutron multiplicity $\nu_{L,H}$ per fragment. That is not available from experiment on the (N, Z) plane yet.

The clear dependence of ν_{tot} on N *and* Z seen in Fig. 6 implies that accurate account of post-scission evaporation in data analysis aiming at recovering primary fragment masses from measured secondary ones requires to perform a correction that depends on both fragment N and Z, at the risk of data mis-interpretation. So far, due to lack of experimental information in the two-dimensional plane, neutron corrections are usually done as function of fragment mass, only (see e.g. Ref. [17] and therein).

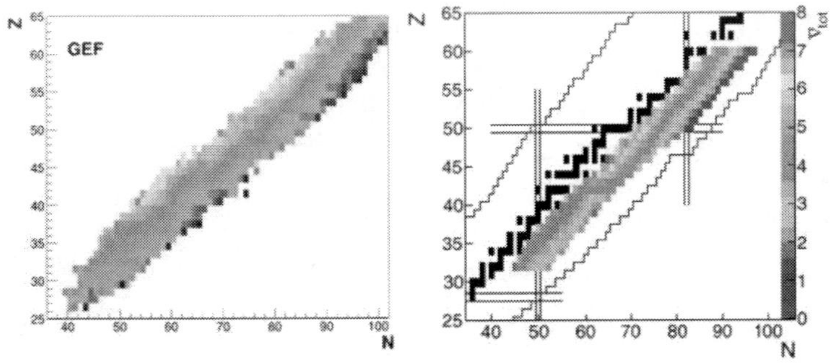

Figure 6. *Right*: total neutron multiplicity ν_{tot} as function of fragment post-neutron (N, Z) partition as deduced at SOFIA for low-energy fission of ^{235}U. Adapted from Ref. [31]. *Left*: identical to the right but for the GEF calculation.

3.4 Importance of correlations

As mentioned in Sect. 2, GEF provides the link between the different fission quantities. That is, it preserves the correlations between these quantities. Calculated correlations can be fruitfully used in interpreting experimental observations, since the inclusive measurements required to extract correlations are rare. As a demonstration of the potential of GEF in this respect, let us consider again the isotopic distributions measured at SOFIA and shown in Fig. 5. The experimental (red) distribution for Sn ($Z = 50$) has a main peak at $A \approx 130$ and exhibits a pronounced shoulder centered at $A \approx 126$. That was interpreted in Ref. [20] as being due to the contribution of both the SL and S1 fssion modes to the production of this element (see also Fig. 22 of Ref. [20]). A left tail is seen in the GEF calculation, too, although less pronounced than in experiment. In order to corroborate the experimental interpretation, knowledge on, either total kinetic energy (TKE), or the aforementioned $\nu_{L,H}$, is best suited. This, because the SL and S1 components are associated to higher and lower intrinsic excitation energy at scission, respectively, i.e., lower/larger (higher/smaller) values for $TKE/\nu_{L,H}$. Since these observables are not available yet from SOFIA, support from model calculations is welcome. It is found (not shown) that GEF expects a contribution from SL and S1 to the Sn isotopic distribution, with distinct TKE and $\nu_{L,H}$.

4 Conclusions

Comparison of recent innovative measurements with calculations by GEF shows the capability of the code to describe reasonably the experimental data over a wide range of fissioning system mass, charge, excitation energy and angular momentum. In addition, the GEF results may help in the discrimination between excitation energy, angular momentum and experimental resolution effects. This study highlights the further need of high-quality experimental information for a better understanding of the fission process.

References

[1] K.-H. Schmidt, B. Jurado, Ch. Amouroux, and C. Schmitt, Nucl. Data Sheets **131**, 107 (2016)
[2] J. Randrup and P. Möller, Phys. Rev. C **88**, 064606 (2013)
[3] M.D. Usang et al., Phys. Rev. C **94**, 044602 (2016)
[4] D. Regnier, N. Dubray, N. Schunck, and M. Verriere, Phys. Rev. C **93**, 054611 (2016)
[5] A.C. Wahl, Report LA-13928 (Los Alamos National Laboratory, 2002) (unpublished)
[6] W.D. Myers and W.J. Swiatecki, Nucl. Phys. A **601**, 14 (1996)
[7] A. Kelić and K.-H. Schmidt, Phys. Lett. B **634**, 362 (2006)
[8] U. Mosel and H.W. Schmitt, Nucl. Phys. A **165**, 73 (1971)
[9] K.-H. Schmidt, A. Kelić, and M.V. Ricciardi, Europh. Lett. **83**, 32001 (2008)
[10] B.D. Wilkins, E.P. Steinberg, and R.R. Chasman, Phys. Rev. C **14**, 1832 (1976)
[11] H. Nifenecker, J. Physique Lett. **41**, 47 (1980)
[12] K.-H. Schmidt and B. Jurado, Phys. Rev. C **86**, 044322 (2012)
[13] K.-H. Schmidt and B. Jurado, Phys. Rev. Lett. **104**, 212501 (2010)
[14] K.-H. Schmidt and B. Jurado, Phys. Rev. C **83**, 014607 (2011)
[15] K.-H. Schmidt and B. Jurado, Phys. Rev. C **83**, 061601 (2011)
[16] B. Jurado and K.-H. Schmidt, J. Phys. G: Nucl. Part. Phys. **42**, 055101 (2015)
[17] D. Duke et al., Phys. Rev. C **94**, 054604 (2016)
[18] W. Younes, Eur. Phys. J. A **53**, 109 (2017)
[19] J. Khuyagbaatar, private communication
[20] E. Pellereau et al., Phys. Rev. C **95**, 054603 (2017)
[21] A. Al-Adili et al., Eur. Phys. J. A **51**, 59 (2015)
[22] S. Goriely et al., Eur. Phys. J. A **51**, 22 (2015)
[23] K.-H. Schmidt, B. Jurado, C. Schmitt, Contribution to the NEA Nuclear Data Week - JEFF Meetings, OECD Nuclear Energy Agency, Paris, France (2017) (unpublished)
[24] R. Leguillon et al., Phys. Lett. B **761**, 125 (2016)
[25] E.M. Kozulin et al., FUSION06 Conference, Venezia, Italy (2006) (unpublished)
[26] M. Caamano et al., Phys. Rev. C **88**, 024605 (2013)
[27] M. Caamano et al., Phys. Rev. C **92**, 034606 (2015)
[28] D. Ramos et al., Acta Pol. B **46**, 329 (2015)
[29] D. Ramos, PhD, University of Santiago de Compostela (2016) (unpublished)
[30] F. Farget, private communication
[31] J.-F. Martin et al., Eur. Phys. J. A **51**, 174 (2015)
[32] J.N. Wilson et al., Phys. Rev. Lett. **118**, 222501 (2017)

EPJ Web of Conferences **169**, 00023 (2018) https://doi.org/10.1051/epjconf/201816900023
Theory-4

Staggering of angular momentum distribution in fission

Pierre Tamagno* and *Olivier* Litaize

CEA, DEN, DER, SPRC, Cadarache, 13108 Saint-Paul-lez-Durance, France

Abstract. We review here the role of angular momentum distributions in the fission process. To do so the algorithm implemented in the FIFRELIN code [1] is detailed with special emphasis on the place of fission fragment angular momenta. The usual Rayleigh distribution used for angular momentum distribution is presented and the related model derivation is recalled. Arguments are given to justify why this distribution should not hold for low excitation energy of the fission fragments. An alternative *ad hoc* expression taking into account low-lying collectiveness is presented as has been implemented in the FIFRELIN code. Yet on observables currently provided by the code, no dramatic impact has been found. To quantify the magnitude of the impact of the low-lying staggering in the angular momentum distribution, a textbook case is considered for the decay of the ^{144}Ba nucleus with low excitation energy.

1 Introduction

The fission process is probably the most complex – and the less accurately understood – phenomenon of nuclear physics. This is mostly due to the fact that many degrees of freedom are involved in the process and that it is extremely difficult to experimentally separate contributions. For instance, kinetic energy of fission fragments are at-best measured with respect to mass number A and fine isotope decompositions are rare. Various isotopes are generated in fission with various "initial" states, and additionally for each of them the decay process is also not perfectly known. As only average experimental quantities are available, it is important to have models that are consistent and as accurate as possible so that experimental data can be properly interpreted. As mentioned above, the decay process of a fission fragment is not perfectly known. Indeed this decay process depends on the nuclear property of the isotope, such as the level scheme and transition branching ratios (or decay widths). These quantities can be obtained experimentally with spectroscopy techniques but the number of accurately measurable levels is limited. The limitations are not only due to statistics or to experimental resolution but can also come from the unstable nature of the isotope (β-decay). Therefore to describe the nuclear structure above some given level, models are used to fulfill gaps. In the present work we highlight the role of a specific degree of freedom in the fission phenomenon: the angular momenta of levels of fission fragments. To illustrate where these distributions can be used in fission models, we present first the model implemented in the FIFRELIN code [1] to describe fission observables. Then we present why we expect that current angular distribution models should be incorrect in the low excitation energy region. We detail then an *ad hoc* modification that we implemented in FIFRELIN

*e-mail: pierre.tamagno@cea.fr

and that should be more accurate. Finally we present a simple case of a decay of a fission fragment with a low excitation energy so that the impact of our modification can be quantified.

2 Fission modeling in FIFRELIN

2.1 Generation of initial fragments

Particle emission in the FIFRELIN code is based on the Bečvář algorithm [2] for the simulation of γ cascades with extension to the emission of prompt neutrons. The decay algorithm and fission fragment (FF) generation are stochastic. Fission events are sampled many times and decay properties of fragments are recorded so that average properties can be provided to compare with experimental data. More detailed quantities can also be provided as well as correlations between observables. The first step in a simulation consists in sampling FF and FF states. The mass fragmentation is sampled in an experimental yield $Y(A)$ such as shown in Fig. 1 for the ^{252}Cf(sf) reaction. The charge fragmentation is sampled according to a gaussian distribution P_Z (see Fig. 1) with a mean value Z_p that is given by the unchanged-charge-density value shifted by $\pm\Delta Z$. For instance, for the light fragment, the most-probable charge is given by $Z_{L,p} = (Z_{CN}/A_{CN})A_L + \Delta Z(A_L)$, where the polarization ΔZ is also obtained from experimental data as illustrated in Fig. 1. Once both fragments are defined, the kinetic energy of the light fragment is sampled from experimental kinetic energy data (mean values $\langle KE\rangle(A)$ and standard deviations $\sigma_{KE}(A)$) and the kinetic energy of the heavy fragment is deduced from momentum conservation.

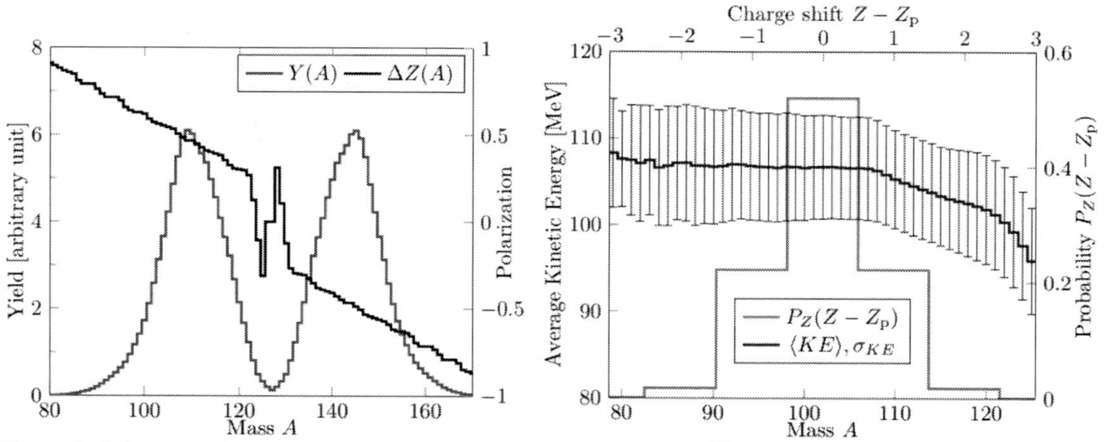

Figure 1. Scission fission yield $Y(A)$ and charge polarization ΔZ for the ^{252}Cf(sf) reaction (left). Related fragments average kinetic energy $\langle KE\rangle(A)$ and standard deviation $\sigma_{KE}(A)$ (right).

From the defined fragments properties, the total excitation energy TXE can be calculated from the total released energy Q by $TXE = Q - KE(A_H) - KE(A_L)$. The total intrinsic excitation energy is given by $E^* = TXE - E_L^{rot} - E_H^{rot} + B_n + E_n^{kin}$, where E_L^{rot} and E_H^{rot} are respectively the rotational energies of the light and heavy fragments. These rotational energies are obtained by: $E^{rot} = J(J+1)/2kI$ where $I = \frac{2}{5}MR^2(1 + 0.31\beta_2)$ is the rigid-body moment of inertia of the fragment and k is an adjustable parameter. To determine the rotational energy of the fragment, angular momentum J is required. It is sampled from a Rayleigh distribution:

$$P(J) = \frac{J+1/2}{\sigma^2} \exp\left[-\frac{(J+1/2)^2}{2\sigma^2}\right], \tag{1}$$

where σ^2 is the so-called spin cutoff parameter. In practice there is one adjustable spin cutoff parameter for each group of fragments ($\sigma_{\rm H}$ and $\sigma_{\rm L}$). They are only used in the initial fragmentation sampling. Finally, the initial parity of both fragments is uniformly sampled and the total intrinsic excitation E^* is dispatched between fragments according to a temperature ratio law [1] involving two additional parameters $R_{\rm T}^{\rm min}$ and $R_{\rm T}^{\rm max}$. We emphasize here that only 5 parameters are adjusted in practice to reproduce average quantities such as prompt neutron multiplicities for the heavy and light fragments ($\bar{\nu}_{\rm H}$ and $\bar{\nu}_{\rm L}$ respectively).

2.2 Fragment decay

At the end of the fragmentation sampling, one has two fragments with well-defined initial states. For instance, for the heavy fragment ($A_{\rm H}, Z_{\rm H}, E_{\rm H}^*, J_{\rm H}, \pi_{\rm H}$). The decay cascade is simulated in a similar manner for both fragments, therefore we only consider an arbitrary fragment state (A, Z, E^*, J, π). The initial state of the fragment is usually at higher energy than the highest known level. Therefore its decay from the initial state down to a long-lived state requires modeling of unknown levels and transitions. This energy region is illustrated in Fig. 2. Below the highest known level ($E_{\rm cut}$) all transitions are experimentally known [3]. Above $E_{\rm cut}$ some levels can be known but additional levels must be sampled to complete the level scheme. Levels are added until a maximum number of levels (or a maximum level density) criteria is matched, corresponding to an energy $E_{\rm bin}$. Above $E_{\rm bin}$ the decay is simulated with equal-width energy bins. The probability of emitting a particle p with properties α from state i to state f is given by the Hauser-Feshbach statistical model:

$$P_p(i \to f, \alpha) = \frac{\Gamma_p(i \to f, \alpha)}{\sum\limits_{f', p', \alpha'} \Gamma_{p'}(i \to f', \alpha')} . \tag{2}$$

In Eq. (2), Γ are the level decay widths. For gamma emission with multipolarity XL, the level decay width is given by

$$\Gamma_\gamma(i \to f, XL) = (E_i - E_f)^{2L+1} \frac{S_\gamma^{XL}(E_i - E_f)}{\rho(E_i, J_i, \pi_i)} . \tag{3}$$

When the excitation energy is higher than the neutron separation energy $S_{\rm n}$, the fragment can emit a neutron. Therefore additional decay channels must be considered. The related decay widths are obtained from optical model calculation with inverse kinematics, yielding

$$\Gamma_{\rm n}(i \to f, \ell j) = \frac{T_{\ell j}(E_{\rm n}^{\rm inv})}{\rho(E_i, J_i, \pi_i)} . \tag{4}$$

Examples of gamma strength functions S_γ^{XL} and neutron transmission coefficients $T_{\ell j}$ can be seen in Fig. 3. In conclusion one can notice from Eqs. (3) and (4) and from the unknown levels sampling that the spin distribution $P(J)$ (that is hidden in ρ in Eqs. (3) and (4)) plays an important role in different places in the decay simulation.

3 Angular momentum distribution

3.1 Rayleigh distribution

Before getting interested in liabilities of Eq. (1), we briefly recall here how it can be obtained [4]. Considering a large number of nucleons (Fermi gas) having their spin projection m_i randomly distributed. The total spin projection $M = \sum_i m_i$ follows a Gaussian distribution (central limit theorem)

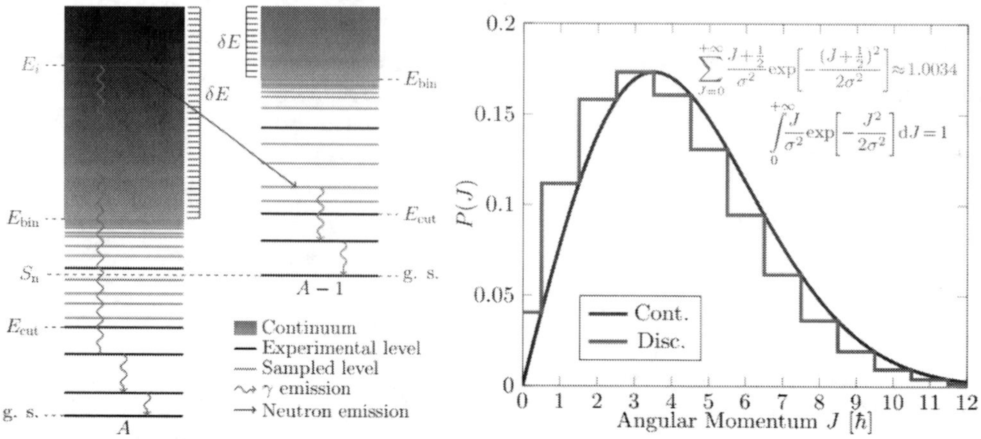

Figure 2. Schematic decay cascade simulated in FIFRELIN (left). Angular momentum distributions obtained with both types shown in Eq. (5) with a spin cutoff $\sigma = 3.5\hbar$ (right).

i.e. $P(M) \propto \exp(-M^2/2\sigma^2)$. The probability that the system has a total angular momentum J is then given by $P(J) = P(M = J) - P(M = J + 1) \approx (J + 1/2)/\sigma^2 \exp[-(J + 1/2)^2/2\sigma^2]$, which is Eq. (1). One can notice that Eq. (1) – although widely used – is not exactly a distribution. Indeed a Rayleigh distribution operates on a continuous variable, whereas Eq. (1) deals with discrete values of J. As a consequence, Eq. (1) is not normalized to unity:

$$\sum_{J=0}^{+\infty} \frac{J+1/2}{\sigma^2} \exp\left[-\frac{(J+1/2)^2}{2\sigma^2}\right] \neq 1 \quad \text{whereas} \quad \int_0^{+\infty} \frac{J}{\sigma^2} \exp\left[-\frac{J^2}{2\sigma^2}\right] \mathrm{d}J = 1\,. \tag{5}$$

One can also notice that Eq. (1) is used indifferently for even-even or for other nuclei. An illustration of both distributions of Eq. (5) is shown in Fig. 2. Beside the normalization issue, the discrete distribution should yield more low angular momenta (red curve is shifted). This is due to the $J + 1/2$ term in the discrete expression [5], otherwise the occurrence probability of $J = 0$ would be zero.

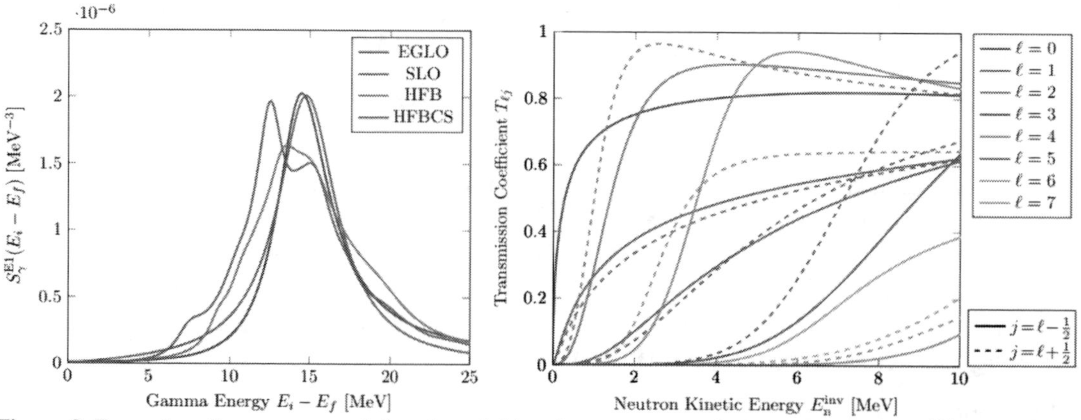

Figure 3. Examples of gamma strength functions (left) and neutron transmission coefficients for ^{144}Ba (right).

3.2 Collective footprint in level density

This normalization and discrete-continuum difference are not the only liabilities of Eq. (1). Indeed one can notice discrepancies between expected results from low-lying levels analysis and the statistical model described in Sect. 3.1. A typical low-lying level scheme for ^{154}Gd is shown in Fig. 4. One can notice the well-known collective structures related to rotational bands built on the ground-state and on β, γ or octupole vibrational states. The $K = 0$ bands have a selective effect on the related band levels. For $K = 0^+$ only even value of J are built, whereas for $K = 0^-$ only odd values of J are present in the band. Additionally, no 0^- states are present in these low-lying levels. From this example one can expect parity-driven structures in the angular momentum distribution. This collective states have a well-known impact on the total level density $\rho(E^*)$ (K_{coll} and K_{vib} enhancement factors [3]). This strong change in behavior between a Fermi gas and a collectiveness-driven system can also explain large differences than can be observed between Fermi-gas spin cutoff σ_{F} and "discrete" spin cutoff σ_{d} extracted from low-lying levels analysis [3].

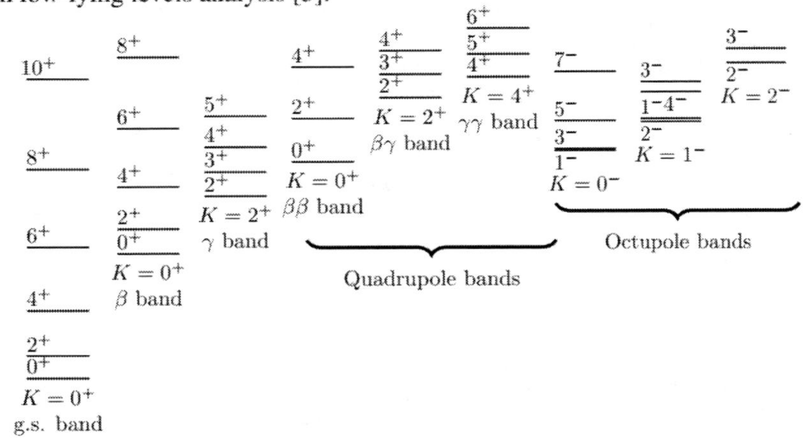

Figure 4. Low-lying collective bands in ^{154}Gd data taken from ENSDF [6].

An additional evidence of the presence of parity-driven enhancements can be obtained from Hartree-Fock-Bogoliubov calculations [7]. In Fig. 5 one can see an example of angular momentum distribution of ^{144}Ba obtained from combinatorial level density based on HFB theory. It can be easily seen that for low energies the staggering effect discussed above enhances even spins for even parity and odd spins for odd parity.

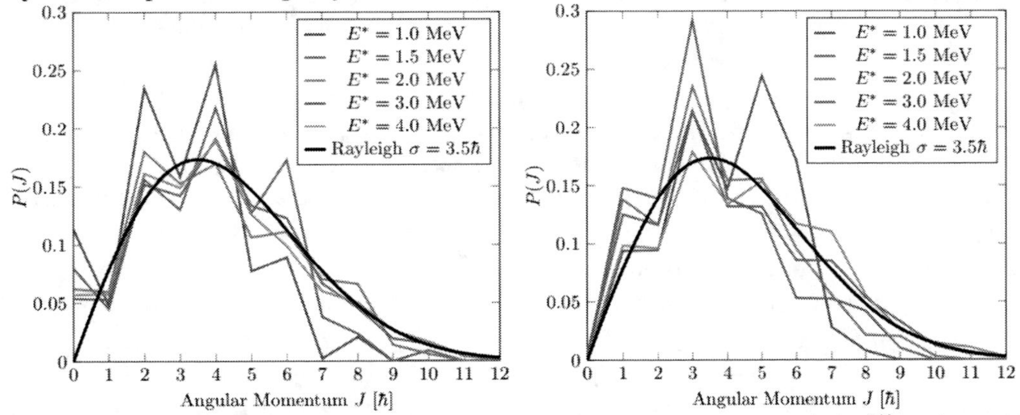

Figure 5. Angular momentum distributions obtained from HFB level density analysis for ^{144}Ba, for positive-parity levels (left) and negative-parity levels (right).

4 Results on textbook case

We investigated the impact of the staggering presented in Sect. 3.2 on fission observables. The usual Eq. (1) has thus been modified in FIFRELIN so that even-odd enhancements can be reproduced. We also had to modify the code so that parity is sampled before angular momentum so that different distributions can be used according to parity. The alternative parity-dependent angular momentum distribution used is:

$$\tilde{P}(J) = \left[\frac{a(1-a)}{2} + 1\right] P(J) + \frac{a(a+1)}{4} \left[P(J+1) + P(J-1)\right] \qquad \begin{matrix} a(\pi = +) = f \\ a(\pi = -) = -f \end{matrix} \tag{6}$$

In Eq. (6), $P(J)$ stands for the usual Rayleigh expression (Eq. (1)). Only even-even nuclei are impacted by this modification as it is expected that single-particle excitations quickly wash out the collective structures. In Eq. (6), f is a parameter controlling the strength of the enhancement. For $f = 0$ one obtains $\tilde{P}(J) = P(J)$, whereas for $f = 1$ one only obtains selected spins. In Fig. 6, Eq. (6) is illustrated for different values of f. To take into account that the staggering disappear with increasing excitation energy, we introduced the energy dependency $f = \exp(-E^*/1.8)$.

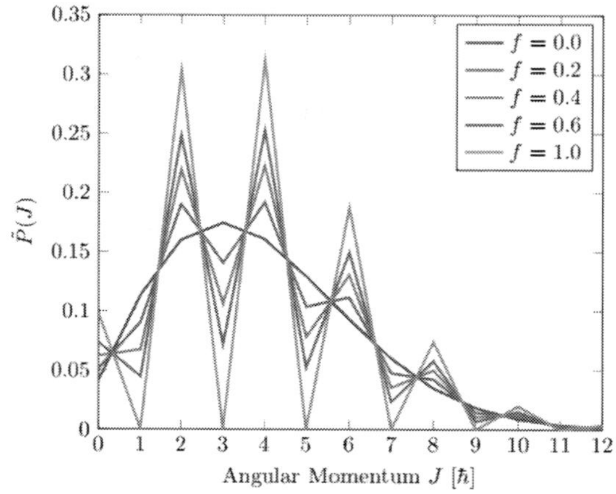

Figure 6. *Ad hoc* angular momentum distributions used to enhance even angular momenta for positive-parity states.

We expected that modifying the functional P that is used in many places in the algorithm would have a significant impact on fission observables provided by the code, for instance gamma and neutron multiplicities and spectra or even isomeric ratios. Yet no such dramatic effect has been found. Two reasons explain this limited impact:

- At high excitation energy, many levels are available as "arrival" states i.e. the level density $\rho(E_f)$ is still high. Therefore if a transition become prohibited to a given level, there will be nearby another level with the proper spin and parity. The decay will then occur with an almost identical probability because the typical energy scale for the variation in gamma strength-function or neutron transmission coefficient is much larger than the average level spacing (see Fig. 3).
- At low excitation energy, experimental transition intensities are known. As they drive the Hauser-Feshbach sampling, spin-parity of experimental levels have limited impact on the decay cascade.

As the code relies on Monte Carlo sampling, quantifying this impact would require outrageous computation time. Therefore we tried to quantify this impact on a simpler case. The decay of a ^{144}Ba isotope exited in a 6^+ state at $E^* = 1.5$ MeV (see Fig. 7) has been simulated with both models P

and $\tilde{P}$. We retained only the first 6 experimentally-known states. To simulate the decay cascade the code samples additional levels between the highest experimentally-known state and the initial energy. The energy and parity samplings are made before spin attributions therefore they are identical in both cases. It can be notice from Fig. 7 that $\tilde{P}$ favor even spins for positive parity and odd spins for negative parity. The simulated gamma intensities can be seen in Fig. 7. One can see that additional rays are observed in the modified case but with intensities about three orders of magnitude smaller than for the main transitions. This explains why differences on main events are not seen on global fission observables.

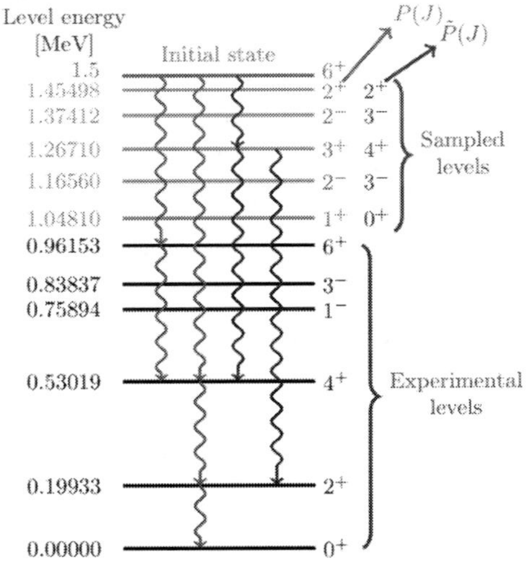

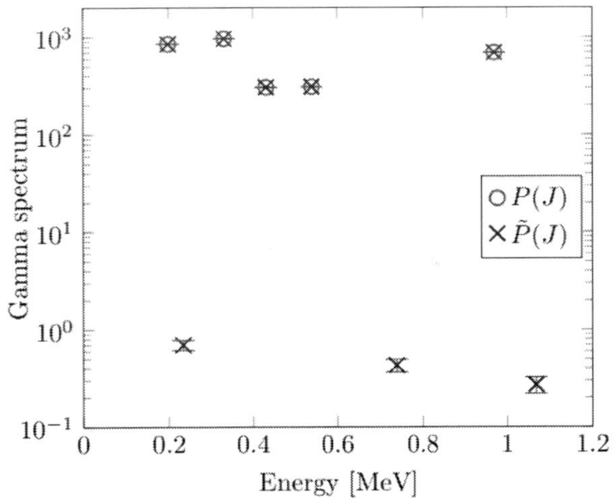

Figure 7. Level scheme obtained in the textbook decay of ^{144}Ba using both models for angular distribution (left). Corresponding calculated gamma intensities (right).

5 Conclusion

The usual Rayleigh distribution used for the description of angular momenta in nuclei is reviewed. It is shown that such a model is expected to be inaccurate at low excitation energies for even-even nuclei. An alternative expression is proposed to simulate the impact of collective states in the distribution. This modification has been shown to have limited impact on average fission observables and its impact has been quantified to be about three orders of magnitude smaller than main decay processes.

References

[1] O. Litaize, O. Serot, Phys. Rev. C **82**, 054616 (2010)
[2] F. Bečvář, Nuclear Instruments and Methods in Physics Research A **417**, 434 (1998)
[3] R. Capote, M. Herman, P. Obložinský, P.G. Young, S. Goriely, T. Belgya, A.V. Ignatyuk, A.J. Koning, S. Hilaire, V.A. Plujko et al., Nuclear Data Sheets **110**, 3107 (2009)
[4] H.A. Bethe, Phys. Rev. **50**, 332 (1936)
[5] A. Gilbert, A.G.W. Cameron, Canadian Journal of Physics **43**, 1446 (1965)
[6] M.R. Bhat, in *Nuclear Data for Science and Technology* (1992), Evaluated Nuclear Structure Data File (ENSDF)
[7] S. Hilaire, personal communication

Fission fragment yields and total kinetic energy release in neutron-induced fission of ^{235,238}U, and ^{239}Pu

F. Tovesson, D. Duke, V. Geppert-Kleinrath, B. Manning, D. Mayorov, S. Mosby, and K. Schmitt

Los Alamos National Laboratory, Los Alamos, New Mexico 77843, USA

Abstract. Different aspects of the nuclear fission process have been studied at Los Alamos Neutron Science Center (LANSCE) using various instruments and experimental techniques. Properties of the fragments emitted in fission have been investigated using Frisch-grid ionization chambers, a Time Projection Chamber (TPC), and the SPIDER instrument which employs the 2v-2E method. These instruments and experimental techniques have been used to determine fission product mass yields, the energy dependent total kinetic energy (TKE) release, and anisotropy in neutron-induced fission of U-235, U-238 and Pu-239.

1 Introduction

Fission nuclear data is important for, and impacts the performance of, several nuclear technologies. Furthermore, the fission process is highly complex and is still, more than 75 years after its discovery, not completely understood. New measurements, particularly of correlated observables, allows us to develop more sophisticated theoretical models and to eventually have truly predictive capabilities for the physics of fission. The technologies that are impacted cover those deployed in service of the nuclear energy, defense, non-proliferation, and nuclear forensics efforts. An experimental program at Los Alamos National Laboratory has been developed to measure several aspects of actinide fission, including product yields and kinetic energy release. These quantities were obtained using different instruments, each designed to capture a measurement of a specific fission observable with high-resolution or with high-statistics. Often, correlations between relevant observables were determined during the measurements. This paper will describe the experiments carried out on U-235, U-238 and Pu-239.

2 Experimental techniques

The experiments were performed at the Los Alamos Neutron Science Center (LANSCE) at Los Alamos National Laboratory. The 800-MeV proton accelerator at LANSCE is used to deliver a pulsed beam to a number of facilities, including the Weapons Neutron Research (WNR) and Lujan Center where multiple flightpaths or instrumental halls are positioned about the source of the neutron beam.

At WNR the primary proton beam hits a water-cooled tungsten target which creates fast neutrons through the spallation process. In the present experiments, we used a 10-m flightpath located at 90 degrees relative to the proton beam at WNR. This is a short flight path, which results in a high neutron flux, and the 90-degree spallation has an average neutron energy of about 2 MeV, which is appropriate for fast neutron-induced fission research. The neutron time-of-flight (TOF) method was used to determine the incident neutron energies.

The Lujan Center also takes advantage of spallation on tungsten for neutron production, but here the target is surrounded by moderating material, resulting in highly thermalized neutron spectra on the various flightpaths. In the current experiments, flightpath 12 was used, which has a cold moderator. Again, the neutron TOF method was used to determine the incident neutron energies.

The different detectors used for fission studies were chosen to be complimentary in the information that they provide. The SPIDER 2v-2E spectrometer provides high-accuracy measurements of fission product mass, but with relatively low efficiency. The ionization chambers provide lower mass resolution, but are highly efficient and have high energy resolution. The time projection chamber (TPC) has high detection efficiency, somewhat lower energy resolution than the ionization chambers, but provides highly accurate particle tracking information.

2.1 Ionization chambers

Frisch-grid ionization chambers were used to study the total kinetic energy release in fission, the gross distributions of fission product masses, and the correlation between the two at difference incident neutron energies.

The ionization chambers and how they are operated are described in reference [1]. The chambers measure the kinetic energy of both of the fragments emitted in binary fission. The sum of the energies is the total kinetic energy release (TKE) of the fragments. Furthermore, the energies of the individual fragments can be used to infer the masses, by using momentum and mass conservation. This is known as the 2E-method, and it relies on previous knowledge of the average number of neutrons emitted for a given fragment mass. The correction limits the mass resolution to 4–5 atomic mass units (amu).

2.2 Spectrometer for ion determination in fission research (SPIDER)

The SPIDER instrument (Fig. 1) measures the velocity and kinetic energy release in the two fragments emitted in binary fission. This allows the fission product masses to be obtained with 1–2 amu resolution, as demonstrated at the ILL in the 1980's. The prototype version of the SPIDER instrument was built with two detector arms, with at total solid angle coverage of about 0.05%. The next version of the instrument is currently being constructed and will have 16 detector arms, bringing the total angular efficiency close to 1%.

2.3 Time projection chamber (TPC)

The Time Projection Chamber (TPC) is described in detail in reference [2]. The TPC was developed mainly for high precision fission cross section measurements, but can also be used to study other aspects of the nuclear fission process. The detector provides full, 3D particle tracking, and is therefore ideal for measuring anisotropy and angular distributions of fission fragments. The tracking is achieved in the TPC with a pixilated readout that provides the (x,y)-projection of a given particle's track, which then is combined with drift time measurements to reconstruct the 3D tracks.

3 Results

3.1 Total kinetic energy release (TKE)

The total kinetic energy release in fission demonstrates a general trend of decreasing with increasing incident neutron energy [3]. There is however, only limited experimental information for the major

Figure 1. Schematic view of the SPIDER instrument.

actinides above 5–10 MeV. Furthermore, some of the fine structures are not well measured, such as the sudden fluctuations in TKE at the thresholds for multi-chance fission channels.

The TKE as a function of incident neutron energy for U-238 [4], U-235, and Pu-239 [5] is shown in Fig. 2. These data sets extend out to 30 MeV, which helps constrain the TKE in the region around 14 MeV. This is of particular interest to applications aspects of nuclear technology. For all systems, the same trends around the opening of multi-channel fission is observed where the TKE increases locally, only to decrease again at higher neutron energies.

3.2 Mass yields

The mass yields for thermal neutron-induced fission were measured with SPIDER for U-235 and Pu-239, and the preliminary result for the former is shown in Fig. 3. Good agreement with the ENDF evaluation is generally observed, although some differences for individual yields are evident. For U-235 there is high confidence in the evaluation, and the difference can likely be attribute to experimental resolution effects and potentially calibration issues. For Pu-239 there is less previous experimental data, and the evaluations are somewhat less reliable.

The Frisch-grid ionization chamber was used to study the change in mass yields as a function of incident neutron energy. The result for U-238 is shown in Fig. 4. While the mass resolution is lower than for the SPIDER instrument, the general trends of the mass yield distributions are clearly observed. As the incident neutron energy increases the asymmetric distribution at low energies changes and a symmetric component is more and more evident. This is a well-known effect, but there is again very limited experimental information above 5–10 MeV.

3.3 Fragment anisotropy

The fission fragment anisotropy in U-235 was measured with the Time Projection Chamber. The results generally agree with literature data, and a manuscript describing the results has been submitted

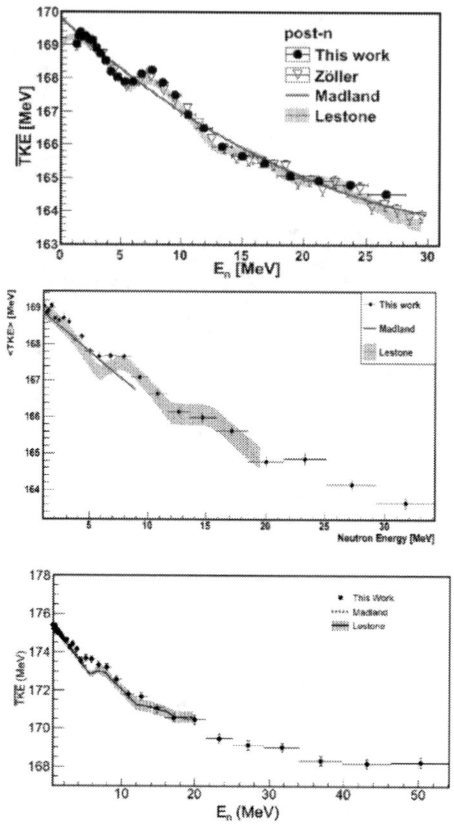

Figure 2. Total kinetic energy (TKE) for U-238 (top), U-235 (middle), and Pu-239 (bottom).

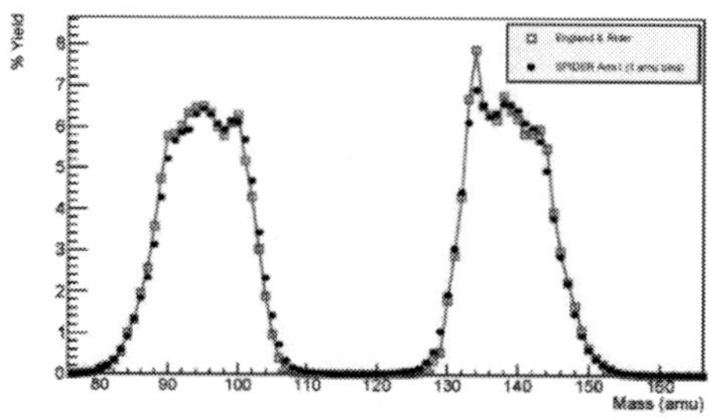

Figure 3. Preliminary mass yields for thermal neutron-induced fission of U-235.

for publication. The TPC is capable of measuring the full angular distributions, but for the initial work the focus was given to the anisotropy, since there is some previous work available for a comparison.

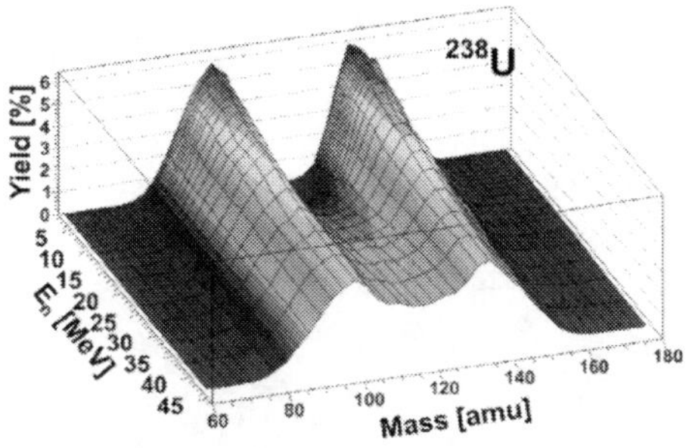

Figure 4. Mass yields of U-235 measured up to 45 MeV of incident neutron energy.

4 Conclusions and outlook

The work presented here shows the current capabilities of studying fission fragment properties at LANSCE. The results provide rather comprehensive information about fission fragments over a wide range of neutron energies for the major actinides. This information can help in benchmarking and guiding advanced fission models, as well as in providing nuclear data relevant to nuclear technology.

One of the missing capabilities for the fission fragment studies is the ability to measure mass yields with high resolution not only for thermal neutrons, but also in the fast region. This is driving the development of MegaSPIDER, which is currently being assembled. This new instrument will be used for experiments at WNR to extend the previous measurements with thermal neutrons to the fast range.

Another further extension of this program is to study the minor actinides. This is already in progress, with a study of U-233. In addition, there are plans for Np-237 and other systems relevant for nuclear reactor fuel cycles such as Pu-240 and Pu-242.

References

[1] S. Mosby, F. Tovesson, A. Couture, D. Duke, V. Kleinrath, R. Meharchand, K. Meierbachtol, J.M. O'Donnell, B. Perdue, D. Richman, D. Shields, Nucl. Instr. and Meth. A **757**, 75 (2014)

[2] M. Heffner, D.M. Asner, R.G. Baker, et al., Nucl. Instr. and Meth. A **759**, 50 (2014)

[3] D.G. Madland, Nucl. Phys. A **772**, 113 (2006)

[4] D.L. Duke, F. Tovesson, A.B. Laptev, S. Mosby, F.-J. Hambsch, T. Brys, and M. Vidali, Phys. Rev. C **94**, 054604 (2016)

[5] K. Meierbachtol, F. Tovesson, D.L. Duke, V. Kleinrath, B. Manning, R. Meharchand, S. Mosby, and D. Shields, Phys. Rev. C **94**, 034611 (2016)

Parameterisation of the residual temperature distribution based on the modelling of successive emission of prompt neutrons

Anabella Tudora[1,*] and *Franz-Josef* Hambsch[2]

[1] *University of Bucharest, Faculty of Physics, 405 Atomistilor Str., Magurele 077125, Romania*
[2] *E.C. Joint Research Centre, Directorate G, Unit G.2., Retieseweg 111, 2440 Geel, Belgium*

Abstract. A new deterministic modelling taking into account the successive emission of prompt neutrons from initial fragments of a fragmentation range {A, Z, TKE} constructed as in the Point-by-Point (PbP) treatment is described. The good agreement of different prompt emission quantities obtained from this modelling (e.g. ν(A), ν(TKE), Eγ(A), Eγ(TKE), etc.) with the experimental data and the results of the PbP model and other Monte-Carlo models validates the present modelling of sequential emission. The distributions of different residual quantities, including the residual temperature distributions P(T) of light and heavy fragments allow to obtain a new parameterisation of P(T) which can be used in the PbP model and the Los Alamos model.

1 Introduction

This paper reports preliminary results of a deterministic treatment of the successive emission of prompt neutrons. The modelling starts with the construction of the initial fragmentation range as in the Point-by-Point (PbP) model of prompt emission [1]. For each initial fragment A, Z at a given TKE covering the range of (A,Z,TKE) an equation for the residual temperature can be solved for each sequence of successive neutron emission. Then different quantities characterizing the residual nucleus and the prompt emission are obtained for each emission sequence (indexed k) corresponding to an initial fragment A, Z at a given TKE, where k is running over the number of prompt neutrons emitted successively (or the number of residual fragments). Using the multiple distributions Y(A,Z,TKE) of initial fragments, different distributions, e.g. of the residual temperature, the residual energy, the average neutron energy in the centre-of-mass frame corresponding to the emission of each neutron as well as the sum of these distributions following the emission of all neutrons can be obtained. Different quantities generically labelled q(A,Z,TKE) corresponding to an initial fragment at a TKE value (e.g. the prompt neutron multiplicity ν(A,Z,TKE), the average prompt γ-ray energy Eγ(A,Z,TKE), the prompt neutron spectrum in the centre-of-mass frame $\varphi(\varepsilon$,A,Z,TKE) etc.) can be obtained by averaging the respective quantities corresponding to each sequence q_k(A,Z,TKE) over the number of sequences (i.e. the number of emitted neutrons).

This modelling was applied to three fissioning nuclei, i.e. ^{252}Cf(SF) and ^{235}U(n_{th},f) for which Y(A,TKE) distributions and prompt neutron data were recently measured at JRC-Geel [2,3] and ^{239}Pu(n_{th},f) for which Y(A,TKE) data measured at JRC-Geel and prompt emission data are available

*e-mail: anabellatudora@hotmail.com

in the EXFOR library. The good agreement of different prompt emission quantities obtained from this modelling (e.g. ν(A), ν(TKE), Eγ(A), <ε>(A) etc.) with the experimental data and the results of other prompt emission models (e.g. PbP, FIFRELIN, etc.) validates the present deterministic modelling of the successive emission of prompt neutrons. The residual temperature distributions P(T) for the light and heavy fragment groups resulting from this modelling offer the possibility to obtain a preliminary parameterisation of P(T). This parameterisation can be used in the PbP model and the Los Alamos model where the sequential emission is taken into account in a global way by a P(T) distribution.

2 Equation of residual temperature following the successive emission of each neutron

For a given fissioning nucleus the initial fragmentation range is constructed as in the PbP treatment [1], i.e. the mass number A of initial fragments covers a large range, from symmetric fission up to a very asymmetric split with a step of one mass unit. For each A three charge numbers Z are taken as the nearest integer values above and below the most probable charge taken as the unchanged charge distribution corrected with the charge polarization $Zp(A)=Z_{UCD}(A)+\Delta Z(A)$. The charge polarization $\Delta Z(A)$ and the rms(A) of the isobaric charge distribution given by the Zp model are used [4]. For each fragmentation TKE values covering a large range (e.g. from 100 to 200 MeV) with a step of 5 MeV are considered.

For each initial fragment A, Z at each TKE (covering the fragmentation range mentioned above) a number of k_{max} prompt neutrons can be successively emitted if the excitation energy of the last residual fragment (with the charge number Z and the mass number A-k_{max}) becomes lower than the neutron separation energy from this nucleus $Sn^{(kmax)}$.

Considering the fragment level densities in the Fermi-gas regime, the temperature of the k-th residual nucleus (following the emission of the k-th prompt neutron) corresponding to each initial fragment at each TKE, is the solution of the following equation:

$$\overline{E_r}^{(k-1)} - S_n^{(k-1)} - <\varepsilon>_k = a_k T_k^2 \tag{1}$$

in which $\overline{E_r}^{(k-1)}$ and $S_n^{(k-1)}$ are the average excitation energy and the neutron separation energy of the precursor, respectively. $<\varepsilon>_k$ is the average energy of the k-th emitted neutron in the center-of-mass frame, a_k and T_k are the level density parameter and the temperature of the k-th residual nucleus, respectively. For the first emitted prompt neutron, k=1, $\overline{E_r}^{(0)} = E^*$ is the excitation energy of the initial fragment (before the first neutron is emitted) resulting from the TXE partition by modelling at scission [5,6].

To solve the residual temperature equations (1) the following approximations are needed:
(a) non-energy dependent level density parameters of initial and residual fragments, e.g. provided by the Egidy-Bucurescu systematic for the back-shift Fermi-gas (BSFG) model [7] or by the Gilbert-Cameron systematic for spherical nuclei [8] and
(b) an analytical expression of the compound nucleus cross-section $\sigma_c(\varepsilon)$ of the inverse process of neutron evaporation which can approximate $\sigma_c(\varepsilon)$ provided by optical model calculations with optical potential parameterisations appropriate for nuclei appearing as fission fragments. Note, the simplest approximation is to consider a constant σ_c. In this case the average prompt neutron energy is $<\varepsilon>_k (T_k) = 2T_k$ and the residual temperature equation (1) has an analytical solution.

In the present calculations $\sigma_c(\varepsilon)$ of each residual nucleus is approximated according to Ref. [9] by the following expression:

$$\sigma_c^{(k)}(\varepsilon) = \sigma_0^{(k)} \left(1 + \alpha_k / \sqrt{\varepsilon}\right) \tag{2}$$

with $\sigma_0^{(k)}$ and $\alpha_k = (\hbar^2/mr_0^2)S_0^{(k)}/A_k^{2/3}$ depending on the mass number A_k and the s-wave neutron strength function of each residual nucleus k.

To verify if the approximation of $\sigma_c(\varepsilon)$ by Eq. (2) is appropriate to be used in the calculation of $<\varepsilon>_k$ entering the transcendent equations (1), the average prompt neutron energy in the centre-of-mass frame obtained with $\sigma_c(\varepsilon)$ of Eq. (2), i.e.

$$< \varepsilon > (T) = T\left(2\sqrt{T} + (3\sqrt{\pi}/4)\alpha\right)\Big/\left(\sqrt{T} + \alpha\sqrt{\pi}/2\right) \tag{3}$$

was compared with $<\varepsilon>$ obtained by using $\sigma_c(\varepsilon)$ from optical model calculations, i.e.

$$< \varepsilon > (T) = \int_0^\infty K(T)\,\varepsilon^2\,\sigma_c(\varepsilon)\,\exp(-\varepsilon/T)\,d\varepsilon,\; K(T)^{-1} = \int_0^\infty \varepsilon\,\sigma_c(\varepsilon)\,\exp(-\varepsilon/T)\,d\varepsilon \tag{4}$$

The average prompt neutron energies $<\varepsilon>$ were calculated according to Eqs. (3) and (4) for temperatures ranging from 0.2 MeV to 2 MeV (which is the usual range of residual temperatures). The calculations were done for many nuclei covering the range of fission fragments. The deviation of $<\varepsilon>$(T) given by Eq. (3) from $<\varepsilon>$(T) based on $\sigma_c(\varepsilon)$ from optical model calculations (with the Becchetti-Greenlees potential) is less than 4% for all fragments. This fact supports the use of the $\sigma_c(\varepsilon)$ given by Eq. (2) in the transcendent equations of residual temperatures.

In the PbP model and other refined models of prompt emission, energy-dependent level density parameters of fragments are used. They are given by the super-fluid model with different shell corrections and different parameterisations of the damping of shell effects and the asymptotic value of the level density parameter. To see if non-energy dependent level density parameters can be used in the residual temperature equations, the variation with energy of the super-fluid level density parameter of many nuclei appearing as fission fragments, for energies ranging up to about 30 MeV, was compared with the constant level density parameter provided by the BSFG systematic of Ref. [7] and the Gilbert-Cameron systematic [8]. The level density parameters of Ref. [7] were used in the present calculations because they approximate well the super-fluid level density parameters for the majority of fission fragments, except the heavy fragments with A around 130 for which the large negative values of the shell corrections (due to the magic numbers N=82, Z=50) lead to a pronounced variation with energy of the super-fluid level density parameters.

Under the approximations (a) and (b) discussed above, for each initial fragment A,Z at each TKE, appearing with a probability given by the fragment distribution Y(A,Z,TKE), the iterative transcendent equations of residual temperatures given by Eq. (1), with k from 1 to k_{max}(A,Z,TKE), are numerically solved, giving the residual temperatures T_k(A,Z,TKE). The energy of each residual nucleus $E_r^{(k)}$(A,Z,TKE) and the average energy in the centre-of-mass frame $<\varepsilon>_k$(A,Z,TKE) of each k neutron emitted sequentially are easily obtained, too.

Examples of distributions of residual temperature and residual energy following the successive emission of neutrons from the heavy and light fragments are illustrated in Fig. 1. The average values of the residual temperatures and residual energies, corresponding to each neutron successively emitted, are given in each frame. Note, the distributions of the last emitted neutrons (k = 6, 7), being very low, were not plotted. Any quantity (generically labelled "q") corresponding to each emitted neutron as a function of initial fragment mass A, or initial TKE can be obtained by averaging q_k(A,Z,TKE) over the Y(A,Z,TKE) distribution, by summing over Z and TKE and over A and Z, respectively. An example of average residual temperatures and energies following the successive emission of each neutron as a function of initial fragment mass A is illustrated in Fig. 2. The saw-tooth shape of these quantities corresponding to the first, second and third emitted neutron is visible. As it can be seen, the successive emission of more than 3 neutrons (i.e. k = 4, 5, 6, 7) is not possible for all initial fragment masses.

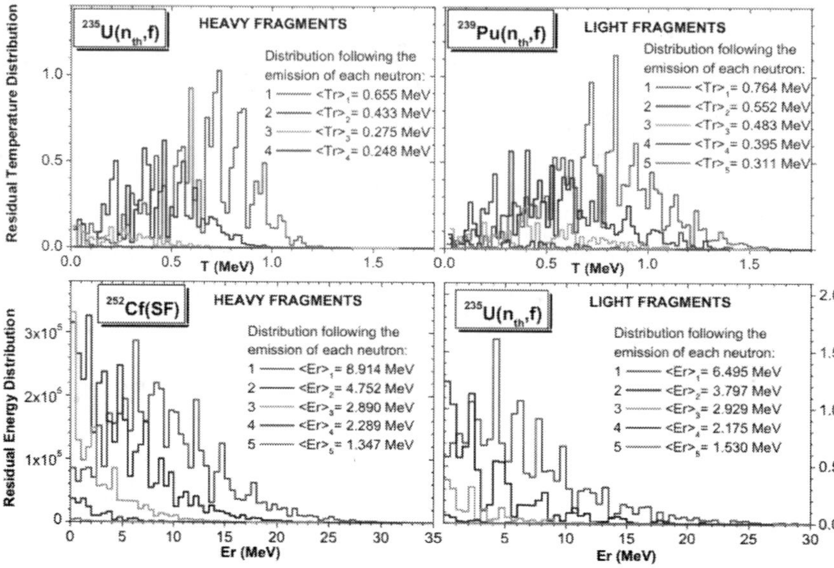

Figure 1. Examples of residual temperature and residual energy distributions following the emission of the first (red lines), second (blue lines), third (green lines), 4-th (wine line) and 5-th neutron (dark yellow) from the heavy and light fragments.

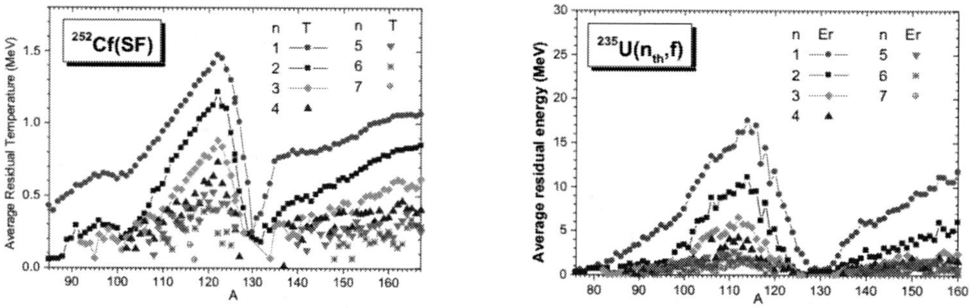

Figure 2. Average residual temperatures (left part) and average residual energies (right part) following the successive emission of each neutron as a function of initial fragment mass A, exemplified for $^{252}Cf(SF)$ and $^{235}U(n_{th},f)$, respectively.

3 Validation of sequential emission calculation by comparison with experimental data and results of other prompt emission models

Any prompt emission quantity corresponding to an initial fragment A, Z at a TKE value, q(A,Z,TKE), can be obtained by averaging $q_k(A,Z,TKE)$ over the number of sequential emissions $k_{max}(A,Z,TKE)$, i.e.:

$$q(A,Z,TKE) = \frac{1}{k_{\max}(A,Z,TKE)} \sum_{k=1}^{K\max} q_k(A,Z,TKE) \tag{5}$$

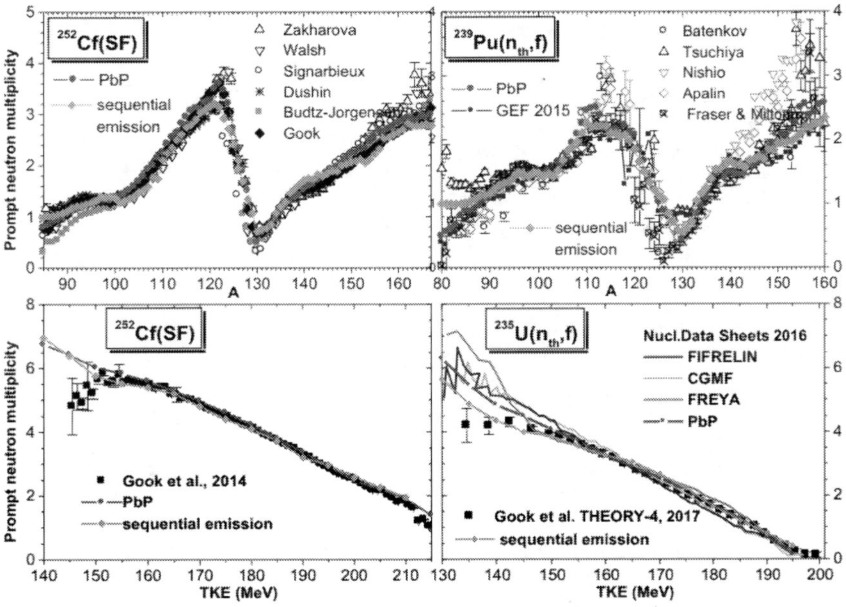

Figure 3. Examples of ν(A) and <ν>(TKE) resulting from sequential emission calculations (green diamonds) in comparison with experimental data (full and open black and grey symbols), the PbP result (red circles) and the results of other prompt emission models.

Different average prompt emission quantities, as a function of initial mass A, $\overline{q}$(A), as a function of TKE, <q>(TKE) are then obtained by averaging q(A,Z,TKE) of Eq. (5) over the Y(A,Z,TKE) distribution (by summing over Z and TKE and over A and Z, respectively). These average quantities are compared with the existing experimental data and also with the results of other prompt emission models. Examples of such comparisons are given in Figs. 3 and 4.

The ν(A) results of sequential emission calculation (green diamonds, upper part of Fig. 3) describe well the experimental data. They are close to the PbP results (red circles) and exhibit a more pronounced structure compared to ν(A) of PbP.

As it can be seen in the lower part of Fig. 3 the PbP results of <ν>(TKE) (red circles connected with lines) give an excellent description of the experimental data of Göök et al. of ^{252}Cf(SF) [2] and especially of ^{235}U(n$_{th}$,f) (this workshop [10]). Note, the PbP result of <ν>(TKE) for ^{235}U(n$_{th}$,f) as well as the results of the computer codes FIFRELIN (blue line), CGMF (cyan line) and FREYA (dark yellow line) were predictions, being reported at the beginning of 2016 [1], before the experimental data of Göök et al. (preliminary results reported in September 2016 [3] and the results presented at this workshop). The <ν>(TKE) results of sequential emission (green diamonds connected with lines) are also in very good agreement with the experimental data and the results of PbP.

Examples of Eγ(A) resulting from sequential emission calculations for ^{235}U(n$_{th}$,f) and ^{239}Pu(n$_{th}$,f) are plotted in the upper part of Fig. 4 with green diamonds in comparison with the experimental data of Pleasonton et al. (black squares) and the results of the PbP model (red circles). Significant differences between the results of PbP and the sequential emission calculation are visible near symmetric fission (where experimental data are missing). Both calculations are in reasonable agreement with the scattered experimental data.

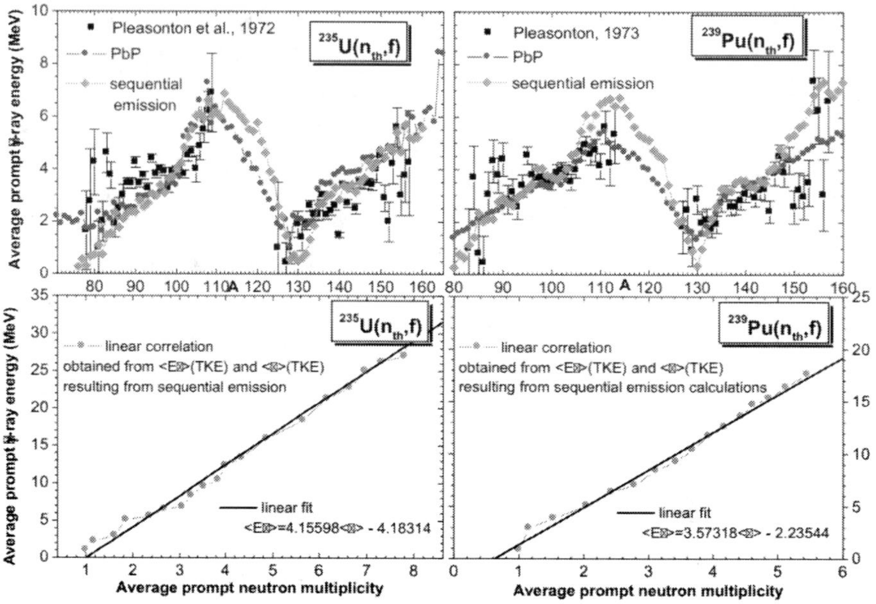

Figure 4. *Upper part:* $E\gamma(A)$ of $^{235}U(n_{th},f)$ and $^{239}Pu(n_{th},f)$: the sequential emission result (green diamonds) in comparison with the experimental data of Pleasonton et al. (black squares) and the PbP result (red circles). *Lower part:* the linear correlation between the prompt γ-ray energy and the prompt neutron multiplicity (green circles) obtained from $<E\gamma>$(TKE) and $<\nu>$(TKE) resulting from sequential emission calculation and the linear fit (black line).

The sequential emission calculations confirm the linear correlation between the average prompt γ-ray energy and the prompt neutron multiplicity, as it can be seen in the lower part of Fig. 4 where the correlation resulting from the sequential emission results of $<E\gamma>$(TKE) and $<\nu>$(TKE) is plotted with green circles and the corresponding liner fit with a black line.

As it can seen in the upper parts of Figs. 3 and 4, ν(A) and $E\gamma$(A) resulting from sequential emission calculations exhibit a more pronounced structure than the PbP results. This fact is not surprising because the modelling of sequential emission is a detailed treatment of the emission of each neutron from each residual fragment formed successively, while the PbP model takes globally into account the sequential emission just by a P(T) distribution with a smooth form. An example is the prompt γ-ray energy corresponding to each initial fragment A, Z at a given TKE. This is calculated in the PbP model as

$$E_{\gamma}(A,Z,TKE) = E^{*}(A,Z,TKE) - \nu(A,Z,TKE)(<\varepsilon>(A,Z,TKE) + \overline{S}n(A,Z,TKE)) \quad (6)$$

in which *E*(A,Z,TKE)* is the excitation energy of the initial fragment (resulting from the TXE partition by modelling at scission), $\nu(A,Z,TKE)$ is the prompt neutron multiplicity, the neutron separation energy is an average value calculated as S_{kn}/k (with k the number of prompt neutrons emitted sequentially). The first order momentum of the prompt neutron spectrum in the centre-of-mass frame $<\varepsilon>$ entering Eq. (6) is obtained by averaging the evaporation spectrum at a given residual temperature over a residual temperature distribution *P(T)* with a smooth form. In the sequential emission treatment $E\gamma$ is calculated at each emission sequence k corresponding to an initial fragment A, Z, at a given TKE

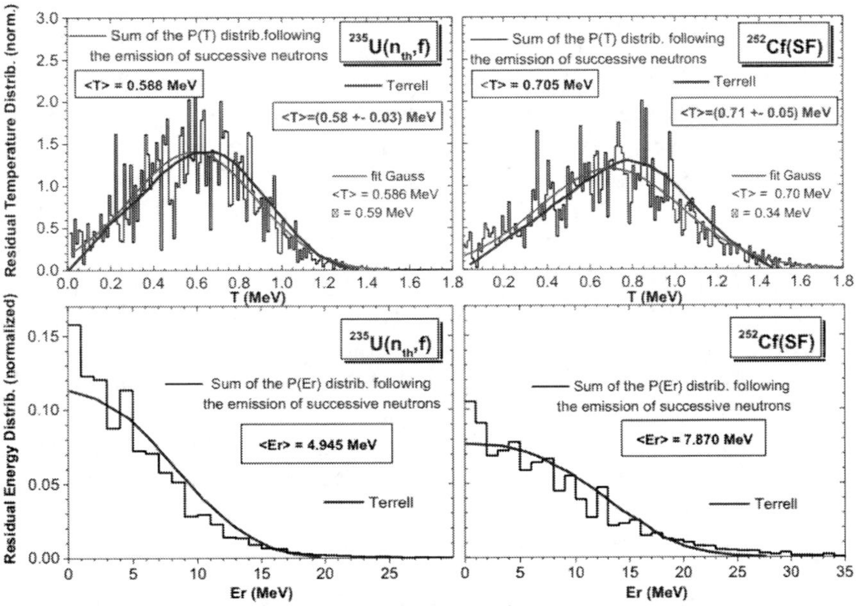

Figure 5. Sum of the residual temperature and residual energy distributions following the emission of successive neutrons from all fragments (histograms plotted black lines) in comparison with the results reported by Terrell [11] (blue lines).

(with k from 1 to k_{max}(A,Z,TKE)), i.e.

$$E_\gamma^{(k)}(A,Z,TKE) = \overline{E}_r^{(k-1)}(A,Z,TKE) - <\varepsilon>_k (A,Z,TKE) - S_n^{(k-1)}(A,Z,TKE) \qquad (7)$$

in which $\overline{E}_r^{(k-1)}$ and $S_n^{(k-1)}$ are the average excitation energy and the neutron separation energy of the residual nucleus Z, A-k+1 and $<\varepsilon>_k$ is the average energy in the centre-of-mass frame of the k-th emitted neutron. These quantities are obtained by solving the transcendent equation of residual temperature (1) at each sequence. The average prompt γ-ray energy corresponding to the initial fragment A, Z, TKE (i.e. the quantity expressed by Eq. (6) in the case of PbP) is obtained by averaging $E_\gamma^{(k)}(A,Z,TKE)$ given by Eq. (7) over the number of sequential emissions according to Eq. (5).

The sum of the residual temperature and residual energy distributions following the emission of successive neutrons from all fragments are given as histograms (black lines) in Fig. 5. As it can be seen the sum of residual temperature distributions is well fitted by Gaussian functions (red lines).

Both distributions, of the residual temperature and the residual energy, are in good agreement with the results of Terrell [11] (blue lines). The average values of the residual temperature (given in each frame of Fig. 5) are very close to the average values reported by Terrell (given in blue).

The good agreement of different prompt emission quantities resulting from the sequential emission calculation with the experimental data and the results of other prompt emission models, as well as the good agreement of the sum of residual temperature and residual energy distributions following the successive neutron emission from all fragments with the results of Terrell, validate the present modelling (with the assumptions and approximations mentioned in Sect. 1).

4 New parameterisation of the residual temperature distribution of light and heavy fragments

In the deterministic models Los Alamos [12,13] and PbP (Ref. [1] and references therein) the sequential emission is taken into account in a global way by a residual temperature distribution P(T) with the following form proposed by Madland and Nix [12]:

$$P(T) = \begin{cases} \left(2/T_m^2\right) T & T \leq T_m \\ 0 & T > T_m \end{cases} \tag{8}$$

In the PbP model the maximum temperature Tm of Eq. (8) is the temperature corresponding to the initial fragment A, Z at a given TKE. In the LA model T_m is the average value of the initial temperature corresponding to the light and heavy fragment groups, i.e. the LA model with non-equal Tm of Ref. [13], or it is the equivalent temperature corresponding to all fragments, i.e. the LA model with equal Tm of Ref. [12]. Consequently the prompt neutron spectrum in the centre-of-mass frame $\Phi(\varepsilon)$ corresponding to an initial fragment at a given TKE is calculated by integrating the prompt neutron spectrum at a given residual temperature $\varphi(\varepsilon,T)$ over the P(T) distribution.

The residual temperature distributions corresponding to the light and heavy fragment groups resulting from the modelling of sequential emission described in Sects. 2 and 3 offer the possibility to obtain a new parameterisation of P(T) as a function of the temperature of initial fragments.

Examples of P(T) distributions resulting from sequential emission calculations are illustrated in Fig. 6 (as histograms plotted with grey lines) for heavy (upper part) and light fragments (lower part).

The P(T) distributions of Madland and Nix, expressed by Eq. (8), are plotted with blue lines. The corresponding Tm (i.e. the average value of the initial temperature) is given in each frame.

The proposed form of P(T) is defined as:

$$P(T) = \begin{cases} (P_{max}/T_1)\, T & T \leq T_1 \\ (P_{max}/(T_2 - T_1))\, (T_2 - T) & T_1 \leq T \leq T_2 \end{cases} \tag{9}$$

with the following conditions:
i) Normalization to 1 from which the relation $P_{max} = 2/T_2$ is obtained,
ii) The average <T> of the P(T) distribution of Eq. (7) must be equal to the <T> value resulting from the sequential emission calculations (which is given in each frame of Fig. 6). From this condition the relation $T_2 = 3 < T > -T_1$ is obtained.

For the studied fissioning systems ^{235}U(n_{th},f), ^{239}Pu(n_{th},f) and ^{252}Cf(SF) we have observed that the values of average residual temperatures of the light and heavy fragment groups resulting from the sequential emission calculations are approximately 0.66 from the average temperature of initial fragments, i.e. <T> = 0.66 $<T_i>$. In other words $< T > = (2/3) < T_i >$ as in the case of P(T) of Madland and Nix. Taking into account this observation and the relations resulting from the conditions i) and ii), the following preliminary parameterisation of P(T) is proposed:

$$P(T) = \begin{cases} (2/T_i^2)\,\alpha T & T \leq cT_i \\ (2/T_i^2)\,(\beta T_i - \gamma T) & cT_i \leq T \leq (2-c)T_i \end{cases} \tag{10}$$

with

$$1/\alpha = c(2-c); \quad 1/\beta = 2(1-c); \quad 1/\gamma = 2(1-c)(2-c) \tag{11}$$

In the PbP treatment, Ti of Eq. (10) denotes the temperature of each initial fragment A, Z at a given TKE. In the case of the most probable fragmentation approach (i.e. the LA model with non-equal T_m) T_i is just T_m (i.e. the average initial temperature of the light and heavy fragment group). The

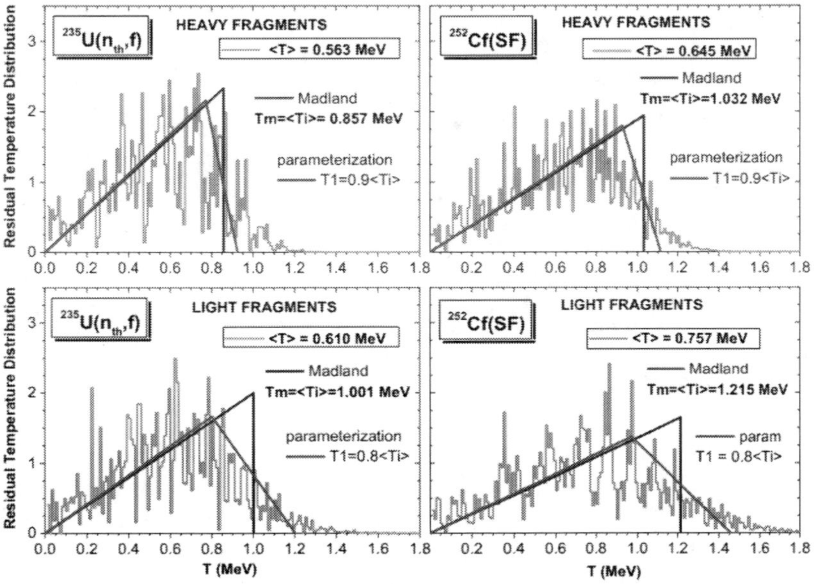

Figure 6. Examples of P(T) for the heavy (upper part) and light fragments (lower part): P(T) from sequential emission calculations (as histograms plotted with grey lines), P(T) of Madland and Nix (blue lines) and the proposed parameterisation of P(T) (red lines).

preliminary parameterisation of Eqs. ((10), (11)) is plotted in Fig. 6 with a red line (i.e. the case $c = 0.9$ for heavy fragments and $c = 0.8$ for light fragments as indicated in the figure).

The use of the preliminary P(T) parameterisation of Eqs. ((8), (9)) in the PbP model calculations leads to results of the prompt neutron multiplicity and prompt γ-ray energy (e.g. ν(A,TKE), ν(A), ν(TKE), $<E\gamma>$(A), $<E\gamma>$(TKE)) which do not differ significantly from the previous PbP results (plotted with red circles in the figures of Sect. 3) obtained with P(T) of Madland and Nix.

The improvement is seen in the prompt neutron spectrum results. An example is given in Fig. 7 where the prompt neutron spectrum in the laboratory frame of ^{252}Cf(SF) calculated with P(T) of Eqs. ((8), (9)) describes better the data of Poenitz and Tamura (left frame) than the calculation done with P(T) of Eq. (6) (right frame).

5 Conclusions

Detailed calculations taking into account the successive emission of prompt neutrons were done for three fissioning systems ^{235}U(n$_{th}$,f), ^{239}Pu(n$_{th}$,f) and ^{252}Cf(SF). The successive transcendent equations of residual temperatures based on the assumption of fragment level densities in the Fermi-gas regime were solved under the approximations:
– non-energy dependent level density parameters of fragments (provided by the Egidy-Bucurescu systematic for BSFG) and
– an analytical expression of the compound nucleus cross-section $\sigma_c(\varepsilon)$ of the inverse process of neutron evaporation, which approximates well the average prompt neutron energy in the centre-of-mass frame $<\varepsilon>$ obtained with $\sigma_c(\varepsilon)$ provided by optical model calculations.

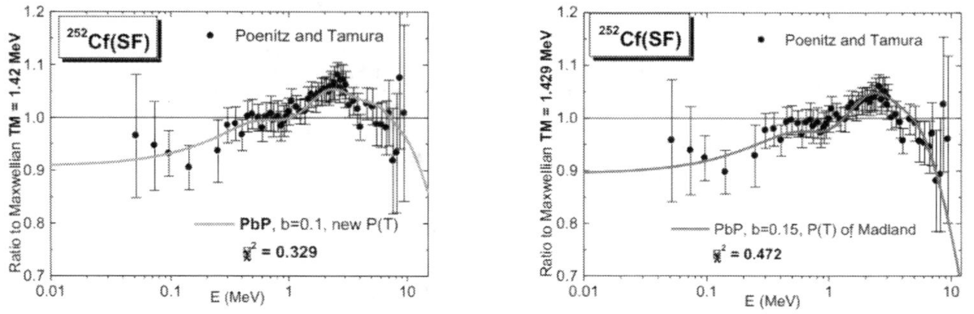

Figure 7. Prompt neutron spectrum of ^{252}Cf(SF) in the laboratory frame calculated with the preliminary parameterisation of P(T) (left part) and the P(T) of Madland and Nix (right part) in comparison with the data of Poenitz and Tamura (renormalized to the calculated spectrum).

This new deterministic modelling of sequential emission is validated by the good agreement of its results (e.g. ν(A), ν(TKE), Eγ(A), Eγ(TKE), $<\varepsilon>$(A), $<\varepsilon>$(TKE)) with the experimental data and the results of other prompt emission models (e.g. PbP, FIFRELIN). Also the sum of the residual temperature and residual energy distributions following the successive neutron emission from all fragments resulting from this modelling are in good agreement with the distributions reported by Terrell [11].

The residual temperature distributions of the light and heavy fragment groups provided by this modelling allow to obtain a preliminary parameterisation of these P(T) distributions which can be used in the deterministic models PbP and Los Alamos with non-equal Tm recently proposed by Madland and Kahler [13].

The use of this preliminary P(T) parameterisation in the PbP model leads to prompt neutron spectrum results describing better the experimental data than the results based on the classical P(T) form of Madland and Nix [12]. The PbP results of prompt neutron multiplicity and prompt γ-ray energy (e.g. ν(A,TKE), ν(A), ν(TKE), $<$E$\gamma>$(A), $<$E$\gamma>$(TKE)) obtained with this preliminary parameterisation of P(T) do not differ significantly from the results obtained with P(T) of Madland and Nix.

The results reported in this paper are preliminary. Sequential emission calculations for other neutron induced fissioning nuclei at higher incident energies (^{234,238}U(n,f), ^{237}Np(n,f) up to about 5 MeV) are in progress, in order to provide an improved parameterisation of P(T). The solving of the successive transcendent equations of residual temperatures by using other prescriptions for the level density parameter of initial and residual fragments and maybe another analytical expression of $\sigma_c(\varepsilon)$ which will approximate better the shape of $\sigma_c(\varepsilon)$ provided by optical model calculations is foreseen.

References

[1] R. Capote, Chen Y.J., F.-J. Hambsch, N.V. Kornilov, J.P. Lestone, O. Litaize, B. Morillon, D. Neudecker, S. Oberstedt, T. Ohsawa, N. Otuka, V.G. Pronyaev, A. Saxena, O. Serot, O.A. Shcherbakov, Shu N.C., D.L. Smith, P. Talou, A. Trkov, A.C. Tudora, R. Vogt, S. Vorobyev, Nucl. Data Sheets **131**, 1 (2016)

[2] A. Göök, F.-J. Hambsch, M. Vidali, Phys. Rev. C **90**, 064611 (2014)

[3] A. Göök, F.-J. Hambsch, S. Oberstedt "Prompt neutron emission and energy balance in 235U(n,f)" ND-2016 Int. Conf. on Nuclear Data for Science and Technology, 11-16 September 2016, Bruges, Belgium

[4] A.C. Wahl, Atomic Data and Nuclear Data Tables **39**, 1 (1988)

[5] C. Morariu, A. Tudora, F.-J. Hambsch, S. Oberstedt, C. Manailescu, J. Phys. G, Nucl. Part. Phys. **39**, 055103 (2012)
[6] A. Tudora, F.-J. Hambsch, I. Visan, G. Giubega, Nucl. Phys. A **940**, 242 (2015)
[7] T. von Egidy and D. Bucurescu, Phys. Rev. C **80**, 054310 (2009)
[8] A. Gilbert, A.C.W. Cameron, Can. J. Phys. **43**, 1446 (1965)
[9] O. Iwamoto, J. Nucl. Sci. Tech. **45**, 910 (2008)
[10] A. Göök, F.-J. Hambsch, S. Oberstedt "Prompt fission neutron emission in the reaction ^{235}U(n,f)", this workshop
[11] J. Terrell, Phys. Rev. **113**, 527 (1959)
[12] D.G. Madland and J.R. Nix, Nucl. Sci. Eng. **81**, 213 (1982)
[13] D.G. Madland, A.C. Kahler, Nucl. Phys. A **957**, 289 (2017)

Measurement of the prompt fission γ-ray spectrum of ^{242}Pu

Sebastian Urlass[1,2,⋆], *Roland* Beyer[1], *Arnd Rudolf* Junghans[1,⋆⋆], *Toni* Kögler[1], *Ronald* Schwengner[1], and *Andreas* Wagner[1]

[1] *Helmholtz-Zentrum Dresden-Rossendorf, Bautzener Landstraße 400, 01328 Dresden, Germany*
[2] *Technische Universität Dresden, 01062 Dresden, Germany*

Abstract. The prompt γ-ray spectrum of fission fragments is important in understanding the dynamics of the fission process, as well as for nuclear engineering in terms of predicting the γ-ray heating in nuclear reactors. The γ-ray spectrum measured from the fission fragments of the spontaneous fission of ^{242}Pu will be presented here. A fission chamber containing in total 37 mg of ^{242}Pu was used as active sample. The γ-quanta were detected with high time- and energy-resolution using $LaBr_3$ and HPGe detectors, respectively, in coincidence with spontaneous fission events detected by the fission chamber. The acquired γ-ray spectra were corrected for the detector response using the spectrum stripping method. About 70 million fission events were detected which results in a very low statistical uncertainty and a wider energy range covered compared to previous measurements. The prompt fission γ-ray spectrum measured with the HPGe detectors shows structures that allow conclusions about the nature of γ-ray transitions in the fission fragments. The average photon multiplicity of 8.2 and the average total energy release by prompt photons per fission event of about 6.8 MeV were determined for both detector types.

1 Introduction

The emission of γ-quanta after nuclear fission plays an important role for the design of future nuclear power plants. The local contribution of the heat generation from γ-rays has to be known very accurately to develop an efficient core-design. Therefore the prompt and delayed γ-emission spectra from the nuclear fission process must be known. In recent years several integral experiments showed discrepancies with simulations up to 30% [1]. For that reason the neutron induced fission prompt γ-ray spectrum (PFGS) for the reactor fuels ^{239}Pu and ^{235}U were taken into the High Priority Request List of the OECD / Nuclear Energy Agency [2].

For future transmutation technologies also the understanding of the burn-up of minor actinides like the longlived ^{242}Pu is from interest. Beside the technological aspect the measurement of the PFGS from spontaneous fission (SF) of ^{242}Pu and the determination of its characteristics is also important for a general parameterization of fission observables. These characteristics are the average energy release $\overline{\epsilon}_\gamma$, the average total energy release $\overline{E}_\gamma$ and average photon multiplicity $\overline{\nu}_\gamma$ per fission. In order

⋆Corresponding author, e-mail: s.urlass@hzdr.de
⋆⋆Principal corresponding author, e-mail: a.junghans@hzdr.de

to separate prompt and delayed γ-rays from fission $LaBr_3$ detectors with a good time resolution are used.

Another aspect for this experiment is the investigation of the dynamics of the fission process. Most fission fragments (FF) are neutron rich nuclei with an initial excitation energy above the neutron separation energy, so that at first prompt neutrons are emitted. A high angular momentum neutron emission is surppressed due to the angular momentum barrier. That means it does not affect the angular momenta of the fission fragments very strongly. If the excitation energy is below the neutron separation energy, prompt γ-ray emission takes place. At a certain point of the deexcitation, when the Yrast-line is reached, the fission fragments are in a collective rotational state. The remaining high angular momentum of the FF can be carried away via electrical quadrupole transitions (E2) [3]. To be able to resolve transitions from known rotational bands in FF, high purity germanium detectors with a high energy resolution were used.

2 Setup, data acquisition and measurement

For the detection of FF a parallel plate fission ionization chamber was used. All details about the PuFC can be found in Ref. [4, 5]. In the following only the most crucial properties the PuFC regarding the PFGS measurement are presented. The PuFC consists of 8 pairs of electrodes, which are disks of 98 mm diameter, mounted with a gap of 10 mm between each other. The cathodes are silicon wafers of 400 µm thickness where the ^{242}Pu was deposited as circular layers of 74 mm diameter using the molecular plating technique [6]. The plutonium content of each deposit is listed in Table 1. In total the fission chamber contains 37.2 mg plutonium.

Table 1. Properties of Pu deposits. The left table a) contains the isotopic vector, which was determined by using mass spectroscopy. The mass of each Pu deposit is shown in the right table b). These quantities were obtained from the SF-rate of ^{242}Pu [4].

a) isotope vector

isotope	mass fraction/%
^{238}Pu	0.002
^{239}Pu	0.005
^{240}Pu	0.022
^{241}Pu	0.002
^{242}Pu	99.967
^{244}Pu	0.002

b) deposit mass

Pu deposit	mass/mg
# 1	5.42 (9)
# 2	4.89 (8)
# 3	4.37 (7)
# 4	4.22 (7)
# 5	4.44 (7)
# 6	4.43 (7)
# 7	4.13 (7)
# 8	5.33 (9)
sum	37.24 (22)

The anodes consist of tantalum with a thickness of 100µm. Due to safety- and radiation protection reasons the housing material is made of 3 mm thick stainless steel, with exception of the front- and end window which are made of 200µm thick stainless steel foil. P10 (90% Argon+10% Methane) is used as counting gas. The total SF-rate in the chamber is about 30 fission events per second, while its corresponding α-activity is about 8 MBq. The quality of the separation of both contributions can be seen in the charge specrum of the first Pu deposit which is depicted in Fig. 1.

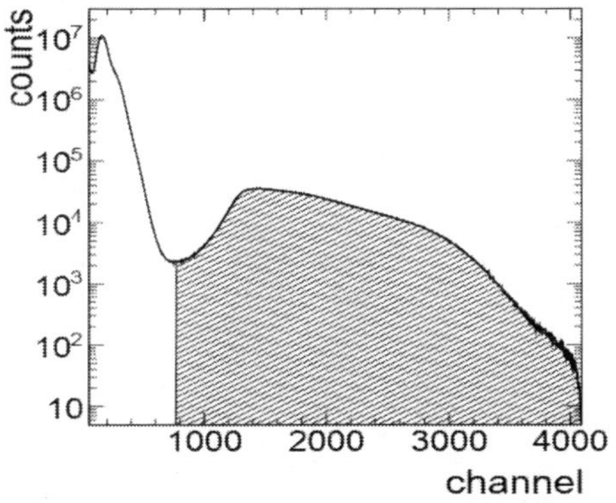

Figure 1. Measured charge spectrum of the first Pu deposit. The charge spectra of the other deposits have a similar spectral shape and are not depicted due to overview reasons. At the channel position of about 800 is a local minimum of the spectrum. Below the minimum the charge spectrum is dominated by α-particles. The induced charge spectrum above the minimum consists of contributions from FF, pile-up of FF with each other and α-particles. The latter one and also multiple pile-up of α-particles with each other is negligible. In order to create the correlation between fission event and γ-ray emission a gate was set for all charge spectra in a way that the defined FF regions contain channel values greater than their corresponding local minima. This region is depicted here for the first deposit as hatched area.

The time resolution of the chamber is about 2 ns (FWHM) and its FF detection probability is nearly 100%. For reasons of pile-up reduction the readout of the Pu signals is done separately for each deposit. These signals are amplified with charge sensitive fast preamplifiers (CSP) which were developed at HZDR. Figure 2 illustrates the entire setup of the experiment and its corresponding electrical scheme.

Two different types of detecors were used to measure the photons in coincidence with a detected fission fragment. These are named in the following as γ-ray detector. As the first type of γ-ray detector three 7.62 cm × 7.62 cm (diameter × length) $LaBr_3$ detectors were used[1]. Their energy resolution[2] is about 4% and the time resolution is approximately 0.8 ns (FWHM). The second type of γ-ray detector are two high purity germanium (HPGe) cluster detectors of the MINIBALL design [7]. Each cluster consists of three capsules containing the HPGe crystal so that all together six HPGe detectors are used. Their energy resolution[2] is about 0.3% while the time resolution is about 10 ns (FWHM). Both time and energy information are measured by the data aquisition and stored in list mode. The triggerlogic of the experiment is based on a logical OR condition between all detectors and is controlled by an FPGA module, which means that each detector can trigger the data acquisition seperately. During the measurement all γ-ray detectors were affected by gain shifts. Recording at first all uncorrelated γ-events shows the advantage for applying a gain correction procedure with high statistics in a short amount of time, which is based on the time dependent shifts of peak positions from the natural background. The initializing and readout of all modules QDC, ADC and TDC is done by RIO4 power PC which communicates via a VME BUS. The RIO4 is operated with a Multibranching System (MBS) which is developed at GSI Darmstadt [8]. The MBS-list-mode data were analysed with the software package Go4 [9]. In the offline analysis a coincidence between PuFC and γ-ray detectors has been created. During the measurement time of about six weeks 70 million fission events were registered. Beside this

[1] Saint-Gobain *BrilLanCe 380 Scintillation Products Technical Note, Performance Summary* (2009).

[2] The energy resolution is here defined as relative quantity $\frac{\Delta E}{E}$ where ΔE corresponds to a measured FWHM at the energy $E = 1333\,keV$ using a ^{60}Co calibration source.

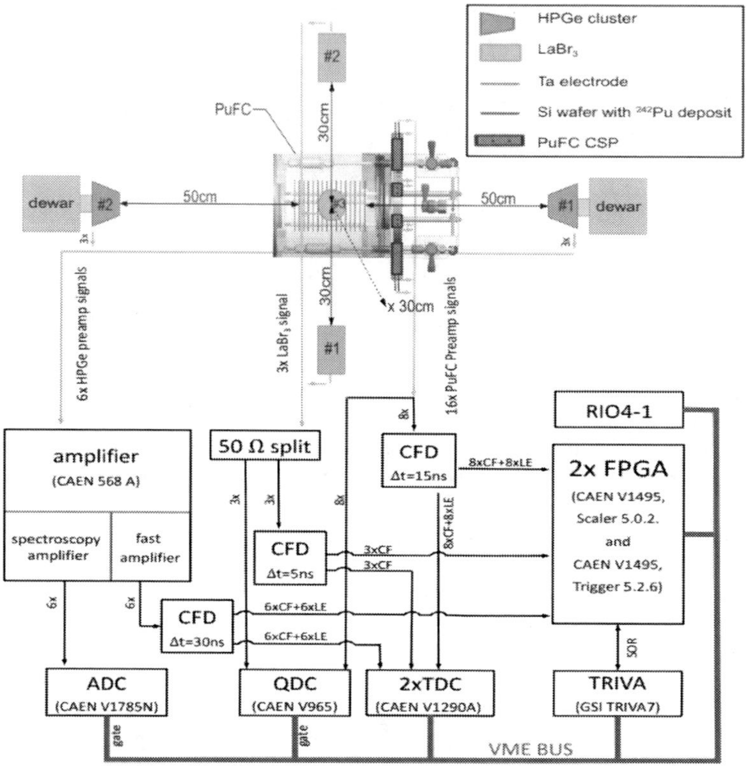

Figure 2. Sketch of experimental setup and electronics. The upper part shows the detector setup. The PuFC with its CSP is in the middle. Around the PuFC the three $LaBr_3$ detectors are placed perpendicular to the FC housing with a distance of 30cm to the center of the FC axis. One of the $LaBr_3$ detectors is mounted perpendicular above the plane of projection, illustrated as light green disk centered on the PuFC. The MINIBALL HPGe detectors are mounted in front of a stainless steel window of the PuFC. The distance between each of these cluster detectors from die closest PuFC deposit is 50cm. A simple scheme of the electronics is illustrated in the lower part. The time information is acquired by a time to digital converter (TDC). The match window for the event trigger for the TDC is $\pm 1\,\mu$s. The charge informations of the PuFC and the $LaBr_3$ signals are recorded by a charge to digital converter (QDC.) The amplitude informations from the HPGe detectors are recorded by an amplitude to digital converter (ADC).

long-term measurement a series of efficiency and energy calibrations with calibrated sources using an identical fission chamber without Pu deposits were performed.

3 Separation of prompt- and delayed γ-rays and Slow rise time correction

In order to get a proper determination of the timing of the signals, constant fraction discriminator (CFD) modules[3], with Amplitude-Risetime-Compensation (ARC) technique [10], are used. This

[3]Inhouse development of HZDR.

ARC-timing property is necessary for the HPGe detectors which have variable rise times because of the different charge collection time inside the HPGe crystal volume. A CFD splits the input signal in an inverted and delayed and an attenuated signal. The superposition of both signals creates a bipolar signal with a zero crossing which is independent from the signal amplitudes, if the rise times of all input signals are the same. To get also a rise time independent zero crossing, the delay Δt between the splitted signals that can be adjusted by an external cable delay were chosen very short compared to the signal rise times, for instance 5 to 30 ns (see Fig. 2). For the application of the ARC-timing, in case for the HPGe detectors, a price has to be paid. The CFD has an internal integrated arming threshold discriminator, which is necessary to prevent to trigger electronic noise. If the zero crossing time occurs before reaching the arming threshold, the discriminator will not be armed and the signal is not registered. This effect occurs very strongly for low amplitude values for the HPGe signals.

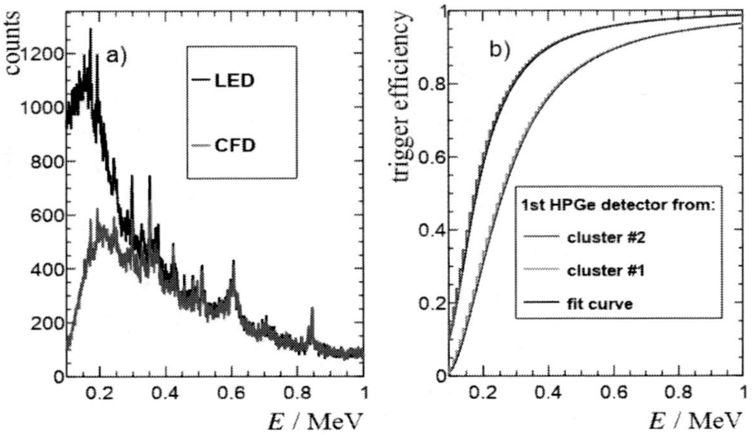

Figure 3. CFD-trigger efficiency for the HPGe detectors. In a) the lack of the CFD arming is illustrated for the first HPGe detector of the second cluster. The integral of the spectrum for CFD triggered events is about 40% less than in comparison to the LED spectrum. In b) the measured trigger efficiencies for the first HPGe detector of the first and second MINIBALL cluster, respectively, are shown with their corresponding fit curves described by Eq. (1). A strong decrease of the trigger efficiencies are visible for energies below 400 keV.

The increase of the delay leads to a better CFD-trigger behaviour but makes the time resolution much worse. The compromise to keep the good time resolution is applying an additional correction in the offline analysis for CFD triggered events at low energies. Therefore a leading edge threshold discriminator (LED) has been used as a reference. The LED is able to register all events with energies greater than 100keV. The ratio of CFD- and LED triggered γ-ray events in coincidence with the PuFC gives a quantity for the lack of the CFD arming and is in the following named as trigger efficiency $\epsilon_{\mathrm{CFD}}(E)$, which is described by an empirical formula. This function for $\epsilon_{\mathrm{CFD}}(E)$ is shown in Eq. (1) and illustrated in Fig. 3b). In Fig. 3a) the difference between the CFD- and LED triggered amplitude spectra is illustrated.

$$\epsilon_{\mathrm{CFD}}(E) = 1 + \frac{1}{a_0 + a_1 \cdot E + a_2 \cdot E^2} \tag{1}$$

The application of ARC timing allows us to get a reasonable time resolution for all the HPGe detectors of about 12 ns in coincidence with the PuFC. The combined time resolution of the coincidence between the $LaBr_3$ detectors and the PuFC was measured to 2.3 ns. In Fig. 4 the energy spectra of both γ-detector types in coincidence with the PuFC are shown. In order to extract only the prompt fission γ-rays a gate for a coincidence window of $\Delta T = \pm 3$ ns for the $LaBr_3$ and $\Delta T = \pm 12$ ns for the

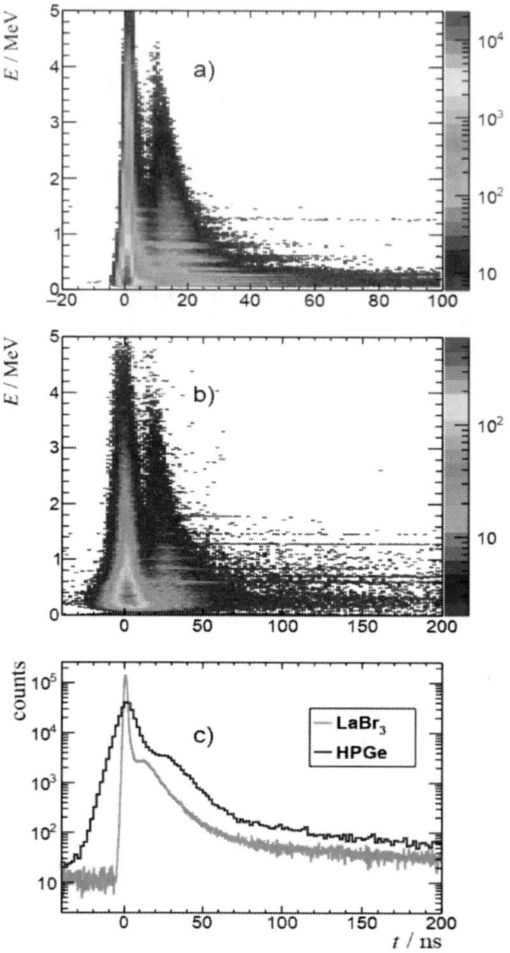

Figure 4. Energy spectra in correlation to prompt- and delayed γ-rays from the fission process. In a) the summed spectrum of all $LaBr_3$ detectors is illustrated. In b) the summed spectrum for all the HPGe detectors is shown. Note the different time scales for both detector types. The time t is calibrated in a way that the time of flight of the prompt γ-rays corresponds to a flight path of 30 cm for the $LaBr_3$ detectors and 50 cm for the HPGe detectors respectively. The time spectra of both γ-ray detector types, projected for all energies, are illustrated in c).

HPGe detectors was set. The delayed γ-rays appearing in this measurement can be categorized in four different contributions:

1. random events:

The γ-rays from the natural background appear randomly in the coincidence time window for the prompt photons and are time independent. Its strongest background contribution comes from ^{40}K (1461 keV) decay and an can be barely seen in Fig. 4b) as horizontal line.

2. (n, n′) on γ-detector material:

In Fig. 4c) the bump structure after the prompt events is due to the (n, n′) reactions, which time delay depends on the distance between the γ-detector and the PuFC. Most of the delayed γ-rays for the $LaBr_3$ detectors could be identified to stem from (n, n′) reactions with the detector material containing the major isotopes ^{139}La, ^{79}Br and ^{81}Br in an energy range of 160–800 keV and can be seen in Fig. 3a) the difference between the CFD- and LED triggered amplitude spectra is illustrated. 4a) and c) about 10 ns after the prompt γ-rays. In case for the HPGe detectors (n, n′) reactions with the most abundant isotopes ^{70}Ge, ^{72}Ge, ^{74}Ge and ^{76}Ge could be identified in an energy range of 600–2000 keV and can be seen in Fig. 4b) and c) about 20 ns after the prompt γ-rays.

3. (n, n′) on the PuFC material:
The large amount of ^{56}Fe in the PuFC material leads to γ-rays with energies of 847 keV from the first excited state and appear due to the short deposit-to-PuFC housing flight path and its corresponding minimal neutron time of flight of 2 ns very close to the time of flight of prompt photons for the $LaBr_3$ detectors. Because of its good time resolution this contribution can be separated from the prompt γ-events. In case for the HPGe detectors (n, n′) reactions on ^{56}Fe can be barely distinguished from prompt γ-quanta but these reactions are also much less expected than for the $LaBr_3$ detectors. Most of the detected photons are emitted in longditudinal direction and pass mainly the 200 μm thick stainless steel windows.

4. Isomeric states from FF:
In this measurement the most significant delayed γ-ray contribution of this type can be identified from an isomeric 6^+ state at 1691 keV of the FF ^{134}Te with a lifetime of about 164 ns (see Ref. [11]). During its deexcitation process also the the short lived first excited 2^+ state (0.64 ps) at 1279 keV) is populated. This photon energy can be seen as horizontal line[4] in Fig. 4a) and b).

4 Response correction, results and discussion

In order to correct the PFGS for their detector response this quantity has to be known. Therefore the response matrices of all γ-ray detectors are simulated, in a energy range from 100 keV to 8 MeV with a 10 keV binning, by Geant4 [12]. In case of the HPGe detectors an additional procedure for the interpolation of the response matrix elements has been done to obtain a matrix elements in a finer energy step size of 2 keV. The detector resolution which is experimentally deduced from the energy calibration measurements is convoluted with the simulated response matrices afterwards. The matrix elements are normalized to the measured efficiencies. For the detector response correction the spectrum stripping method has been used [13]. The energy steps of the obtained emission spectra are equidistant with the same binning as the response matrices. The maximum energy limit is set to 5 MeV. The PFGS are extrapolated above 4.5 MeV for the $LaBr_3$ and above 2 MeV for the HPGe detectors. The minimum energy of the $LaBr_3$ detector was set to 100 keV and for the HPGe detectors to 150 keV due to the discriminator threshold behaviour. The final reponse corrected PFGS of our measurement are depicted in Fig. 5 in comparison to the measurement from Oberstedt et al. [14] and the calculation by the CGMF code [15], whose acquired PFGS also refers to the fission reaction ^{242}Pu(SF). The conditions and characteristics of these PFGS are shown in Table 2.

In our case the integral from the $LaBr_3$ spectrum with its lower energy threshold and lower amplitudes in comparison to the HPGe is about the same as the integrated HPGe spectrum which leads the the same multiplicity of $\bar{\nu}_\gamma \approx 8.1$. In comparison to the spectra from Oberstedt et al. and the CGMF calculation their values are lower than our measurement but agree for the most part of the spectrum within the errorbars to our results.

For the low energy region, as it is depicted in Fig. 5b), it can be seen that the PFGS of our $LaBr_3$ spectrum agrees quite well with the CGMF calculation. They differ from our HPGe spectrum and the result from Oberstedt et al. while these latter two spectra show a good agreement with each other. In the energy region for the PFGS which is shown in Fig. 5c) the CGMF calculation has

[4]The intensity should decrease exponentially in dependency of the time but cannot be seen here due to the short time window compared to the life time.

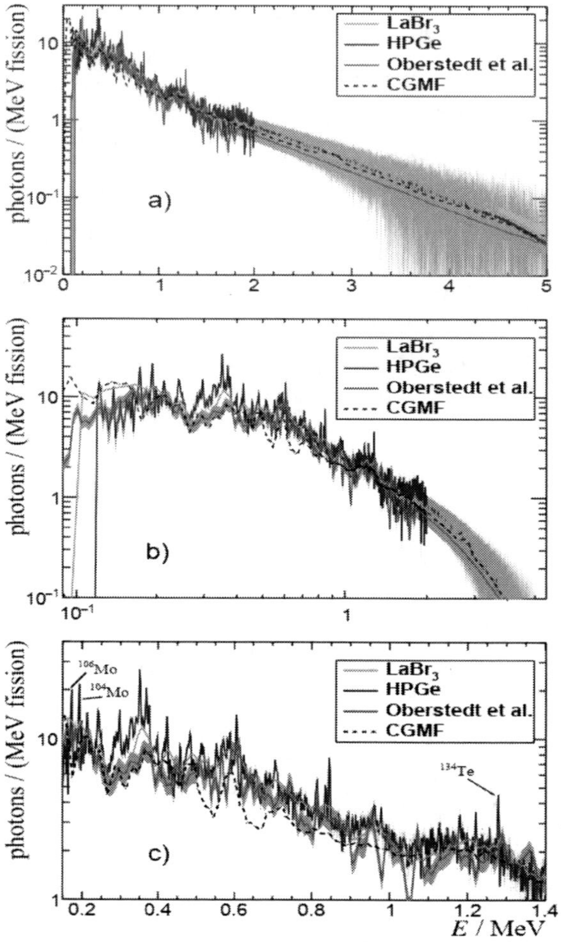

Figure 5. PFGS for ^{242}Pu from our measurement, averaged for all $LaBr_3$ and HPGe detectors, in comparison with the results from Oberstedt et.al [14] and the calculation from Stetcu [16] by the CGMF code [15]. The errorbars of the statistical uncertainties are depicted in a lighter color in comparison to its corresponding line color. In a) the PFGS with the entire energy range, in b) the PFGS in logarithmic energy scale are illustrated. The identified transitions from FF are pointed out in c) where a detail of the PFGS is shown.

significant lower values in comparison to the other measured spectra. The multiplicity of the spectrum of Oberstedt et al. and the CGMF calculation have approximately the same value of $\overline{\nu}_\gamma \approx 6.7$ due to the lower height of amplitude. This results also in a lower average total energy in comparison to our measurement whereas the mean energies agree better with each other[5].

The main contribution from the uncertainties of our PFGS characteristics result from systematic effects. These are efficiency- and response correction and the additional slow rise time trigger efficiency correction for the HPGe. The γ-ray energy from ($2^+ \rightarrow 0^+$) quadrupole transitions of major FF ^{134}Te (1279 keV), ^{104}Mo (193 keV) ^{106}Mo (172 keV) could be identified (see Fig. 5c). Their life times are in the order of ns or less [11]. So it can be assumed that they completely contribute inside

[5] In case for the HPGe the mean energy is expected to be higher due to the higher threshold.

Table 2. Parameter of the PFGS from ^{242}Pu from this measurement in comparison with the results from Oberstedt et al. and the calculation by the CGMF code. The conditions low energy threshold E_{thr} and coincidence time window ΔT of the PFGS are shown with the characteristics average multiplicity per fission $\overline{\nu}_\gamma$, total energy $\overline{E}_\gamma$ and mean energy $\overline{\epsilon}_\gamma$.

	This work		Oberstedt et al. [14]	CGMF [16]
	$LaBr_3$	HPGe		
ΔT / ns	± 3	± 12	± 10	± 10
E_{thr} / keV	100	150	100	100
$\overline{\nu}_\gamma$	8.1 ± 0.2	8.2 ± 0.5	6.72 ± 0.07	6.7
$\overline{E}_\gamma$ / MeV	6.8 ± 0.2	6.9 ± 0.4	5.66 ± 0.06	5.87
$\overline{\epsilon}_\gamma$ / keV	839 ± 10	870 ± 20	843 ± 12	870

the coincidence time window. For ^{134}Te the counts at the energy 1279 keV are enhanced by an additional contribution due to the ^{134m}Te deexcitiation which are occurring in the defined coincidence time window.

5 Conclusion

The PFGS measurement of ^{242}Pu was done with unprecedented statistics and for the first time HPGe detectors were used. Particularly for this type of detector a lot of effort has expended in terms of the the slow rise time rejection. The statistical uncertainty could be kept sufficiently low for very small energy steps which allows us to resolve E2 transitions from abundant FF. An isomeric state of the FF ^{134}Te (1279 keV) could be identified. Furthermore, the measured spectrum from the $LaBr_3$ detectors could be extended to 4.5 keV because of the high counting statistics. In comparison to the recent measurement from Oberstedt et al. and the CGMF calculation the spectral shape agrees well whereas the amplitudes differs from each other. With this setup, using $LaBr_3$ detectors and this type of fission chamber, also a neutron induced PFGS measurement is conceivable. The HPGe detectors are suited to study delayed γ-rays from isomeric states of FF, preferably from SF sources.

This work has been supported by the German Federal Ministry for Education and Science within the TRAKULA project (02NUK13A). The authors want to thank Andreas Hartman for the maintenance service of the HPGe detectors and Manfred Sobiella who made technical drawings of the MINICLUSTER which could be directly embedded in the Geant4 simulation. We want to thank A. and S. Oberstedt for their valuable advices about the PFGS measurement and also for the provision of their recent data of the PFGS of ^{242}Pu. Furthermore we want to thank I. Stetcu who made the model calculation for ^{242}Pu with the CGMF code.

References

[1] G. Rimpault, D. Bernaud, D. Blanchet, C. Vaglio-Gaudard, S. Ravaux and A. Santamarine Needs of Accurate Prompt- and delayed γ-spectrum und Multiplicity for Nuclear Reactor Designs. Phys.Proc. **31**, 3-12 (2012)

[2] WPEC. *The High Priority Request List for Nuclear Data (HPRL).* Working Party on International Evaluation Co-Operation, OECD / NEA (2011)

[3] R. Vandenbosch und J.R. Huizenga *Nuclear Fission* Academic Press Inc. (1973)

[4] T. Kögler, *Bestimmung des neutronen-induzierten Spaltquerschnitts von ^{242}Pu*. Dissertation, Technische Univeristät Dresden, Germany (2016)

[5] T. Kögler, R. Beyer, R. Hannaske, A.R. Junghans, R. Massarczyk, R. Schwengner and A. Wagner. *The nELBE (n, fis)*, Experiment, CERN-Proceedings-**2014-002**, 25-30 (2014)

[6] A. Vascon, J. Runke, N. Trautmann, B. Cremer, K. Eberhardt and C. Düllmann. Quantitative molecular plating of large-area ^{242}Pu targets with improved layer properties. Appl. Radiat. Isot. **95**, 36–43 (2015)

[7] D.W. Weißhaar, *MINIBALL: Ein neuartiges Gamma-Spektrometer mit ortsauflösenden Germaniumdetektoren* PhD thesis, Universität Köln (2003)

[8] H.G. Essel und N. Kurz. *GSI Multi-Branch System - Reference Manual, Version 5.1.* GSI, Gesellschaft für Schwerionenforschung mbH, Planckstraße 1, 64291 Darmstadt, Germany (2010)

[9] J. Adamczewski-Musch, M. Al-Turany, D. Bertini, H. G. Essel and S. Linev. *The Go4 Analysis Framework, Version 5* GSI, Gesellschaft für Schwerionenforschung mbH, Planckstraße 1, 64291 Darmstadt, Germany (2015)

[10] G.F. Knoll. *Radiation detection and measurement* John Wiley & Sons Inc., ISBN: 978-0-470-13148-0

[11] Brookhaven National Laboratory. *Nuclear Data Center* (2017)

[12] J. Allison, S. Agostinelli, K. Amako, J. Apostolakis, H. Araulo, P. Arce, M. Asai, D. Axen, S. Banerjee, G. Barrand, F. Behner, L. Bellagamba, J. Boudreau, L. Broglia, A. Brunengo, H. Burkhardt, S. Chauvie, J. Chuma, R. Chytracek, G. Cooperman et al. Geant 4- a simulation toolkit. Nucl. Instr. and Meth. A **506**, 250-303 (2003)

[13] A.K. Furr, E.L. Robinson and C. Robins. A Spectrum Stripping technique for qualitative activation analysis using monoenergetic gamma spectra. Nucl. Instr. and Meth. A **63**, 205-209 (1968)

[14] S. Oberstedt, A. Oberstedt, A. Gatera, A. Göök, F.-J. Hambsch, A. Moens, G. Sibbens, D. Vanleeuw und M. Vivaldi. Prompt fission γ-ray spectrum characteristics from ^{240}Pu(SF) and ^{242}Pu(SF) Phys. Rev. C **93** (2016)

[15] I. Stetcu, P. Talou, T. Kawano and M. Jandel. Properties of prompt-fission-γ-rays Phys. Rev. C **90** (2014)

[16] I. Stetcu. Private Communication (2017) Los Alamos National Laboratory (LANL)

Fission observables from 4D Langevin calculations with macroscopic transport coefficients

Mark D. Usang[1,4,⋆], *Fedir A.* Ivanyuk[1,2], *Chikako* Ishizuka[1], *Satoshi* Chiba[1,3], and *Joachim A.* Maruhn[5]

[1] *Tokyo Institute of Technology, Tokyo, Japan*
[2] *Institute for Nuclear Research, Kiev, Ukraine*
[3] *Astronomical Observatory of Japan, Tokyo, Japan*
[4] *Malaysia Nuclear Agency, Bangi, Malaysia*
[5] *Institut für Theoretische Physik, Frankfurt, Germany*

Abstract. We have extended the Langevin equations to 4 dimensions (4D) by allowing the independent deformation for the left (δ_1) and right fragments (δ_2) of the fissioning nucleus. At the moment we are only able to use them in conjunction with the macroscopic transport coefficients. Nevertheless, we can see a considerable improvement in the preliminary results for the fission observables, especially those related to the total kinetic energy (TKE) of fission fragments. By plotting the TKE distributions we have revealed the super-long fission modes in ^{236}U and super-short fission modes in ^{257}Fm. By plotting the distribution of δ against the fragment's TKE we have noted a correlation between the values of δ and Brosa's fission modes. We have found that the standard fission modes correspond to prolate tips of the light fragments while the complementary heavy fragments have oblate fission tips. On the other hand, if both fragments were prolate at the tips, we get super-long fission modes. If both fragments were oblate at the tips, we get super-short fission modes.

1 Introduction

The calculation of fission observables using Langevin equations has recently progressed from using 3D calculations to 4D. Previously [1], we constructed the 3D deformation potential within the two center shell model (TCSM) [2] and then allow the shape of the nucleus to evolve along the trajectories calculated by Langevin equations. The elongation of the nucleus was described by the normalized distance z_0/R_0 between the centers of the left and right potentials. Another collective coordinate, the mass asymmetry $\alpha = (A_1 - A_2)/(A_1 + A_2)$, was approximated by the asymmetry of the fragment's volume, $\alpha = (V_1 - V_2)/(V_1 + V_2)$. The fragments deformation is regulated mainly by the parameters δ_i. In 3D calculations, we put in the $\delta_1 = \delta_2$ restriction but for the current work on 4D calculations, we allow δ_1 to change independently of δ_2. Thus, in the current 4D calculations, the collective coordinates used in the Langevin calculations with macroscopic transport coefficients are $q_i = \{z_0/R_0, \delta_1, \delta_2, \alpha\}$.

usang.m.aa@titech.ac.jp;mark_dennis@nm.gov.myt
⋆e-mail: usang.m.aa@titech.ac.jp; mark_dennis@nm.gov.my

The Langevin equations [3] describe the changes of q_i and their conjugate momenta p_i with respect to time,

$$\frac{dq_i}{dt} = \left(m^{-1}\right)_{ij} p_j \; ; \quad \frac{dp_i}{dt} = -\frac{\partial V(q, E_{int})}{\partial q_i} - \frac{1}{2}\frac{\partial}{\partial q_i}\left(m^{-1}\right)_{jk} p_j p_k - \gamma_{ij}\left(m^{-1}\right)_{jk} p_k + g_{ij} R_j(t). \quad (1)$$

The inertia tensor m_{ij} is calculated by the Werner-Wheeler method [4] and friction tensor γ_{ij} is calculated by using the wall-and-window approach [5–9] with the reduction factor $k_s = 0.27$. In the text, we refer to this inertia and friction tensor as macroscopic transport coefficients as it was inspired by hydrodynamical assumptions.

The thermal intrinsic energy, E_{int} is estimated at each time step from the energy balance,

$$E_{int} = E_x - \frac{1}{2}\left(m^{-1}\right)_{ij} p_i p_j - V(q, T = 0) = aT^2 \,, \quad (2)$$

where E_x is the initial excitation energy of the system. In the case of neutron induced fission we assume $E_x = S_n + E_n$, where S_n and E_n are respectively the neutron separation energy and incident neutron kinetic energy. The temperature of the system is related to E_{int} by the Fermi gas relation, $E_{int} = aT^2$, where a is the level density parameter [10].

The strength of the random force g_{ij} is related to the diffusion tensor, $D_{ij} = g_{ik} g_{jk}$ and for the diffusion tensor we used the modified Einstein relation, $D_{ij} = T^* \gamma_{ij}$. Here T^* is the effective temperature [11, 12],

$$T^* = \frac{\hbar\varpi}{2} \coth \frac{\hbar\varpi}{2T} \,, \qquad \text{with } \hbar\varpi = 2 \text{ MeV} \,. \quad (3)$$

In the high temperature limit T^* coincides with T. At low temperature the quantal aspects became dominant and the magnitude of diffusion coefficient becomes larger than those obtained from the classical Einstein relation. The parameter ϖ is the local frequency of the collective motion and in principle it could be calculated at each deformation point. Unfortunately this would consume too much time. Hence, we simply use the constant value $\hbar\varpi = 2$ MeV. This value is close to the result from zero-th energy calculations [13]. In the limit of $T \to 0$, we have $T^* \to 1$ MeV, but at high temperature limit, $T^* \to T$.

The potential energy surface $V(q, E_{int})$ is calculated by the macroscopic-microscopic method [14–16],

$$V(q, E_{int}) = V_{\text{LD}}^{def}(q) + [V_{\text{shell}}(q) + V_{\text{pair}}(q)]\, \Phi(E_{int}) \,, \quad \text{with} \quad \Phi(E_{int}) = e^{-E_{int}/E_d} . \quad (4)$$

The macroscopic part of the potential energy $V_{\text{LD}}^{def}(q)$ was calculated within the finite-range liquid drop model [17]. The energy and pairing shell corrections $V_{shell}(q)$ and $V_{pair}(q)$ at $T = 0$ were calculated with the finite-depth deformed Woods-Saxon potential. For this the TCSM shapes were expressed in terms of deformed Cassini ovaloids and the code by Pashkevich [18, 19] was used to calculate the single-particle energies. The damping factors $\Phi(E_{int})$ in present work was calculated by Ignatyuk prescription [20] with the shell damping energy $E_d = 20$ MeV.

We start the Langevin calculations as far inside the saddle as possible in order to account for the stochastic fluctuation in the equilibrated medium of nuclear collective motion inside the saddle. As the nucleus evolves, we check the radius of the neck, R_{neck}. Once the condition $R_{\text{neck}} = 0$ is fulfilled, we consider such event as a fission event. The value of α at this configuration is extracted and the pre-neutron emission fragments mass are obtained from the formula, $A_i^{\text{pre}} = 0.5 A_{CN}(1 + (-1)^{i-1}\alpha)$ for $i = \{1, 2\}$ in reference to the left and right fission fragment, respectively. A_{CN} is the compound nucleus mass, which is here either 236 or 257. Hence, we are able to calculate the post-neutron emission mass by,

$$A_i^{\text{post}} = A_i^{\text{pre}} - \bar{\nu}(A^{\text{pre}}). \quad (5)$$

In present work we simply use the experimental results for $\nu(A^{\mathrm{pre}})$ from the reactions with the incident neutron energy $E_n = 0.5$ MeV [21].

2 Numerical Results

The preliminary results of our calculations for ^{236}U and ^{257}Fm are shown in Fig. 1 and Fig. 2, respectively. In each figure, the fission fragment mass yield and total kinetic energy (TKE) distribution are shown. In each of these figures, we plot the pre-neutron emission fission fragment yield from our calculation. In the case of ^{236}U, we plot also the estimated post-neutron emission fission fragment yield from the calculated pre-neutron yield.

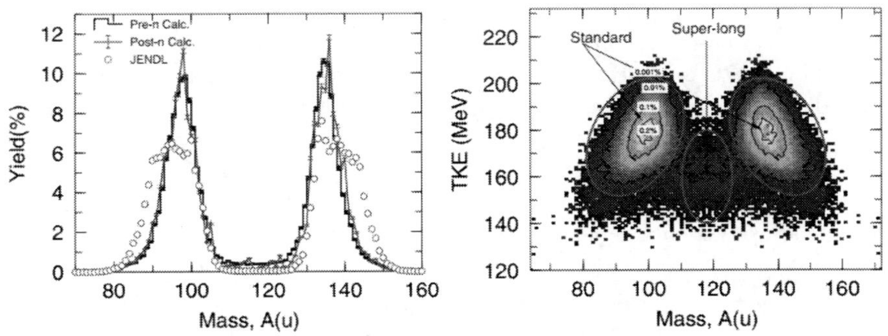

Figure 1. The calculated fission fragment yield of ^{236}U at $E_x = 7$ MeV on the left, and TKE on the right. On the left figure, pre-neutron emission fission fragment (Pre-n Calc.) and post-neutron emission fission fragment (Post-n Calc.) are compared with the JENDL thermal fission product yield. The post-neutron emission fission fragment mass is calculated according to Eq. (5). On the right figure, we mark Brosa's standard fission modes (Standard) and super-long fission modes (Super-long).

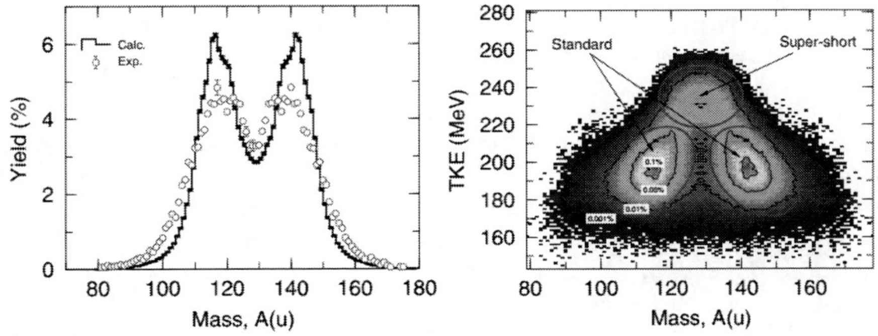

Figure 2. The calculated fission fragment yield of ^{257}Fm at $E_x = 7$ MeV on the left, and TKE on the right. The pre-neutron emission fission fragment yield (Calc.) is compared with the experimental [22] spontaneous fission data (Exp.). We mark the presence of standard and super-short fission modes on the TKE figure.

Both calculations for ^{236}U and ^{257}Fm are done at $E_x = 7$ MeV and we see that the average mass of the light and heavy fragments align with the estimates from either evaluated or experimental yields.

More precisely, in the case of ^{236}U, the pre-neutron emission light fission fragment mass average is 99.17 and the heavy fission fragment mass average is 136.83. After neutron emission, the light and heavy fragment mass average are 97.14 and 135.59 each. This should be compared with the evaluated data from JENDL [23] – the average mass of 94.75 and 138.68 for the light and heavy fission fragment, respectively. In the case of ^{257}Fm, the average mass for light and heavy fragments are 116.40 and 140.60. The experimental mass averages [22], 115.41 and 141.40 each for light and heavy fission fragments, are very close to our calculations. This clearly implies that the potential surface is correct or at least the position of the potential minimum or saddle is correct.

However, the peaks are thinner and too tall. It seems the macroscopic transport coefficients used in present work are not so good.

2.1 Distributions of TKE

Regarding the TKE, on the other hand, the general shape and distribution are quite good. The TKE distribution on both ^{236}U and ^{257}Fm on Figs. 1 and 2 clearly reveals the standard fission modes. In ^{236}U, the standard fission modes are prevalent regardless of the fission fragments TKE. According to Brosa et al. [24] these fission modes are formed by the prescission shapes of the nucleus. The standard fission modes are described to be *slightly asymmetric and of "standard" length.* Indeed, in the case of ^{236}U, the average length measured in terms of center of mass distance $\langle R_{12} \rangle$ is around 2.5 to 2.6 times larger than the radius of spherical nucleus R_0 for the fragments with $A \leqslant 110$ and $A \geqslant 130$.

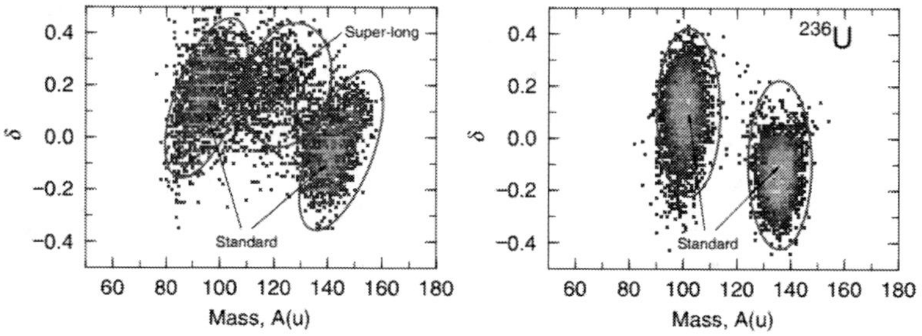

Figure 3. The distribution of deformation parameter δ_i calculated at $E_x = 7$ MeV for ^{236}U at the scission configuration with low TKE (160 MeV ≤ TKE ≤ 165 MeV, left) and high TKE (185 MeV < TKE ⩽ 190 MeV, right). We indicate the emerging Brosa's fission modes.

In Fig. 3 we show the distribution of parameter δ at the scission point. One can see the clusters of fission fragments with an asymmetric mass around $\delta \approx 0.2$ for the lighter fragments and $\delta \approx -0.1$ for the heavy fragments. The value of δ describes the shapes of tips of the fissioning nuclei. Positive δ corresponds to a prolate tip, negative δ corresponds to oblate tip shape and $\delta = 0$ means that the tips are spherical. It is easy to infer from here, that if the tips of both fragments are prolate, we will get a longer prescission nucleus as it was stretched apart. If they are both oblate, the prescission fragments would be closer to each other in order to conserve the volume. The long prescission shapes are described by Brosa as the super-long fission modes and the short prescission shapes are called super-short fission modes.

We can observe the super-long fission modes in ^{236}U from Figs. 1 and 3 at symmetric mass around $110 \leqslant A \leqslant 130$. The fragments at these symmetric mass mostly appear at TKE lower than 180 MeV. In Fig. 3, it is clear that it clusters around $\delta \approx 0.2$ also, but because it was of symmetric mass, both complementary fragments would be similarly prolate at the tips. The final confirmation of this fission modes came from the length of the prescission shapes, yielding $\langle R_{12}\rangle$ of approximately $2.8R_0$ to $2.9R_0$. This is far longer than the prescission shape we see for the standard fission modes of ^{236}U.

In ^{257}Fm the standard fission modes have prescission shapes that are slightly longer than that for ^{236}U, $\langle R_{12}\rangle \approx 2.7R_0$. The TKE of the standard fission modes however only gets as large as 220 MeV, as we can see from Fig. 2. In fact, as we can see from Fig. 4, the standard fission modes of ^{257}Fm are present only at low TKE and completely absent from contributing to high TKE fission fragments. Similar to ^{236}U standard fission modes, in ^{257}Fm the light fragments are prolate and the heavy fragments are oblate at the tips. At high TKE of around 235 MeV, only fission fragments from the super-short fission modes can be seen. At such high TKE, we see that the fission fragments clustered around symmetric fragment mass with very oblate fissioning shapes at $\delta \approx -0.2$. As such, the length of the prescission shape, $\langle R_{12}\rangle$ is less than $2.45R_0$ at the symmetric mass of A=128.

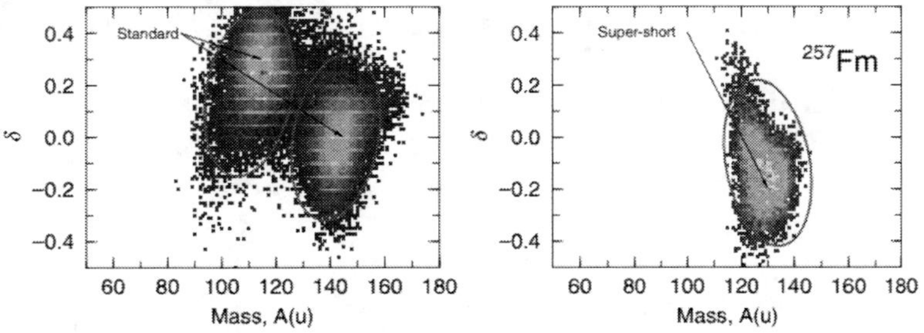

Figure 4. The distribution of deformation parameter δ_i calculated at E_x = 7 MeV for ^{257}Fm at scission that has low TKE (190 MeV ≤ TKE ≤ 195 MeV, left) and high TKE (235 MeV < TKE ⩽ 240 MeV, right). We indicate also the emerging Brosa's fission modes.

3 Conclusion

The systematics by Viola [25] predicts the TKE for ^{236}U and ^{257}Fm at 170.15 MeV and 194.31 MeV, respectively. Both these estimates deviate from the corresponding experimental TKE. In the case of ^{236}U, the systematics slightly overestimates the TKE for thermal fission which gave the value around 169.71 MeV, according to JENDL. On the other hand, Viola systematics underestimates the TKE for ^{257}Fm experimental TKE [26] of 197.6 MeV. This deviation can be caused by the presence of either super-long or super-short fission modes, if the Brosa systematic only tracks TKE from standard fission modes. Our calculations, however, are unable to further clarify this problem at the moment because the average TKE in our calculation is too high at 176.59 MeV and 202.55 MeV, respectively, for ^{236}U and ^{257}Fm due to the slightly poor mass yield. If we were able to match either the evaluated or experimental average TKE by fixing the yield, we would probably be able to examine the validity of this question by removing the contribution from symmetric yield.

In the present study, the shape of the tips seems to be inherent in allowing for further lengthening or shortening of prescission shape. If the tips are prolate, the shape at the scission point is more elongated and the Coulomb interaction energy between fragments – smaller, and oblate shapes lead to a larger Coulomb interaction energy.

It would also be interesting to expand the current work to consider the fission observables from ^{258}Fm and other nuclides of interest.

Recently we have submitted a more detailed paper [27] on the fission observables from ^{236}U using the 4D Langevin approach with macroscopic transport coefficients and some other improvements.

The present work included the results for *Comprehensive study of delayed-neutron yields for accurate evaluation of kinetics of high-burn up reactors* entrusted to Tokyo Institute of Technology by the Ministry of Education, Culture, Sports, Science and Technology of Japan (MEXT).

References

[1] M.D. Usang, F.A. Ivanyuk, C. Ishizuka, S. Chiba, Phys. Rev. C **94**, 044602 (2016)
[2] J. Maruhn, W. Greiner, Z. Phys. **251**, 431 (1972)
[3] Y. Abe, S. Ayik, P.G. Reinhard, E. Suraud, Phys. Rep. **275**, 49 (1996)
[4] K.T.R. Davies, A.J. Sierk, J.R. Nix, Phys. Rev. C **13**, 2385 (1976)
[5] J. Blocki, Y. Boneh, J. Nix, J. Randrup, M. Robel, A. Sierk, W. Swiatecki, Ann. Phys. **113**, 330 (1978)
[6] G. Adeev, A. Karpov, P. Nadtochii, D. Vanin, Phys. Part. Nucl. **36**, 378 (2005)
[7] H.J. Krappe, K. Pomorski, *Theory of Nuclear Fission: A Textbook*, Vol. **838** (Springer Science & Business Media, 2012)
[8] W. Swiatecki, Nucl. Phys. A **428**, 199 (1984)
[9] J.R. Nix, A.J. Sierk, Nucl. Phys. A **428**, 161 (1984)
[10] J. Tōke, W. Światecki, Nucl. Phys. A **372**, 141 (1981)
[11] H. Hofmann, D. Kiderlen, Int. J. Mod. Phys. E **07**, 243 (1998)
[12] H. Hofmann, C. Grégoire, R. Lucas, C. Ngô, Z. Physik A **293**, 229 (1979)
[13] D.L. Hill, J.A. Wheeler, Phys. Rev. **89**, 1102 (1953)
[14] V. Strutinsky, Nucl. Phys. A **95**, 420 (1967)
[15] V. Strutinsky, Nucl. Phys. A **122**, 1 (1968)
[16] M. Brack, J. Damgaard, A.S. Jensen, H.C. Pauli, V.M. Strutinsky, C.Y. Wong, Rev. Mod. Phys. **44**, 320 (1972)
[17] H.J. Krappe, J.R. Nix, A.J. Sierk, Phys. Rev. C **20**, 992 (1979)
[18] V. Pashkevich, Nucl. Phys. A **169**, 275 (1971)
[19] V. Pashkevich, A. Rusanov, Nucl. Phys. A **810**, 77 (2008)
[20] A. Ignatyuk, K. Istekov, G. Smirenkin, Sov. J. Nucl. Phys. **29**, 450 (1979)
[21] R. Müller, A.A. Naqvi, F. Käppeler, F. Dickmann, Phys. Rev. C **29**, 885 (1984)
[22] J.P. Balagna, G.P. Ford, D.C. Hoffman, J.D. Knight, Phys. Rev. Lett. **26**, 145 (1971)
[23] K. Shibata, O. Iwamoto, T. Nakagawa, N. Iwamoto, A. Ichihara, S. Kunieda, S. Chiba, K. Furutaka, N. Otuka, T. Ohasawa et al., J. Nucl. Sci. Technol. **48**, 1 (2011)
[24] U. Brosa, S. Grossmann, A. Müller, Phys. Rep. **197**, 167 (1990)
[25] V.E. Viola, K. Kwiatkowski, M. Walker, Phys. Rev. C **31**, 1550 (1985)
[26] D.C. Hoffman, M.R. Lane, Radiochim. Acta **70**, 135 (1995)
[27] C. Ishizuka, M.D. Usang, F.A. Ivanyuk, J.A. Maruhn, K. Nishio, S. Chiba (submitted to Phys. Rev. C)

EPJ Web of Conferences **169**, 00028 (2018) https://doi.org/10.1051/epjconf/201816900028
Theory-4

Study of photon emission with the fission event generator FREYA

Ramona Vogt[1,2,⋆] and *Jørgen* Randrup[3,⋆⋆]

[1] *Nuclear and Chemical Sciences Division, Lawrence Livermore National Laboratory, Livermore, CA 94551, USA*
[2] *Physics Department, University of California at Davis, Davis, CA 95616, USA*
[3] *Nuclear Science Division, Lawrence Berkeley National Laboratory, Berkeley, CA 94720, USA*

Abstract. The event-by-event fission model FREYA is employed to study photon observables. The model has been expanded beyond the simple statistical photon emission reported previously to include the discrete RIPL-3 lines. We update the earlier results and discuss the sensitivity of the results to the FREYA input parameters.

1 Introduction

The computational model FREYA [1–7] generates complete fission events, i.e. it provides the full kinematic information on the two product nuclei as well as all the emitted neutrons and photons. In its development, an emphasis had been put on speed so that large event samples can be generated fast. To make FREYA events as realistic as possible, FREYA relies on experimental data supplemented by simple physics-based modeling. In its standard version, to treat a given fission case, FREYA needs the fission fragment mass distribution $Y(A)$ and the total kinetic energy TKE(A) for the relevant energy. $Y(A)$ is taken either directly from the measured yields or as a five-Gaussian fit to the data which makes it possible to parameterize its energy dependence [3].

In order to generate an event, FREYA first selects the mass split based on $Y(A)$. The fragment charges are then sampled from the normal distributions suggested by experiment [3]. The linear and angular momenta of the two fragments and their internal excitations are subsequently sampled. After their formation, the fully accelerated fragments de-excite first by neutron evaporation and then by photon emission. In addition to spontaneous fission, FREYA treats neutron-induced fission up to $E_n = 20\,\mathrm{MeV}$; the possibility of pre-fission evaporation is considered as well as pre-equilibrium neutron emission (which plays a role at the highest energies). For full details, see Refs. [1–7].

In the new version of FREYA [7], we modulate the statistical emission of photons by the giant dipole resonance (GDR). Each photon emission reduces the total fragment angular momentum by $1\hbar$. Statistical photons are emitted isotropically with an energy sampled from a black-body spectrum with the GDR form factor,

$$\frac{dM_\gamma}{dE_\gamma} \sim \frac{\Gamma^2_{\rm GDR}E^2_\gamma}{(E^2_\gamma - E^2_{\rm GDR})^2 - \Gamma^2_{\rm GDR}E^2_\gamma}\, E^2_\gamma\, e^{-E_\gamma/T} \qquad (1)$$

⋆e-mail: vogt2@llnl.gov
⋆⋆e-mail: jrandrup@lbl.gov

where T is the temperature. The position of the resonance is taken as $E_{GDR}/\text{MeV} = 31.2/A^{1/3} + 20.6/A^{1/6}$ [8], while its width is $\Gamma_{GDR} = 5\,\text{MeV}$. Note that in Ref. [4], photon emission assumed black-body radiation only with a spectrum of the form $dM_\gamma/dE_\gamma \sim E_\gamma^2\,e^{-E_\gamma/T}$. The GDR has the effect of making the photon spectrum harder.

In addition, we now include the low energy photon lines from the RIPL-3 database [9]. The RIPL-3 library tabulates a large number of discrete electromagnetic transitions for nuclei throughout the nuclear chart but complete information is available for only relatively few of them. However, as we will discuss, by invoking certain assumptions it is possible to construct, for each product species, a table of the possible decays from the lowest discrete levels, i.e. the level energies $\{\varepsilon_\ell\}$, their half-lives $\{t_\ell\}$, and the branching ratios of their various decays. Then whenever the fragment de-excitation leads to an excitation energy below any of the tabulated levels, FREYA switches to a discrete cascade based on the RIPL-3 data. The discrete cascade is continued until the half-life t_ℓ exceeds the specified value of t_{max} or the ground state is reached. These lines have the most immediate impact on the low energy end of the photon spectrum, $E_\gamma < 1$ MeV.

The RIPL-3 data files are organized by element, with one file for each Z value. Each such file contains similarly structured listings for those isotopes of that element for which data exist. For any tabulated nucleus, we seek to include all levels "in a complete level scheme", as indicated in the isotope header line on the data file. Each listed level ℓ may decay into a total of $n(\ell)$ lower levels $\{\ell'\}$ and the associated relative transition rate $P(\ell \to \ell')$ is indicated for each one, if available. Often the rate for a listed transition is not given and the corresponding transition is then ignored.

However, if all the decay rates from a given level ℓ are missing we assign decay rates from that level to all of the lower levels ℓ' based on a simple phase-space consideration:

$$P(\ell \to \ell') \sim [E_\ell - E_{\ell'}]^2 \exp[-(J_\ell - J_{\ell'})^2/2d_J^2]$$

Here E_ℓ is the energy of level ℓ, J_ℓ is its listed spin, and we use $d_J = 1$. It should be recognized that there are many more added transitions (201568) than tabulated transitions (75809). That is primarily because a level for which there are tabulated decay rates tends to decay to only some of the levels below it, whereas a level without tabulated decay rates is allowed to decay to *any* level below it. Furthermore, levels without tabulated decay rates tend to be high-lying and thus have many lower levels.

In the previous gamma decay procedure, the product nucleus first made statistical radiation (while conserving the angular momentum) until the yrast line was reached and then proceeded towards the ground state by collective emission. In order to incorporate the discrete decays, we take note of the highest tabulated discrete energy E_ℓ and then follow the earlier procedure until the total excitation has fallen below that value. (It should be noted that the statistical decay procedure was modified so that the angular momentum is reduced by one unit for each emission in order to better emulate the predominantly $E1$ and $M1$ character of those transitions.) In order to ensure that the nucleus arrives at a tabulated level, the energy of the last statistical transition is increased (slightly) so that the final state is the closest lower-lying level. Further de-excitation is then carried out by means of the discrete rates, as described above.

If there are no RIPL-3 transitions available for a given fragment, the de-excitation occurs by emission of "collective" photons along the yrast line. Each emitted photon reduces the angular momentum by $2\hbar$, as in Ref. [4].

In either case, photon emission continues until the energy is below some specified minimum energy, g_{min}, which is generally detector dependent.

FREYA contains a number of adjustable parameters that control various physics aspects:

dTKE, an overall shift of $\overline{\text{TKE}}(A)$ relative to the input TKE(A), used to adjust to the average neutron multiplicity $\overline{\nu}$;

e_0, the overall scale of the Fermi-gas level density parameters;

x, the advantage in excitation energy given to the light fragment;

c_T, the relative statistical fluctuation in the fragment excitations;

c_S, the ratio of the "spin temperature" to the "scission temperature";

g_{min}, the minimum energy photon registered and counted by FREYA;

t_{max}, the maximum half-life of a level during the photon decay process which stops when the level half-life exceeds t_{max}.

The three parameters that most affect photon observables are c_S, g_{min} and t_{max}. We give some examples here of how changing these parameters affects the photon results. The value of c_S governs the amount of rotational energy in the system and thus strongly affects most observables. The values of g_{min} and t_{max} are detector dependent, related to the energy threshold for photon detection and the time gate for real detectors so that the results for a particular experimental configuration depend on this information to make a comparison. We will show the effect of changing c_S and g_{min} on the total photon energy and multiplicity as a function of fragment mass. However, the effect of changing t_{max} on the photon observables is more subtle. Therefore, we will show the changes in the photon multiplicity relative to a maximum time of $\sim$5 μs for several different values of g_{min} and c_S.

2 Parameter sensitivity

We first discuss the effect on photon observables caused by changing the parameter c_S, effectively changing the rotational energy in the system. The other parameters are kept fixed while c_S is varied. We choose to show results for $c_S = 0.2$, 0.8, 1.4 and 2.0. The value of 0.8 is rather close to the best fit value of 0.87 for ^{252}Cf(sf) [10] but here we take equal spacings between values instead of using a particular fitted value. The value 0.2 is taken as a lower bound, reducing the rotational energy in the system without removing it altogether. The two values larger than unity serve to illustrate the range of the effect.

Figure 1 presents the total photon energy and photon multiplicity as functions of fragment mass for the four values of c_S. The statistical uncertainties on the one million FREYA events used to obtain the results for each value of c_S are shown. They are typically very small. For example, the uncertainties on the energy in Fig. 1a) are typically negligible, only appearing far into the tails of the fragment distributions, $A < 90$ and $A > 160$, where only a few hundred events are sampled. Because the dip in the mass yield at symmetry is reduced for ^{252}Cf(sf) relative to other asymmetrically fissioning isotopes, the uncertainties near $A = 126$ are reduced relative to those in the tails.

In Fig. 1a), it can be seen that the photon energy simply grows by approximately 0.5 MeV when c_S increases by 0.6. There is no visible mass dependence on changing c_S. However, the photon multiplicity does show some modification of the fragment mass dependence with increasing c_S, see Fig. 1b). While the general trend is the same for all values of c_S, there is a larger increase in the multiplicity for the heavy fragment while the dip at the $A \sim 132$ doubly-closed shell deepens with increasing c_S. In addition, there appears to be a rather flat plateau for the light fragment masses $100 < A < 120$ for $c_S = 0.2$ that acquires a positive slope as c_S increases.

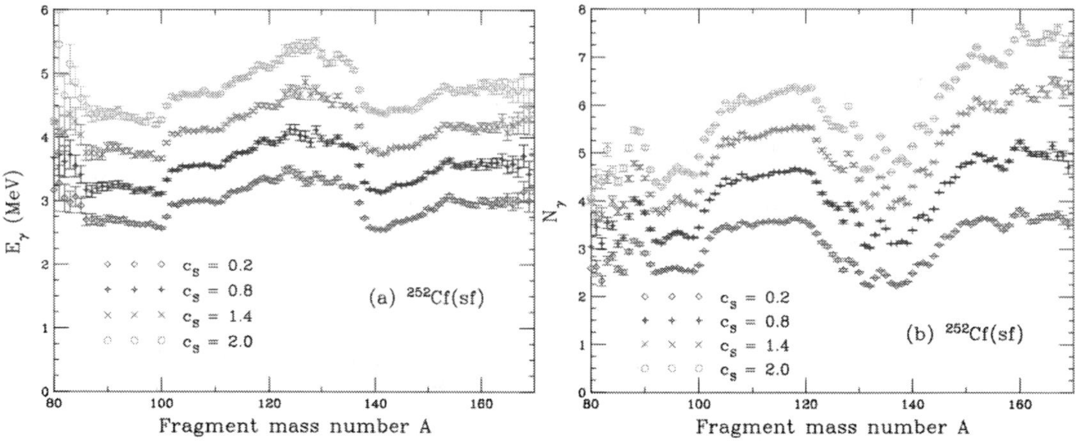

Figure 1. The total photon energy (a) and photon multiplicity (b) as functions of fragment mass for ^{252}Cf(sf) from FREYA calculated for four different values of the parameter c_S which sets the amount of the fragment spin. Results are shown for $c_S = 0.2$ (red), 0.8 (blue), 1.4 (magenta) and 2.0 (green). Each calculation is based on one million FREYA events. The statistical uncertainties in the calculations are shown.

We note, however, that these results are calculated assuming that no other parameter value changes. Thus, if the total excitation energy is held fixed, increasing the rotational energy, as happens for increased c_S, then less energy is available for neutron emission. Thus increasing c_S while keeping the other parameters fixed will decrease the average neutron multiplicity. Although this could be compensated for by changing the value of *d*TKE, one has to be careful to adjust it within reasonable physics limits to not spoil the agreement of the calculation with other measured observables.

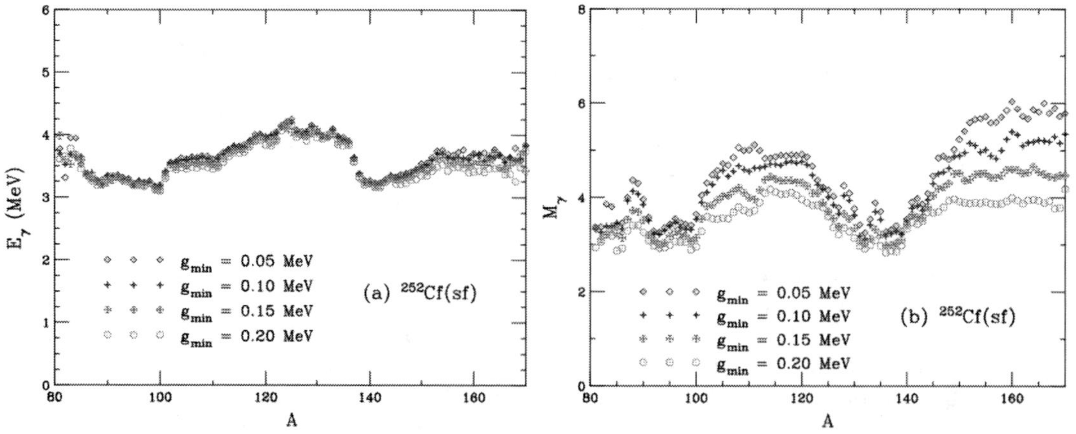

Figure 2. The total photon energy (a) and photon multiplicity (b) as functions of fragment mass for ^{252}Cf(sf) from FREYA calculated for four different values of the parameter g_{min}, the minimum photon energy detected. Results are shown for $g_{\text{min}} = 0.05$ (red), 0.10 (blue), 0.15 (magenta) and 0.20 (green). Each calculation is based on one million FREYA events.

Figure 2 shows the result of changing $g_{\min}$ between 0.05 and 0.2 MeV. Since the results overlap significantly here, we do not show the statistical uncertainties. However, we note that they would be similar to those shown in Fig. 1. The change in total photon energy, shown in Fig. 2a), is negligible since the largest value, $g_{\min} = 0.2$ MeV, makes, on average, a $\sim 5\%$ change in the photon energy emitted from a given fragment. The effect on the photon multiplicity is significant albeit not as large as changing c_S by a factor of 10, between 0.2 and 2, as shown in Fig. 1b). As $g_{\min} \rightarrow 0$, more and more soft photons could be emitted, increasing the total multiplicity.

In Fig. 2b), we can see that the largest change in M_γ with A occurs for $A < 100$ and $125 < A < 140$. Both are in the vicinity of closed nuclear shells where one expects reduced emissions due to the lower total excitation energy imparted to these fragments. The photon multiplicity from the heavy fragment appears to be the most affected by changing $g_{\min}$ with a $\sim 50\%$ change in M_γ at $A \sim 160$ relative to a $\sim 20\%$ change near $A \sim 110$. The difference arises from the higher spin of the heavy fragment for the best fit value of c_S, see Ref. [7]. The higher spin implies larger rotational energy, a great probability of photon emission and thus higher photon multiplicity from the heavy fragments. Because the multiplicity dependence on A is reduced as $g_{\min}$ increases, M_γ may be expected to become approximately independent of A at higher values of $g_{\min}$ where only statistical photon emission, generally higher energy, is possible. For the effect of increasing $g_{\min}$ up to 2 MeV on the average values of E_γ and M_γ, see Ref. [7].

With the addition of the RIPL-3 library discrete photon lines, the time window in which the detector operates can affect the measured photon spectrum, particularly at low energies. For example, if the photon cascade from a fission fragment lands on a long-lived isomeric state, the decay will not proceed further during the measurement time and no more photons will be emitted during the process. FREYA can populate such isomeric states based on the initial fragment distributions and neutron emission.

To show how these ratios could change with inputs, we look first at the cumulative photon multiplicity and total photon energy as a function of $t_{\max}$ for different values of $g_{\min}$ in Fig. 3a). We choose relatively low values of $g_{\min}$, from 0.05 to 0.20 MeV. Given the rather small differences between

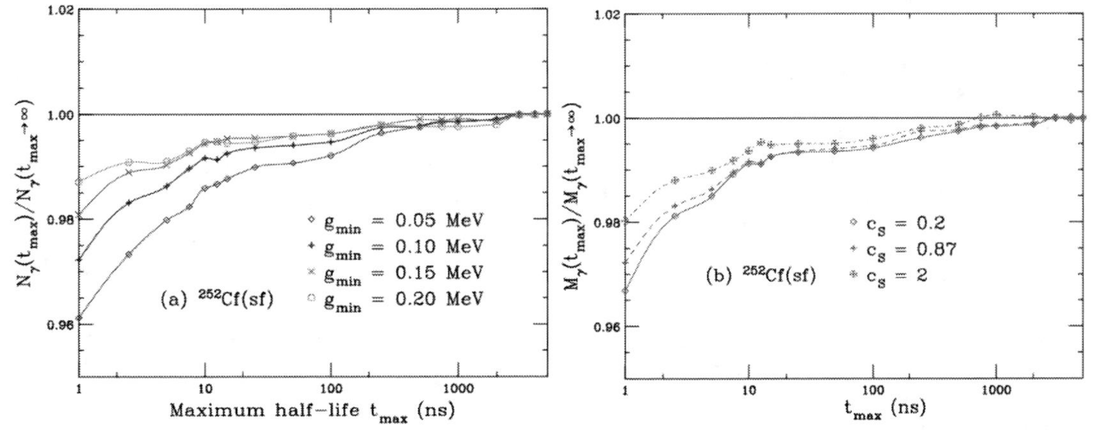

Figure 3. The total photon multiplicity relative to the maximum value at $t_{\max} \rightarrow \infty$ (effectively $t_{\max} = 5\ \mu$s) for ^{252}Cf(sf) as a function of photon half-life $t_{\max}$ for (a) changing $g_{\min}$ ($g_{\min} = 0.05$ (solid), 0.10 (dashed), 0.15 (dot-dashed) and 0.20 (dotted) MeV) and (b) changing c_S ($c_S = 0.2$ (solid), 0.87, the default value (dashed) and 2.0 (dot-dashed)). Each point is from a calculation of one million FREYA events.

$g_{\min} = 0.15$ and 0.20 MeV, the results for higher values of $g_{\min}$ would become indistinguishable, especially for higher $t_{\max}$. The largest effect is on the multiplicity, shown in Fig. 3a), and for $t_{\max} \leq 100$ ns. The greatest difference is between $g_{\min} = 0.05$ MeV and 0.10 MeV. Higher values of $g_{\min}$ have a reduced effect because the discrete levels tend to emit rather soft photons. The effect on the cumulative total photon energy is minimal.

Figure 3b) shows the dependence on the cumulative photon multiplicity on $t_{\max}$ for several values of c_S. The middle ratio is the best fit value from the ^{252}Cf(sf) fit [10], $c_S = 0.87$ and is the same as the results shown in Fig. 3a) for $g_{\min} = 0.10$ MeV. The other two values, $c_S = 0.2$ and 2 are the upper and lower limits used in the calculations shown in Fig. 1. The dependence on c_S is weaker than that on $g_{\min}$.

We do not show uncertainties here, the uncertainty on the average photon multiplicity over one million FREYA events is on the sub-percent level. However, some of the non-smooth behavior of the ratios may be due to such sub-percent fluctuations in the total multiplicity between FREYA runs.

The results presented here are only a relatively small subset of the observables we have studied. To see all the results as well as comparison to available data, see Ref. [7].

The work of R.V. was performed under the auspices of the U.S. Department of Energy by Lawrence Livermore National Laboratory under Contract DE-AC52-07NA27344. The work of J.R. was performed under the auspices of the U.S. Department of Energy by Lawrence Berkeley National Laboratory under Contract DE-AC02-05CH11231. The authors also thank the United States Department of Energy National Nuclear Security Administration Office of Defense Nuclear Nonproliferation Research and Development for support for this work.

References

[1] J. Randrup and R. Vogt, Phys. Rev. C **80**, 024601 (2009)
[2] R. Vogt and J. Randrup, Phys. Rev. C **84**, 044621 (2011)
[3] R. Vogt, J. Randrup, D.A. Brown, M.A. Descalle, and W.E. Ormand, Phys. Rev. C **85**, 024608 (2012)
[4] R. Vogt and J. Randrup, Phys. Rev. C **87**, 044602 (2013)
[5] R. Vogt and J. Randrup, Phys. Rev. C **90**, 064623 (2014)
[6] J. Randrup and R. Vogt, Phys. Rev. C **89**, 044601 (2014)
[7] J. Randrup and R. Vogt, submitted to Phys. Rev. C [arXiv:1708.07312]
[8] B.L. Berman and S.C. Fultz, Rev. Mod. Phys. **47**, 713 (1975)
[9] R. Capote *et al.*, Nucl. Data Sheets **110**, 3107 (2009)
[10] R. Vogt, A. Nicholson, J. Randrup, I. Gauld and S. Croft, Proc. 1st ANS Advances in Nuc. Nonpro. Tech. and Policy, Santa Fe, NM, 2016, LLNL-CONF-690741

Angular distribution of scission neutrons studied with time-dependent Schrödinger equation

Takahiro Wada[1,★], *Tomomasa* Asano[1], and *Nicolae* Carjan[2]

[1] *Department of Pure and Applied Physics, Kansai University, 3-3-35 Yamate-cho, Suita, 564-8680, Japan*
[2] *Flerov Laboratory of Nuclear Reactions, Dubna, Moscow region, Russia*

Abstract. We investigate the angular distribution of scission neutrons taking account of the effects of fission fragments. The time evolution of the wave function of the scission neutron is obtained by integrating the time-dependent Schrodinger equation numerically. The effects of the fission fragments are taken into account by means of the optical potentials. The angular distribution is strongly modified by the presence of the fragments. In the case of asymmetric fission, it is found that the heavy fragment has stronger effects. Dependence on the initial distribution and on the properties of fission fragments is discussed. We also discuss on the treatment of the boundary to avoid artificial reflections.

1 Introduction

At the moment of scission, the neck that has connected the two fission fragments ruptures. This neck rupture can be very quick because the number of nucleons involved in the neck region is relatively small. After the scission, the neck disappears being absorbed by the fragments leaving some nucleons behind in the neck region and these nucleons are observed as particle emission. The number of scission particles carries the information on the scission process. There have been attempts to estimate the number of scission neutrons theoretically [1,2]. The results depend on the nuclear shape such as the neck radius before scission. If we extract the reliable number of scission neutrons from experiments, we can get information on the nuclear shape at the time of scission.

In measuring the scission neutrons experimentally, the information on their angular distribution is essential. The emission of scission neutron is normally supposed to be isotropic in the rest frame of the mother nucleus in the lowest order approximation. However, since they are emitted in the close vicinity of the fission fragments, the final angular distribution of the scission neutron is modified by the fragments through scattering and re-absorption. In the previous study, we proposed a framework to calculate the angular distribution of scission neutrons by solving a time-dependent Schrödinger equation [3]. It was shown that the angular distribution of the scission neutron is strongly modified by the presence of the fragments, namely, the attractive nuclear potential enhances the yields around 0 and 180 degrees, while the absorptive potential diminishes them. We also investigated the angular distribution of scission protons and discussed the effect of the Coulomb field of the fragments [4]. It was shown that the angular distribution of the scission proton is also strongly modified by the fragments. The Coulomb field of the fragments tends to focus the yields around 90 degrees, while the

★e-mail: wadataka@kansai-u.ac.jp

attractive nuclear potential enhances the yields around 0 and 180 degrees. In this paper, we examine the dependence of the angular distribution of scission neutrons on the initial distribution of neutron and on the properties of fission fragments, such as deformation of the fragments. In the next section, the formulation to calculate the angular distribution of the scission particles is revisited. In Sect. 3, results are presented for the case of the fission of ^{236}U for both symmetric and asymmetric divisions. Next, we discuss on the treatment of the boundary to avoid artificial reflection. Finally, a summary is given.

2 Framework

We start with a time-dependent Schrödinger equation (TDSE),

$$i\frac{\partial\psi}{\partial t} = H\psi, \quad H = -\frac{1}{2m}\nabla^2 + U, \quad \hbar = 1, \tag{1}$$

where ψ denotes the neutron wave function, H is the Hamiltonian, and U is the potential that represents the effect of the fission fragments. Assuming the axial symmetry, we solve the TDSE in two-dimensional grid space (ρ, z). The time development is obtained with the use of the mid-point integration,

$$\psi(t + \Delta t) = \psi(t) - i\Delta t\, H\psi(t + \Delta t/2). \tag{2}$$

By decomposing ψ into the real and the imaginary part, $\psi = P + iQ$, the numerical solution is obtained using the leap-frog method [3]. The potential U has two centers and is parameterized in Woods-Saxon form centered at the position of each fragment. For the case of spherical fragments, U is expressed as,

$$U(\rho, z) = \frac{V_0 + iW_0}{1 + \exp\left(\frac{\sqrt{\rho^2+(z+B_2)^2}-R_{F2}}{a}\right)} + \frac{V_0 + iW_0}{1 + \exp\left(\frac{\sqrt{\rho^2+(z-B_1)^2}-R_{F1}}{a}\right)}, \tag{3}$$

where B_i is the distance between the center of scission neutron distribution and the i-th fragment, a is the diffuseness, and R_{Fi} is the radius of the potential of i-th fragment. The distance Bi is given as $B_i = \eta R_{Fi}$, where η is the elongation parameter which is to be deduced from the systematics of the total kinetic energy.

We adopt a Gaussian wave packet for the initial wave function of the scission neutron, $\psi(t = 0) = C\exp(-(\alpha_\rho\rho^2 + \alpha_z z^2)))$, where C is a normalization constant and α's denotes the width of the wave packet. We now set a sphere of radius R in the grid space and calculate the neutron flux on this spherical surface,

$$\mathbf{j}(r, t) = \frac{1}{2im}\left(\psi\,\nabla\psi * -\psi * \nabla\psi\right). \tag{4}$$

We then calculate the number of outgoing scission particles per unit solid angle per unit time and integrate it with time. The angular distribution is given by the number of particles per unit solid angle that passed the surface up to time t,

$$\frac{dN(\theta, t)}{d\Omega} = \int_0^t \frac{d^2N(\theta, t')}{d\Omega dt}dt' = \int_0^t \mathbf{j}(R, \theta, t')\cdot\mathbf{n}(R, \theta)R^2dt', \quad \mathbf{n} = e_r. \tag{5}$$

The energy density and the energy flow are defined respectively as

$$\varepsilon(\mathbf{r}, t) = \frac{\hbar^2}{2m}\frac{\partial\psi*}{\partial\mathbf{r}}\cdot\frac{\partial\psi}{\partial\mathbf{r}} + U(\mathbf{r}, t)\,\psi * \psi\,, \tag{6}$$

and

$$\mathbf{j}_\varepsilon(\mathbf{r},t) = \frac{\hbar}{2im}\left\{\frac{\hbar^2}{2m}\left(\frac{\partial\psi}{\partial\mathbf{r}}\cdot\nabla\frac{\partial\psi*}{\partial\mathbf{r}} - \frac{\partial\psi*}{\partial\mathbf{r}}\cdot\nabla\frac{\partial\psi}{\partial\mathbf{r}}\right)\right\} + U(\mathbf{r},t)\mathbf{j}(\mathbf{r},t). \tag{7}$$

The energy that is brought by scission neutrons to a certain angle θ is calculated in the same manner as in Eq. (5),

$$\frac{dE(\theta,t)}{d\Omega} = \int_0^t \mathbf{j}_\varepsilon(R,\theta,t')\cdot\mathbf{n}(R,\theta)R^2dt', \quad \mathbf{n} = e_r, \tag{8}$$

and the average energy of scission neutron for each direction is given by

$$\langle\varepsilon(\theta)\rangle = \frac{dE(\theta,t)}{d\Omega}\Bigg/\frac{dN(\theta,t)}{d\Omega}. \tag{9}$$

3 Results and discussions

As an example, we consider the fission of ^{236}U that corresponds to the neutron induced fission of ^{235}U. From a systematic study of the average total kinetic energy (TKE) of the fragments, Zhao et al. derived the elongation parameter h which is the ratio between the average distance between the fragments to the contact distance $r_0(A_1^{1/3} + A_2^{1/3})$, where A_1 and A_2 are the mass numbers of the fragments and $r_0 = 1.17$ fm [5]. The average distance between the fragments is determined so that the corresponding point charge Coulomb energy is equal to the average TKE. They obtained $\eta = 1.53$ for the asymmetric fission in U region. In the calculation, we simply use the same elongation parameter for B_i as $B_i = \eta r_0 A_i^{1/3}$. The distance between the grid points is typically 0.1 fm in both z- and ρ-directions. The time step for the integration is typically $\Delta t = 0.02$ fm/c. The radius R is set to 60 fm and the integration is performed up to $t = 5\times10^{-21}$ s. In calculating the time development, we take account of the motion of the fragments due to the Coulomb repulsion between the fragments. In the following, $V_0 = -40$ MeV and $W_0 = -1$ MeV are used for the strength of the potential unless otherwise is mentioned.

3.1 Dependence on the initial wave function

First, we investigate the dependence on the initial wave function for the symmetric fission. The initial wave function is assumed to be of the form of $\psi(t=0) = C\exp(-(\alpha_\rho\rho^2 + \alpha_z z^2))$. The initial average kinetic energy of scission neutron is given as,

$$\langle\varepsilon\rangle = \frac{\hbar^2}{2m}\left(2\alpha_\rho + \alpha_z\right). \tag{10}$$

The left panel of Fig. 1 shows the angular distribution of scission neutrons for the isotropic initial wave function ($\alpha_\rho = \alpha_z$) with $<\varepsilon> = 0.5$, 1.0, 1.5, and 2.0 MeV. The yields along fission axis, 0 and 180 degrees, increase as $<\varepsilon>$, while the yield around 90 degrees is almost independent of $<\varepsilon>$.

In the right panel of Fig. 1, we show the results for the cases of deformed initial wave function. The value of α_ρ is fixed and α_z is given as $\alpha_\rho/\alpha_z = 1, 9/4, 4, 25/4$. When we increase the deformation of the initial wave function, we obtain less yields around 90 degrees, while the yields around 0 and 180 degrees do not change as much. As a whole, the effect of the attractive nuclear potential is to increase the yeilds around 0 and 180 degrees.

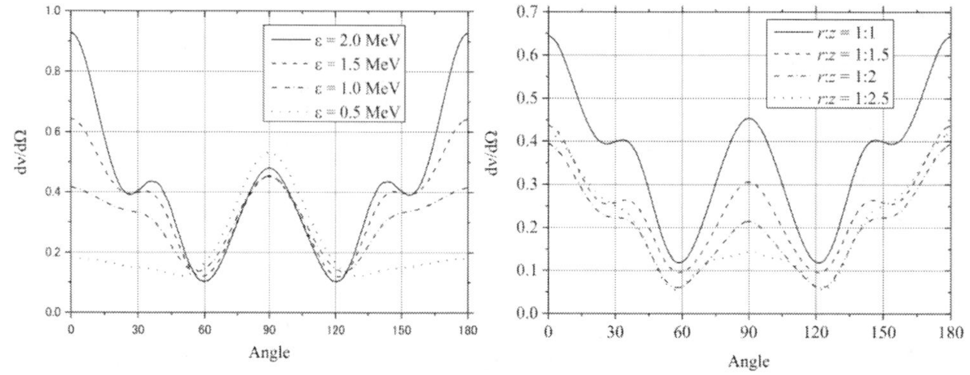

Figure 1. Angular distribution of scission neutron. *Left panel*: dependence on the width of the wave packet, *Right panel*: dependence on the deformation of the wave packet.

3.2 Dependence on the motion and the deformation of the fragments

Next, we investigate the case of asymmetric fission. The mass number of the heavy fragment is assumed to be $A_1 = 140$. The left panel of Fig. 2 displays the dependence on the motion of the fragments. Three cases are shown; the dashed line shows the result when the fragments are fixed at the initial position, the solid line corresponds to the case that the fragments start from the same position as the first case at rest then accelerated by the Coulomb repulsion, and the dotted-dash line is the case with the pre-scission kinetic energy of 15 MeV. In the second and the third case, we keep the TKE unchanged. Note that heavier fragment is on 0 degree and the lighter one is on 180 degrees. As is seen in Fig. 2, contrary to the experimental observation of the angular distribution of prompt neutrons, we observe more neutrons in the direction of heavier fragment. This indicates either that there is a significant contribution of the post scission emission of neutrons from fission fragments or that we need to employ more sophisticated initial condition of scission neutrons [2]. By comparing the solid and the dashed lines, we see that the change of the yield is smaller on the heavier fragment side than that on the light fragment side. The oscillation around 90 degrees is more enhanced for the case of the fixed fragments. That means that the yield around 90 degrees is sensitive to the position of the fragments. This tendency is confirmed when we compare the solid line with the dotted-dash line.

We have assumed spherical fragments so far. Now we examine how the angular distribution changes when the fragments are deformed. We assume that the fragments are deformed axial-symmetrically along the fission axis and the optical potential $U(\rho, z)$ is calculated accordingly. The surface of a fragment is expressed as,

$$R(\theta) = R_\delta(1 + \delta \cos\theta) \ . \tag{11}$$

where R_δ is determined to conserve the volume of the fragment. We show the results in the right panel of Fig. 2 for several cases. It is seen that the angular distribution on the heavier fragment side is sensitive to the deformation of heavier fragment and vice versa.

3.3 Angular distribution of the average energy

Before discussing the average energy of scission neutrons, we will briefly remark on the effects of absorption by the fragments. The left panel of Fig. 3 shows the comparison of the angular distribution

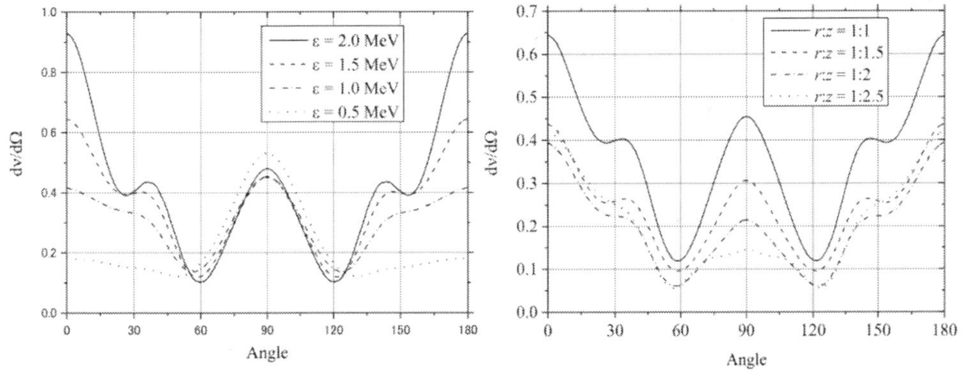

Figure 2. Angular distribution of scission neutron for asymmetric fission. *Left panel*: dependence on the motion of the fragments, *Right panel*: dependence on the deformation of the fragments.

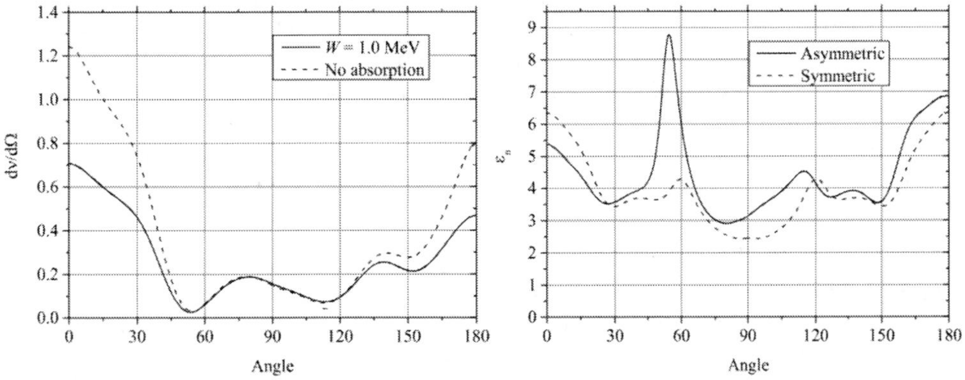

Figure 3. *Left panel:* comparison of the angular distribution of scission neutrons for the cases with and without absorption. *Right panel:* angular distribution of average energy of scission neutrons for symmetric and asymmetric fission.

calculated with ($W_0 = -1$ MeV) and without ($W_0 = 0$) the absorption. It is seen that the yields between 60 and 120 degrees do not change with the absorption, while the yields around 0 and 180 degrees are reduced significantly by the absorption.

The average energy of emitted neutron is calculated according to Eqs. (6)–(9). The right panel of Fig. 3 shows the angular distribution of the average energy for the symmetric and the asymmetric division. No precision kinetic energy is included and the absorption is set to $W_0 = -1$ MeV. It is seen that the average energy of neutron is high compared with the average kinetic energy of the initial wave function, $<\varepsilon> = 1.5$ MeV in this case. This is partly due to the motion of the fragments and is partly due to the finite duration of the calculation. Qualitatively speaking, slow component stays inside of the boundary for a long time, while the components with large kinetic energy arrive at the boundary faster. It is seen that the yields around 0 and 180 degrees have relatively high kinetic energy due to the attractioin of the fragments in both symmetric and asymmetric cases.

3.4 Treatment of boundaries

In order to avoid unrealistic reflections of the wave function at the border of the grid, there are essentially two methods; one is to prepare a large calculation grid and put a week absorbing potential outside of the sphere of radius R, on which we calculate the outgoing flux, the other is to use a smaller calculation grid and to use a proper artificial boundary condition to avoid the reflection. The former method has been employed in our calculation so far. One drawback of this method is that we need extra CPU time for the calculation outside of the physical region.

One example of the latter is a so-called "transparent boundary condition" (TBC) [6]. For the case of one-dimensional problem, it assumes that the wave function can be approximated by a single energy plane wave, i.e. $\psi(x) \propto e^{ikx}$. Because of the second differential in the kinetic energy term of the Hamiltonian, we need to extrapolate the wave function beyond the boundary. With TBC, the extrapolation is done as,

$$\psi(x_M + \Delta x) = \frac{\psi(x_M)^2}{\psi(x_M - \Delta x)} \tag{12}$$

where x_M denotes the boundary point. When we denote $f(x) = \log(\psi(x))$, Eq. (11) corresponds to the linear approximation, $f(x+\Delta x) = 2f(x) - f(x - \Delta x)$. We now propose to use the second-order approximation, $f(x + \Delta x) = 3f(x) - 3f(x - \Delta x) + f(x - 2\Delta x)$ to improve the accuracy of the extrapolation. We call this new method "second-order TBC" and applied it to a one-dimensional problem for free particle for which we know the exact solution. It was found that the error due to the extrapolation was much smaller compared with that with the original TBC. We also applied this method to the calculation of scission neutron. Though some fine tuning was necessary in the application, we obtained essentially the same accuracy as the one with the imaginary potential method but with less CPU time.

4 Summary

The effects of the scattering and re-absorption by the fission fragments on the angular distribution of scission particles have been investigated in the framework of the time-dependent Schrödinger equation. The effects of the fragments are taken account in terms of the nuclear optical potentials. We solve the time-dependent Schrödinger equation of the emitted neutron and the angular distribution is calculated by counting the outgoing flux. It has been demonstrated that the attractive nuclear potential enhances the yields around 0 and 180 degrees, while the absorptive potential diminishes them. In the case of asymmetric fission, the heavy fragment which is larger in size causes stronger effect to the angular distribution of scission neutrons. In order to elucidate the essential features of the effects of the fragments, we have examined the angular distribution by changing the initial wave function. We have also examined dependence on the position and the deformation of the fragments. The angular distribution of the average energy of emitted neutrons has been discussed. Finally, we have proposed a new boundary condition to minimize the reflection on artificial boundaries due to finiteness of the grid.

References

[1] N. Carjan, P. Talou, O. Serot, Nucl. Phys. A **792**, 102 (2007)
[2] M. Rizea, N. Carjan, T. Wada, Physica Procedia **47**, 27 (2013)
[3] T. Wada, T. Asano, *Proc. 5th Int. Conf. on Fission and Properties of Neutron-Rich Nuclei* (World Scientific, 2014), 409
[4] T. Wada, T. Asano, N. Carjan, Physica Procedia **47**, 33 (2013)
[5] Y.L. Zhao et al., Phys. Rev. C **62**, 14612 (2000)
[6] N. Carjan, M. Rizea, D. Storottman, Comp. Phys. Comm. **173**, 41 (2005)

Studies of fission fragment yields via high-resolution γ-ray spectroscopy

J.N. Wilson[1], *M.* Lebois[1], *L.* Qi[1], *P.* Amador-Celdran[2], *D.* Bleuel[3], *J.A.* Briz[4], *R.* Carroll[5], *W.* Catford[5], *H.* De Witte[6], *D.T.* Doherty[7], *R.* Eloirdi[2], *G.* Georgiev[8], *A.* Gottardo[1], *A.* Goasduff[8], *K.* Hadyñska-Klek[9], *K.* Hauschild[8], *H.* Hess[10], *V.* Ingeberg[11], *T.* Konstantinopoulos[8], *J.* Ljungvall[8], *A.* Lopez-Martens[8], *G.* Lorusso[12], *R.* Lozeva[8], *R.* Lutter[13], *P.* Marini[14], *I.* Matea[1], *T.* Materna[7], *L.* Mathieu[15], *A.* Oberstedt[16], *S.* Oberstedt[17], *S.* Panebianco[7], *Zs.* Podolyak[5], *A.* Porta[4], *P.H.* Regan[5,12], *P.* Reiter[10], *K.* Rezynkina[6], *S.J.* Rose[11], *E.* Sahin[11], *M.* Seidlitz[10], *O.* Serot[18], *R.* Shearman[512], *B.* Siebeck[10], *S.* Siem[11], *A.G.* Smith[19], *G.M.* Tveten[11], *D.* Verney[1], *N.* Warr[10], *F.* Zeiser[11], and *M.* Zielinska[7]

[1]*Institut de Physique Nucléaire, CNRS-IN2P3, Univ. Paris-Sud, Université Paris-Saclay, 91406 Orsay Cedex, France*
[2]*European Commission, Joint Research Centre, Directorate G for Nuclear Safety and Security,Unit G.I.5, Postfach 2340, 76215 Karlsruhe, Germany*
[3]*Lawrence Livermore National Laboratory, 7000 East Ave, Livermore, CA 94550, USA*
[4]*Subatech/Ecole des Mines, 4 rue Alfred Kastler, 44307 Nantes, France*
[5]*Department of Physics, University of Surrey, Guildford, GU2 7XH, UK*
[6]*Institute for Nuclear and Radiation Physics, KU Leuven, 3000 Leuven, Belgium*
[7]*Irfu, CEA, Université Paris Saclay, 91191 Gif-sur-Yvette Cedex, France*
[8]*CSNSM Orsay, CNRS-IN2P3, Univ. Paris-Sud, Université Paris-Saclay, 91405 Orsay Cedex, France*
[9]*INFN Laboratori Nazionali di Legnaro, Viale dell'Universita, 2, 35020 Legnaro, Italy*
[10]*Institut für Kernphysik, Zülpicher Strasse 77, 50937 Köln, Germany*
[11]*University of Oslo, Department of Physics, P.O. Box 1048, Blindern 0316 Oslo, Norway*
[12]*National Physical Laboratory, Hampton Road, Teddington, TW11 0LW, UK*
[13]*LMU München, Schellingstrasse 4, 80799, Munich, Germany*
[14]*CEA/DAM, DIF, Bruyères-le-châtel, 91297 Arpajon, France*
[15]*CENBG Bordeaux, Chemin du Solarium, Le Haut Vigneau, BP. 120, 33175 Gradignan, France*
[16]*Extreme Light Infrastructure - Nuclear Physics (ELI-NP), 077125 Bucharest-Magurele, Romania*
[17]*European Commission, Joint Research Centre, Directorate G for Nuclear Safety and Security, Unit G.2, 2440 Geel, Belgium*
[18]*CEA Cadarache, 13108 Saint-Paul-lez-Durance, France*
[19]*University of Manchester, Dept. Of Physics, Oxford Road, Manchester, M13 9PL, UK*

Abstract. Precise spectroscopic information on the fast neutron induced fission of the ^{238}U(n,f) reaction was recently gained using a new technique which involved coupling of the Miniball high resolution γ-ray spectrometer and the LICORNE directional neutron source. The experiment allowed measurement of the isotopic fission yields for around 40 even-even nuclei at an incident neutron energy of around 2 MeV where yield data are very sparse. In addition spectroscopic information on very neutron-rich fission products was obtained. Results were compared to models, both the JEFF-3.1.1 data base and the GEF code, and large discrepancies for the S1 fission mode in the Sn/Mo isotope pair

https://doi.org/10.1051/epjconf/201816900030

were discovered. This suggests that current models are overestimating the role played by spherical shell effects in fast neutron induced fission. In late 2017 and 2018 the nu-ball hybrid spectrometer will be constructed at the IPN Orsay to perform further experimental investigations with directional neutrons coupled to a powerful hybrid Ge/LaBr3 detector array. This will open up new possibilities for measurements of fission yields for fast-neutron-induced fission using the spectroscopic technique and will be complimentary to other methods being developed.

1 Introduction

The measurement of the isotopic distribution of fission fragments or fission yields and how they vary as a function of the mass, charge, excitation energy and angular momentum of the fissioning system is very important for a deeper understanding of nuclear fission. The precise details of the mass and charge splits are determined by the complex shell effects in the nascent fragments and parent nucleus. Fission models must be capable of reproducing these fragment distributions with a high degree of accuracy. Furthermore, for neutron induced fission in particular the knowledge of fission yields has some important consequences for energy applications and the functioning of reactors. These are the production of neutron poisons (e.g. ^{135}Xe), the production of beta-delayed neutron emitters (e.g. ^{137}I), the decay heat, the production of long-lived waste components (e.g. ^{137}Cs) and the emitted reactor anti-neutrino flux and spectra. This latter point is important for both non-invasive core monitoring and anti-proliferation and short base-line anti-neutrino experiments and short range neutrino oscillations.

There are several methods of obtaining experimental information on fission yields that have been developed over the last few decades. Simple mass yield curves can be obtained via radiochemical separation and spectroscopy after an actinide sample is irradiated with neutrons [1]. However, obtaining the fully characterised yields in mass and charge of the fragments is much more experimentally challenging. For thermal neutron induced fission extensive measurements have been performed over the last two decades with recoil mass spectrometers such as Lohengrin [2]. For fast neutron induced fission yield measurements exist for 14 MeV neutrons thanks to the very high neutron fluxes possible from neutron generators using the t(d,n)^{4}He reaction. However, for energies in between these two extremes which are most relevant for fast reactor applications (1–4 MeV) the current experimental data are very sparse.

1.1 Surrogate reaction yield measurements

Two recent techniques have been able to provide complete fission yield data in this intermediate energy region using surrogate reactions in inverse kinematics. This requires acceleration of actinide nuclei to hundreds of MeV per nucleon. The SOFIA experiment [3] at GSI uses photofission as the reaction mechanism which achieves complete characterisation of both fission fragments and the emitted neutrons on an event-by-event basis. Fragment distributions for fissioning nuclei over a wide range of A and Z can be measured leading to a wealth of new and important experimental information. However, the experiment is currently unable to control for compound nucleus excitation energy, which is on average high (11 MeV) and has a broad distribution. This makes direct comparison with fast-neutron-induced fission difficult.

The Vamos/Spider [4] setup at GANIL uses multi-nucleon transfer reactions in inverse kinematics to induce fission. The main advantage is that the detection of the outgoing ejectiles from the transfer reaction in the spider charged particle detector allows for reconstruction of the compound nucleus excitation energy if no neutron emission is assumed. However, only one of the two fragments is measured in the VAMOS mass spectrometer and the multi-nucleon transfer reaction can provide a large angular momentum transfer to the compound system (beyond 20 hbar). The angular momentum distribution of the fissioning system is thus radically different from that of the equivalent neutron-induced reaction for similar excitation energies. It is not at all clear what effect such differences would have on resultant isotopic yields. However, nuclear shell effects are known to change as a function of spin, and thus it would be surprising if large differences in initial angular momenta would not have some effect on the resulting yield distributions.

1.2 2E-2v measurements

The most direct and potentially accurate method to perform yield measurements is using a double armed 2E-2v spectrometer which directly measures the kinetic energy of each fragment, and the precision time-of-flight (TOF) over the flight path for each arm. This gives access to pre and post neutron masses, and charges through energy loss measurements. At least four of these devices are under development (e.g. STEFF [5], SPIDER [6], FALSTAFF [7] and VERDI [8]). Eventually the devices will be used with fast neutron beams available at new facilities such as NFS. However, the actinide samples used must be extremely thin to avoid energy losses and hence limited to less than 1 mg of material. In the best case fission rates of a few per second seems feasible at the highest neutron fluxes. However, the intrinsic efficiencies for detecting fission fragments is very low (typically 1–3%) due to the small solid angle coverage of the detectors in the spectrometer arms. This implies very long experiments will be needed to gather sufficient statistics, even with state-of-the-art intense neutron beams from new high-flux facilities.

1.3 Prompt spectroscopic methods

One rarely-used method for measuring fission yields, whose potential has possibly been overlooked, is the use of high-resolution gamma-ray spectroscopy to measure the prompt gamma rays de-exciting the fission fragments. Each fragment has a unique gamma decay signature dependent on its particular nuclear structure and the pathways via which its excited states decay. To uniquely identify fragments requires double or triple gamma ray coincidence spectroscopy, which in turn needs a high efficiency, high resolution gamma ray spectrometer using Germanium detectors.

The spectroscopic method was first used to measure yields of a ^{252}Cf spontaneous source [9] at Oak ridge in the 1990's with an array of 12 Ge detectors. The method was subsequently deployed to measure fission yields in the heavy ion induced ^{208}Pb(^{18}O,f) reaction with the Euroball spectrometer [10]. However, until recently this method has been essentially limited to yield measurements of spontaneous fission since the technique has particular problems correcting for isomers (underproduction) and fragment population via beta-decay (overproduction).

However, with the recent coupling of the MINIBALL spectrometer [11] to the LICORNE directional neutron source [12,13], this method can now be extended to direct measurements of yields for fast-neutron-induced fission.

The spectroscopic method has certain limitations so it currently cannot achieve the same level of precision as other techniques. However, relative errors of less than 20% seem feasible. The measurement precision is limited by at least four factors:

(i) The detection efficiency of the spectrometer must be known as accurately as possible. At the lowest gamma ray energies (<200 keV) this can prove difficult, and introduce systematic errors to the measurement.

(ii) A sum of all the gamma intensity feeding the first excited states must be carried out to determine the absolute yield. This can be particularly challenging for odd nuclei where the intensity is fragmented and there are many weakly populated decay paths feeding the first excited state.

(iii) Since the technique is dependent on gamma coincidence measurements it is impossible to determine the proportion of total intensity directly feeding the ground state for any particular fragment. Although this is expected to be quite small in most cases, the ground state feeding may vary between nuclei and hence introduce some small systematic errors.

(iv) Large corrections may be necessary for missing intensity due to the presence of isomeric states or beta decay.

In our experiments we have shown that the latter point can be mitigated by use of a pulsed beam which exploits time correlations to allow measurement of missing intensity from isomeric states and simultaneously strongly selects against uncorrelated beta-decays.

2 Experimental details

In February 2015, the LICORNE directional neutron source was successfully coupled to the Miniball spectrometer at the IPN Orsay. The full details of the LICORNE/MINIBALL experiment can be found here [14]. LICORNE is an inverse kinematic neutron source based on the p(^{7}Li,n)^{7}Be reaction which can achieve high fluxes of up to 10^7 n/s/cm^2 for samples placed a few cm away from the source.

The MINIBALL spectrometer consists of 24 Ge crystals packaged in 8 triple-clusters. Each cluster was placed approximately 14 cm from the fission target, and the array covered one third of the total solid angle, providing a total photopeak efficiency of around 6% at 1 MeV. The LICORNE directional neutrons had an average energy of 1.7 MeV just above the ^{238}U fission threshold. The maximum neutron cone angle was 19.5 degrees and the beam was pulsed with a period of 400 ns and pulse width of 2 ns. A sample of 31.35 g of depleted ^{238}U was irradiated over an effective beam time of 11 days with fissions induced at an estimated rate of 25 kHz.

The MINIBALL array was designed as a high-efficiency, low-multiplicity device for radioactive beam experiments and is not Compton suppressed. The peak-to-total is fairly low 22%, and thus limits the resolving power of the device when used to select nuclei using standard double or triple gamma coincidence techniques. The Compton background in MINIBALL is particularly large at energies in the 200–500 keV range. Furthermore, self-shielding of low energy gamma rays by the compact metal target limited the sensitivity of the experiment below 500 keV. Measurements of yields for odd nuclei were thus particularly challenging and hence the published results focused on the 40 most strongly populated even-even nuclei in the fission of ^{238}U.

3 Results

The absolute yields measured in the experiment can again be found in our recent publication [14], but are seen to be reasonably well-reproduced by the JEFF3.3.1 and ENDFB7 evaluations and the GEF fission code apart from the yields of Sn/Mo isotopes which showed significant deviations. This is interpreted as a quenching of the S1 fission mode and an overestimation of the role of spherical shell effects in fission for this particular nucleus.

A gamma ray coincidence spectrum of a typical gate on the $2^+ \rightarrow 0^+$ transition in a strongly populated even-even fragment can be seen in Fig. 1. The spectrum shows both the higher-lying transitions in ^{96}Sr and also low-lying transitions from all its binary partners ^{142}Xe, ^{141}Xn, ^{140}Xe, ^{139}Xe, etc. corresponding to evaporation of 1,2,3,4 etc. neutrons.

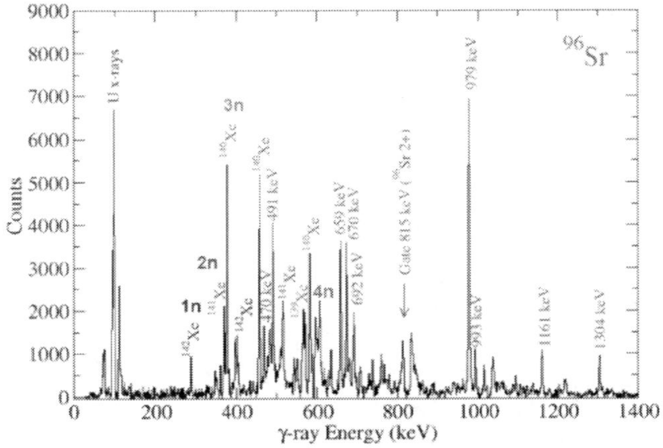

Figure 1. A typical gamma ray coincidence spectrum obtained from gating on the $2^+ \rightarrow 0^+$ transition in ^{96}Sr at 815 keV. Coincidences from gamma rays emitted from higher lying states in ^{96}Sr can be observed along with the corresponding binary partners of $^{142-139}$Xe.

An interesting possibility for a more complete analysis of the gamma-ray coincidence data, arises from the collection of correlated fragment information. It is possible to deduce the relative probabilities for the number of neutrons emitted conditional on the detection of ^{96}Sr by measuring the relative intensities of its binary Xe partners. If combined with similar data from all the other Sr fragments populated in the experiment, then a neutron multiplicity distribution can be deduced for the Sr/Xe charge split.

It has been relatively straight forward to extract this information from the LICORNE/MINIBALL data for Sr and Xe because the first excited states of both odd and even Sr and Xe nuclei are relatively high in energy, hence intensity measurements didn't suffer from the sample self-shielding problem which normally occurs for odd nuclei with low energy gamma transitions. Figure 2 shows graphically the summary of these absolute yield measurements for the observed Xe partners correlated with every observed Sr isotope. (It should be noted that the measurement could be performed again independently by gating on all observed Xe isotopes and measuring the corresponding Sr binary partners).

The first observation from this figure is that the more stable the selected Sr isotope, the more neutrons on average are evaporated and vice-versa. For example, the 8n neutron channel becomes just visible when gating on the ^{94}Sr closest to stability, and the 2n and 3n channels dominate when selecting the most exotic Sr isotopes. This correlation is clearly as it ought to be. The second observation is that the distribution of emitted neutrons appears to become broader when moving closer to stability. This also appears reasonable since there are many more available combinations for the number of neutrons emitted by each Sr and Xe fragment when the sum is larger on average.

Once the complete correlated yield matrix has been extracted for the particular charge split (in this case Se/Xe) it is possible to sum the results for each neutron multiplicity. Hence the complete neutron multiplicity distribution for the isotope pair can be deduced. This can be seen in Fig. 3.

3.1 The nu-ball array and improvements to the spectroscopic technique

In 2017 and 2018 an experimental campaign with the nu-ball spectrometer is planned. Nu-ball consists of 34 Ge detectors 24 clovers and 10 coaxial coupled to 20 LaBr3 detectors from the FATIMA

 https://doi.org/10.1051/epjconf/201816900030

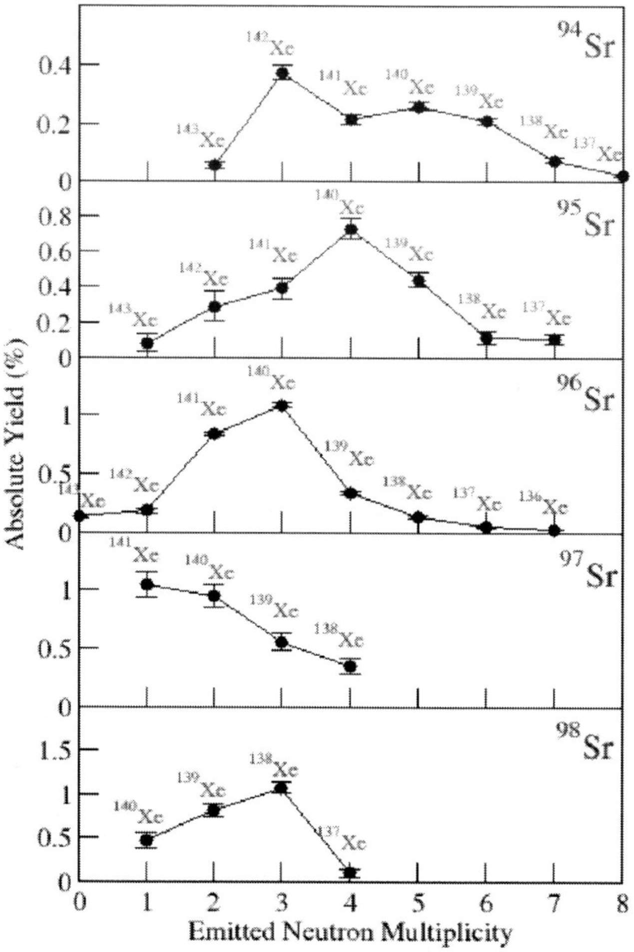

Figure 2. Absolute yields of correlated Sr-Xe nuclide pairs.

collaboration. It is currently under construction at the IPN Orsay and will be commissioned in October 2017. The photopeak efficiency at 1 MeV is expected to be 5.5%. Nu-ball aims to improve on the resolving power of MINIBALL by at least an order of magnitude since the Ge detectors have anti-Compton BGO shields and will be Compton suppressed. This will raise the peak-to-total ratio of the device at 1 MeV from 20% to around 50%. In addition, the self-shielding of the massive actinide samples to be irradiated with neutrons at the centre of the spectrometer (tens of grams of ^{238}U and ^{232}Th) will be hugely diminished. The newly developed metal samples will have average densities of between 0.5 and 1.0 g/cm^3 compared to a density of 19.1 g/cm^3 for uranium metal. The reduction in effective thickness of material that gamma rays must travel through to escape the surface can be shown to be proportional to a power of the ratio of densities $(\rho^{new}/\rho^{old})^{2/3}$.

Therefore, reducing the density of the sample by a factor of 40 in our case will produce a corresponding reduction in the gamma-ray attenuation of the sample by over a factor of 10 and hugely increase the efficiency of the spectrometer in the low energy range (100–500 keV). Lowering the

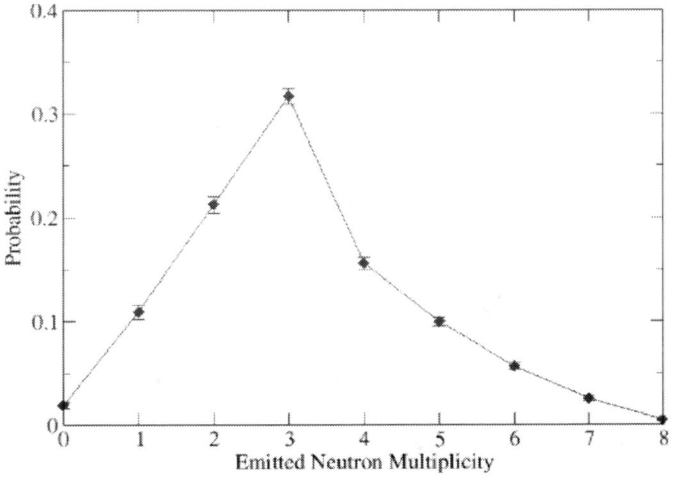

Figure 3. Deduced neutron multiplicity distribution for the Sr-Xe isotope charge split.

average density of the actinide samples is achieved by extending the sample in space over several centimetres, either by stacking spaced out thin foils, or by encapsulating metallic turnings.

4 Conclusion

The combination of a fast directional, neutron source coupled to a high-resolution gamma-ray spectrometer has been used to demonstrate the feasibility of making direct fission-yield measurements for fast-neutron induced fission in 1–4 MeV range. This energy range is very important for energy applications for both current and future (Gen IV) nuclear reactors yet current yield data are very sparse.

The recent LICORNE/MINIBALL experiment has also shown the importance of a pulsed neutron beam for accurate yield measurements with gamma spectroscopy which is essential to suppress beta decay and correct for the presence of isomers. Furthermore, the spectroscopic technique produces correlated fragment data which allows extraction of neutron multiplicity distributions for different fragment charge splits. This is an additional observable in fission that can now be compared with theory and hence may allow for further refinement of fission models and a deeper understanding of the fission process.

Improvements are currently being made to the spectroscopic technique to measure fragment yields for fast-neutron induced fission. The nu-ball spectrometer is currently under construction at the IPN Orsay which will improve the resolving power by an order of magnitude and samples with a much lower average density are being designed to solve the self-shielding problem. These improvements should allow higher precision yield measurements of ^{238}U(n,f) and ^{232}Th(n,f) in the near future. The samples to be used in these cases will contain tens of grams of actinide material. However, with the improved resolving power of the spectrometer and the potential coupling to active (ionization chamber) targets, it ought to be possible to perform yield measurements with only hundreds of mg of material and lower fission rates of the order of $\sim$1 kHz. Hence yield measurements for the key fissile isotopes of ^{235}U, ^{233}U and ^{239}Pu, ^{241}Pu may soon be within reach with the gamma spectroscopic technique for fast-neutron-induced fission. These will hopefully provide a useful compliment to measurements with the different 2E-2v techniques planned for future fast-neutron facilities such as NFS.

References

[1] S. Nagy, K.F. Flynn, J.E. Gindler, J.W. Meadows, and L.E. Glendenin, Phys. Rev. C **17**, 163 (1978)

[2] H.G. Börner and F. Gönnenwein, The Neutron: A Tool and an Object in Nuclear and Particle Physics, World Scientific, Singapore (2012)

[3] L. Grente et al., EPJ Web of Conferences **122**, 01006 (2016)

[4] F. Farget et al., Eur. Phys. J. A **51**, 175 (2015)

[5] A.J. Pollitt, A.G. Smith, I. Tsekhanovich, J.A. Dare and L. Murray, EPJ Web of Conferences **93**, 02018 (2015)

[6] F. Tovesson, et al., EPJ Web of Conferences **62**, 05002 (2013)

[7] D. Doré et al., Nuclear Data Sheets **119**, 346–348 (2014)

[8] M.O. Frégeau et al., EPJ Web of Conferences **62**, 05003 (2013)

[9] G.M. Ter-Akopian et al., Phys. Rev. C **55**, 3 (1997)

[10] A. Bogachev et al., Eur. Phys. J. A **34**, 23–28 (2007)

[11] N. Warr et al., Eur. Phys. J. A **49**, 40 (2013)

[12] M. Lebois, J.N. Wilson, P. Halipré, B. Leniau, I. Matea, A. Oberstedt, S. Oberstedt, and D. Verney, Nucl. Instrum. Methods Phys. Res., Sect. A **735**, 145 (2014)

[13] J.N. Wilson, M. Lebois, P. Halipré, S. Oberstedt, and A. Oberstedt, Phys. Procedia **64**, 107 (2015)

[14] J.N. Wilson et al., Phys. Rev. Lett. **118**, 222501 (2017)

AUTHOR INDEX

Åberg Sven 133
Adamian G. G. 103
Ahmadov Gadir....65
Al-Adili A....127
Al-Adili Ali 1, 119
Amador-Celdran P 219
Andreev A. V....103
Andreyev Andrei....87
Antonenko N. V....103
Aritomo Yoshihiro 87
Asai Masato 87
Asano Tomomasa....213
Bartel J....109
Berikov Daniyar....65
Bernard David....49
Beyer Roland....57, 191
Billnert Robert 97
Blanc Aurélien....49
Bleuel D....219
Briz J. A....219
Carjan N. 143
Carjan Nicolae....213
Carlsson Gillis 133
Carroll R....219
Catford W....219
Chatillon A. 127
Chebboubi Abdelaziz....49, 77
Chiba S....27
Chiba Satoshi 87, 201
Choudhury D....127
Cong Vu D. 43
Danilyan Gevorg 65
Doherty D. T....219
Døssing Thomas 133
Dragic Aleksandar 43
Duke D. 173
Eloirdi R. 219
Eronen Tommi 119
Furman Walter 149
Gagarsky Alexei 65
Gatera A....127
Gatera Angélique 7, 97
Georgiev G....127, 219
Geppert-Kleinrath V. 173
Goasduff A....219
Göök A. 127
Göök Alf....1, 7, 15, 97
Gorelov Dimitri....119
Gottardo A....219
Goverdovski Andrei....149
Goverdovskii Andrei....71
Gustavsson Cecilia 1
Hadyñska-Klek K. 219
Hambsch Franz-Josef....1, 7, 15, 43, 149, 179
Hauschild K....219
Hess H. 219
Hirose Kentaro 87
Hutanu Vladimir 65
Ingeberg V. 219
Ishizuka C. 27
Ishizuka Chikako 201
Ivanyuk F. A. 27
Ivanyuk Fedir A. 201
Jaffke Patrick 35
Jansson Kaj....1, 119
Jokinen Ari 119
Jovancevic Nikola....43
Julien-Laferrière Sylvain....49
Junghans Arnd R. 57
Junghans Arnd Rudolf....191
Jurado Beatriz 155
Kankainen Anu....119
Kessedjian Grégoire....49
Ketlerov Vladimir 71
Khryachkov Vitaly....71
Klenke Jens....65
Knezevic David 43
Kögler Toni....57, 191
Konstantinopoulos T. 219
Kopatch Yuri....65
Kopatch Yury 149
Köster Ulli....49
Krmar Miodrag....43
Lantz Mattias....1, 119
Laurent B. 127
Lebois M....127, 219
Litaize Olivier....49, 77, 165
Ljungvall J. 219
Lopez-Martens A....219
Lorusso G....219
Lozeva R....219
Lutter R....219
Maj A....127
Makii Hiroyuki....87
Manning B. 173
Marini P. 219
Maruhn Joachim A. 201
Masalovich Sergey....65
Matea I. 127, 219
Materna T. 219
Materna Thomas 49
Mathieu L....219
Mattera Andrea....1, 119
Mayorov D. 173
Meplan Olivier....49
Mitrofanov Vecheslav 71
Mitsyna Ludmila V. 43
Moens André....7
Möller Peter....35, 133
Moore Iain D....119

AUTHOR INDEX

Mosby S. 173
Nagame Yuichiro 87
Nerlo-Pomorska B. 109
Nishio Katsuhisa 87
Novitsky Vadim 65
Oberstedt A. 127, 219
Oberstedt Andreas 7, 97, 149
Oberstedt S. 127, 219
Oberstedt Stephan 1, 7, 15, 43, 97, 149
Ohtsuki Tsutomu 87
Olivier Frégeau Marc 1
Orlandi Riccardo 87
Österlund Michael 1
Panebianco S. 219
Pasca H. 103
Penttilä Heikki 119
Petrache Costel M. 87
Podolyak Zs. 219
Pomorski K. 109
Pomp Stephan 1, 119
Porta A. 219
Prokofiev Alexander V. 1, 119
Qi L. 127, 219
Rakopoulos Vasileios 1, 119
Randrup Jørgen 133, 207
Rapala Michal 49
Regan P. H. 219
Reiter P. 219
Revay Zsolt 43
Rezynkina K. 219
Rinta-Antila Sami 119
Rizea M. 143
Rose S. J. 127, 219
Ruskov Ivan 149
Sage Christophe 49
Sahin E. 219
Sato Tetsuya K. 87
Schmidt Karl-Heinz 155
Schmitt C. 109, 127
Schmitt Christelle 35, 155
Schmitt K. 173
Schwengner Ronald 57, 191
Seidlitz M. 219
Sergachev Alexei 71
Serot O. 219
Serot Olivier 49, 77
Shcherbakov Oleg 149
Shearman R. 219
Sibbens Goedele 7
Siebeck B. 219
Siem S. 219
Smith A. G. 219
Solders Andreas 1, 119
Stetcu Ionel 35
Stieghorst Christian 43
Sukhovoj Anatoly M. 43
Talou Patrick 35
Tamagno Pierre 77, 165
Tanaka Shouya 87
Tarrío Diego 1
Thulliez Loïc 77
Tovesson F. 173
Toyoshima Atsushi 87
Tsekhanovich Igor 87
Tsukada Kazuaki 87
Tudora Anabella 179
Tveten G. M. 219
Urlass Sebastian 191
Usang M. D. 27
Usang Mark D. 201
Vanleeuw David 7
Vermeulen Mark 87
Verney D. 219
Vidali Marzio 1, 7
Vogt Ramona 207
Wada Takahiro 213
Wagner Andreas 57, 191
Ward Daniel 133
Warr N. 219
Wasilewska B. 127
Wilson J. N. 127, 219
Witte H. De 219
Zeiser F. 127, 219
Zielinska M. 219